Y0-BYK-327

DEVELOPMENTAL AND CELL BIOLOGY SERIES

EDITORS
M. ABERCROMBIE D. R. NEWTH
J. G. TORREY

NUCLEAR CYTOLOGY IN RELATION TO DEVELOPMENT

NUCLEAR CYTOLOGY IN RELATION TO DEVELOPMENT

F. D'AMATO
Professor of Genetics, University of Pisa

CAMBRIDGE UNIVERSITY PRESS
CAMBRIDGE
LONDON · NEW YORK · MELBOURNE

Published by the Syndics of the Cambridge University Press
The Pitt Building, Trumpington Street, Cambridge CB2 1RP
Bentley House, 200 Euston Road, London NW1 2DB
32 East 57th Street, New York, NY 10022, USA
296 Beaconsfield Parade, Middle Park, Melbourne 3206, Australia

First published 1977

Printed in Great Britain at the
University Press, Cambridge

Library of Congress Cataloging in Publication Data

D'Amato, Francesco, 1916–
Nuclear cytology in relation to development.

(Developmental and cell biology series; 6)
Bibliography: p.
Includes index.
1. Cell nuclei. 2. Developmental cytology.
I. Title. [DNLM: 1. Cell nucleus. QH595 D155n]
QH595.D33 574.8′732 76-46045
ISBN 0 521 21508 0

To Maria Grazia

Contents

Preface

When the advisory editors of the Developmental and Cell Biology series of the Cambridge University Press asked me to write a monograph on 'Nuclear Cytology in Relation to Development', I accepted their proposal with great interest. It seemed to me that such a monograph could offer an excellent opportunity to discuss comparatively many aspects of development and growth in plants and animals, trying to properly emphasize peculiar features in plants which do not commonly find due appreciation in discussion of the long-standing problems of development.

When writing the book, I have made all efforts to achieve a balanced treatment of animal and plant systems. Whether my endeavour has been successful, at least partially, can only be judged by the reader. I am aware of the vast fields I have left unexplored (among these, important aspects of abnormal and pathological development) and of the many papers which I have not cited. I apologize to all who feel that their contribution should have been included. They may, however, contemplate that the length assigned to the text demanded many omissions. I have tried to compensate for them, where possible, by giving references to review articles. The citation of literature ends about December 1975 (papers dating 1976–7 were known to me as preprints).

Professor John G. Torrey of Harvard University read all of the manuscript and made many minor corrections and some useful suggestions for which I am very grateful to him. I am also grateful to Mrs M. Velia Bellani and Mr P. L. Leoni of our Institute of Genetics at the University in Pisa, for typing the manuscript, and to Mr Leoni for preparing the drawings for the original figures.

Pisa
June 1976

F. D'AMATO

1

Life cycles

General

An essential feature of living organisms is reproduction, the ability to produce other individuals of the same species. Although the continuous flow of generations in any species of animals or plants presumably did not escape man's observation for centuries, the material basis for the transition from one generation to the next remained obscure until about a hundred years ago. It was in the second half of the last century that Hertwig and Strasburger, working on sea urchins and plants respectively, showed that the key event in fertilization is the fusion of the nucleus of the male gamete with the nucleus of the female gamete. Further observations led, within a few decades, to the demonstration that:

(i) during cell division, the nucleus produces by mitosis, or karyokinesis, two daughter nuclei each of which is identical in its chromosome complement to the other and to the original nucleus;

(ii) the chromosome number of gametes (n)* is half the chromosome number of the zygotes and of the somatic cells derived from it (2n);

(iii) the production of gametes – and, in many plants, of cells with n chromosome number (spores) which give rise to gamete bearing individuals (gametophytes) – is the result of meiosis, consisting of two successive nuclear divisions which lead to reduction of the chromosome number from 2n to n.

The above observations have laid the foundations for a correct interpretation of life cycles in sexually reproducing eukaryotes, i.e. organisms, both unicellular and multicellular, which have typical nuclei with a nuclear envelope and chromosomes, as well as nuclear divisions in the form of mitosis and meiosis. In eukaryotes, fertilization involves the fusion of the two gametes to form a zygote containing two complete genomes. These zygotes are called holozygotes in contrast with the merozygotes (Wollman, Jacob & Hayes, 1956) in bacteria (prokaryotes), in which only a part of the genome of a donor cell is transferred to a recipient cell which contributes both its cytoplasm and its entire genome.

* In diploids, n corresponds to the basic chromosome number (x): 2n = 2x. In polyploids, the gametic or sporic (n) and the zygotic or somatic (2n) chromosome numbers are further specified by adding the level of ploidy. Thus, e.g. 2n = 4x = 24 designates a tetraploid individual or species having 6 as its basic chromosome number; in case of a regular meiosis, the resulting cells (gametes or spores) have n = 2x = 12 chromosomes.

In most plants and in some groups of animals, besides sexual reproduction, different methods of asexual (= agamic) reproduction may operate: these are often much more efficient than sexual reproduction in the dispersal of the species. During asexual reproduction, new individuals are formed by mitosis from single cells (agamogony), groups of cells or even organs (vegetative reproduction or propagation) which are detached from the individual. The population of individuals derived asexually from a common ancestor is said to form a clone. Clones are either haploid or diploid depending on the chromosome complement of the ancestor.

In analyzing the life cycle of sexually reproducing species it is essential to ascertain the relative positions and the extent of the temporal separation of the two events which alternate in the cycle: fusion of the gametes (amphimixis) and reduction in chromosome number (meiosis). In the following pages, some patterns of life cycles will be discussed. For more extensive treatments of life cycles, including terminological questions and genetic and evolutionary aspects, reference is made to Martens (1954*a*, *b*), Wetmore, De Maggio & Morel (1963), Alston (1967), Resende (1967), Raper & Flexer (1970) and Feldmann (1972).

Diplontic organisms

In all metazoans and in most protozoans development normally starts from a zygote, which undergoes a series of mitoses to produce a diploid organism. At maturity, this organism reproduces sexually by forming haploid gametes from diploid cells by means of meiosis. Since meiosis forms gametes, it is called gametogenic or, less properly, gametic. Some authors use the term terminal meiosis to emphasize that it occurs at the end of a life cycle, which is wholly occupied by diploid individuals, the haploid state being represented by single cells: the gametes (Fig. 1.1). In describing diplontic as well as haplontic life cycles (see below), several authors use the term 'phase' to denote the haploid and the diploid state. This would mean that in a diplontic life cycle, although no alternation of generations occurs, there is an alternation of two nuclear phases, one haploid and the other diploid. As pointed out by Martens (1954*a*) and subscribed to by others (e.g. Feldmann, 1972), the concept of phase should imply a process of development by mitosis, the minimum requirement for the establishment of a phase being one mitosis. Martens' distinction between nuclear state (the state of single cells as gametes, spores or zygotes) and nuclear phase (a generation, reduced as it may be, which originates by mitosis from either a zygote or a spore: diplophase and haplophase respectively) acquires special significance when the evolution of the haplo-diplontic life cycle in phanerogams, the seed plants (see below), is considered.

Some animals actually show alternation of two or more morphological types due to the occurrence, in addition to sexual reproduction, of asexual propagation processes including pedogenesis (asexual propagation of immature diploid ani-

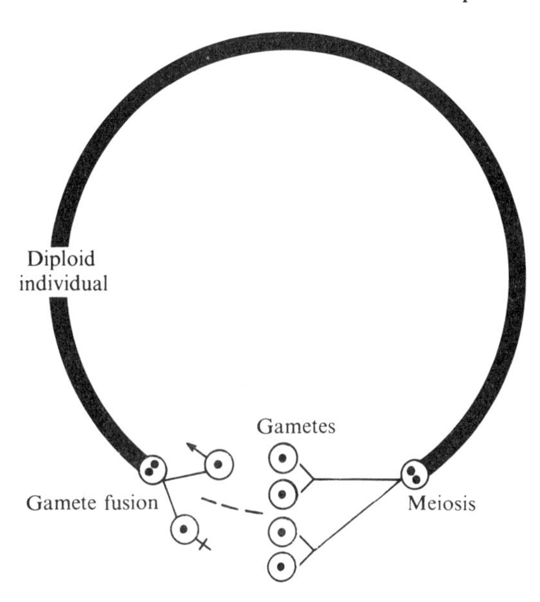

Fig. 1.1. Diagram of a diplontic life cycle. The diploid phase is represented by a thick line and the outline of a nucleus with two nucleoli; the haploid state, comprising single cells (gametes), is represented by the outline of a nucleus with one nucleolus.

mals), viviparity, parthenogenesis, predominance or exclusiveness of the female sex. Other factors contributing to the complexity of life histories in some animals are migration to different hosts and metamorphosis. As an example, consider the complex life history of the liver flukes (trematodes). The adult sheep fluke (*Fasciola hepatica*) inhabiting the liver releases fertilized eggs which are excreted by the host and develop into a miracidium, a ciliated free-swimming larva. The miracidium bores into a certain type of snail where it forms a cyst (sporocyst) from which many larvae of a second type, the rediae, are produced. A third type of larva, the cercariae, develops from the rediae within the tissues of the snail: the cercariae leave the snail and swim actively to reach vegetation on which they encyst. If consumed by a sheep, the cyst resumes development into an hermaphroditic adult fluke. The sequence of pedogenetic developments, which precedes the attainment of the adult state in the sheep liver fluke and in other trematodes, appears most striking if it is considered that the three different larval types and the adult animal all have the same genetic endowment, namely that of the original zygote.

Another example of the ability of the same set of genes to express alternative forms of the organism in the course of a life cycle is offered by the complete metamorphosis in holometabolous insects. The larva, which originates from the egg, may differ completely from the adult; it contains, in addition to its larval structures, a number of primordia, the imaginal disks, which give rise during

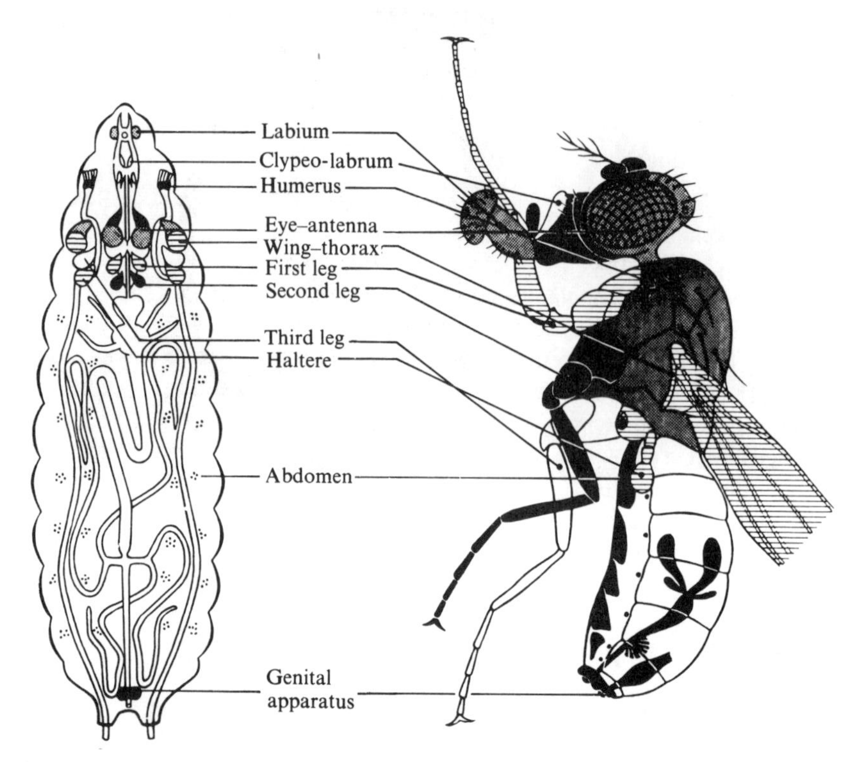

Fig. 1.2. Schematic representation of the larval organization and the location of different imaginal disks in Diptera. Disks and their corresponding adult derivatives are connected by lines and given the same hatching or shading (after Nöthiger, 1972).

metamorphosis to specific structures of the adult. When the larva enters the pupal stage, most of the larval tissues are destroyed by histolysis and the adult is practically formed anew from the imaginal disks, whose cells undergo dramatic differentiative changes. In the intensively studied *Drosophila*, most, if not all, imaginal disks first appear as invaginations or thickenings of the epidermis in the later embryonic stages; moreover, the imaginal disks responsible for the formation in the adult (imago) of the head, the thorax and its appendages, the integument of the abdomen and several internal organs (gut, salivary glands, gonads) have been identified (Nöthiger, 1972) (Fig. 1.2). Since the imaginal disks have provided outstanding material for studies on developmental processes, we will refer to them on other occasions.

If we now turn our attention to the plant kingdom, we see that the diplontic life cycle is rare. It is found in some diatoms, in some brown algae (Fucales) and in species of lower fungi. In some species, the diploid individuals are monoecious, because they produce male and female gametes on different portions of their body (thallus); in other species, male and female individuals occur (dioecism). Besides sexual reproduction, asexual propagation by means of propagules, either unicellular (conidia) or multicellular, is very common.

Haplontic organisms

In relatively few species of Protozoa and in species of green and red algae and of fungi, development starts from a single haploid cell, a spore, which results from meiosis in a zygote. In general, from one zygote four spores (tetrad) are formed; in most species of ascomycetes, however, the zygote undergoes a third mitosis after two meiotic divisions so that eight haploid nuclei and spores are produced. In exceptional cases, more than eight spores are formed, due to the occurrence of two or more mitoses after the meiotic process. In unicellular species, each spore produces a sometimes very large progeny of identical individuals (colony, clone); within the colony, the fusion of two cells behaving as gametes (conjugation) forms a diploid zygote which undergoes meiosis (meiocyte or gonotokont). In multicellular organisms, each spore forms one individual which reproduces sexually; here again, monoecy or dioecy may occur. Since meiosis forms spores, it is called sporogenic; some authors speak of zygotic meiosis, others of initial meiosis, because it takes place at the beginning of a life cycle which is wholly

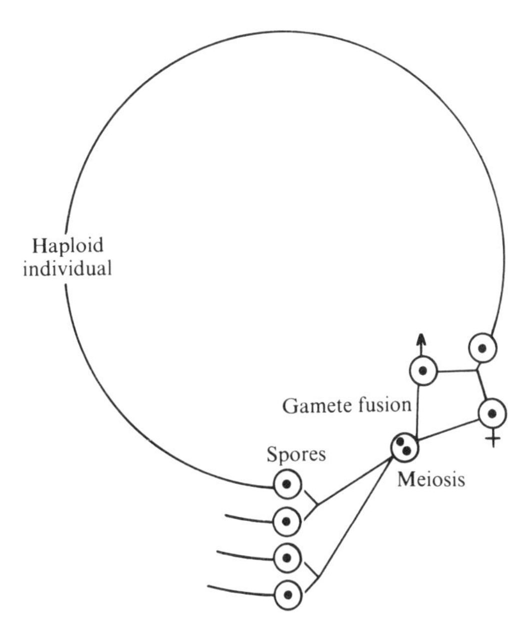

Fig. 1.3. Diagram of a haplontic life cycle. The haploid phase is represented by a thin line and the outline of a nucleus with one nucleolus the diploid state, comprising single cells (zygotes), is represented by the outline of a nucleus with two nucleoli.

occupied by haploid individuals (haplontic cycle), the diploid state being represented by single cells, the zygotes (Fig. 1.3). In haplontic organisms, asexual reproduction may occur by means of single vegetative cells, the conidia, which by mitosis form new haploid individuals identical to those which originated from spores. Conidia and spores, then, share a common mode of development (Haeckel's monogony as opposed to amphigony or development by fusion of gametes); they are, however, quite distinct in origin, the former being vegetative cells and the latter products of meiosis. In view of the significance of this distinction (Battaglia, 1955*a*), the indiscriminate use of the term spore in some text books, and even in scientific papers, is much to be regretted, because it may render the description of some life cycles incomprehensible to a non-specialist. This is the more so when the term spore is also applied to diploid conidia which are released from asexually reproducing diploid individuals both in diplontic and diplo-haplontic life cycles.

In a few species of green algae (e.g. *Acrosiphonia spinescens*), the zygote undergoes extensive growth in the absence of mitosis and has a much longer life-span than the haploid individuals which develop from the spores.

Diplo-haplontic organisms

The diplo-haplontic life cycle is by far the most widespread life cycle among plants. It is characterized by an intermediate meiosis and consequently by an alternation of generations, one haploid and one diploid. The diploid zygote produces by mitosis a diploid individual, a diplophyte, also called a sporophyte because it forms spores by meiosis. Spores undergo mitotic divisions and develop into haploid individuals (haplophytes) which eventually contain the gametes (gametophytes) (Fig. 1.4). Since gametophytes are haploid as their gametes are, they are correctly interpreted as gametiferous (gamete bearing) individuals, in opposition to the gametogenic individuals such as the diplonts, which form gametes by meiosis (D'Amato, 1971).

The diplo-haplontic life cycle shows great variation in different taxonomic groups in relation to specific adaptations, phylogenetic position of the taxa etc. A developmentally interesting type of the diplo-haplontic life cycle is that found in some species of algae belonging to the genera *Ulva, Cladophora, Zanardinia, Dictyota, Rhodochorton*, which show an alternation between morphologically indistinguishable (homomorphic) and sometimes equally sized (equivalent) gametophytes and sporophytes. In dioecious species (*Dictyota dichotoma*), the life cycle comprises three identical individuals: a sporophyte, a male gametophyte and a female gametophyte, whose distinction is only possible on cytological grounds. As shown by a now classical tetrad analysis (Schreiber, 1935), of the four spores which are produced meiotically in each sporangium in *Dictyota*, two give rise to female gametophytes and two to male. In many other species of algae (see list in Feldmann, 1972), there is an alternation of morphologically different (hetero-

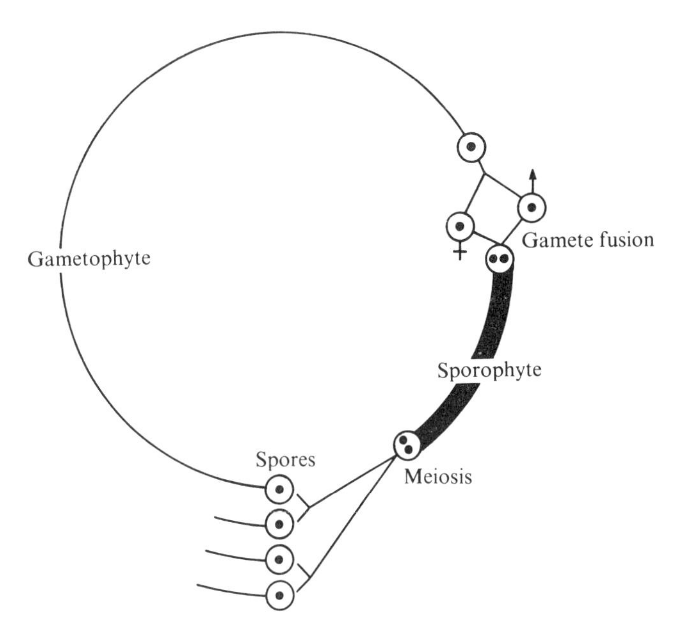

Fig. 1.4. Diagram of a haplo-diplontic life cycle with predominance of the haplophase (e.g. mosses). The haploid phase is represented by a thin line and the outline of a nucleus with one nucleolus; the diploid phase is represented by a thick line and the outline of a nucleus with two nucleoli.

morphic) generations, generally with one generation dominating the other. A more complex life cycle which occurs in red algae (e.g. *Polysiphonia*) comprises one gametophyte and two separate diplophytes. The zygote develops inside the carpogonium (female organ) of the gametophyte to form a diplophyte (carposporophyte) which lives as a parasite of the gametophyte; this diplophyte produces diploid conidia ('carpospores') which form an independent diplophyte (tetrasporophyte) reproducing by typical haploid spores (four spores in each sporangium). Gametophytes and tetrasporophytes are practically indistinguishable: however, carposporophytes and tetrasporophytes, both diploid, are completely different in structure.

Among higher plants, i.e. the Embryophyta comprising the Bryophyta (mosses and liverworts), the Pteridophyta (ferns, lycopods, horsetails, etc.) and the Phanerogamae (seed plants), the diplo-haploid life cycle shows a progressive decrease in the extent and size of the haplophase in the series from the mosses to the seed plants.

In the bryophytes, the leafy plant represents the haplophyte, which originates from a spore and reproduces sexually by means of egg cells (one in each archegonium) and motile sperm cells contained in great numbers in each antheridium. After fertlization, the zygote grows *in situ* into a diploid embryo and afterwards a non-leafy sporophyte in whose sporangia spores are produced by meiosis. Spores are morphologically alike; in dioecious species, they give rise to male and female gametophytes in a 1:1 ratio.

In the pteridophytes, the trend towards a progressive morphological and physiological dominance of the diplophase over the haplophase, which reaches its end-point in the angiosperms, is very clear. The diplophyte (e.g. the fern or the horsetail plant) produces sporangia, generally grouped in sori, on either vegetative or modified leaves (sporophylls): in each sporangium, the spore mother cells undergo meiosis to form tetrads of spores. In the isosporous species (e.g. isosporous ferns, horsetails), the spores give rise to minute free-living monoecious gametophytes (prothalli). From the fertilized egg a diploid embryo is formed, which further develops into the leafy plant (sporophyte). In the heterosporous species (e.g. heterosporous ferns, *Selaginella*, *Isoëtes*), the sporophyte produces sporangia of two different types (megasporangia and microsporangia) borne on different sporophylls (mega- and microsporophylls). From the two types of spore produced (megaspores and microspores) female and male gametophytes (mega- and microgametophyte respectively) develop: these gametophytes are even smaller than the gametophytes of isosporous pteridophytes. Indeed, they generally occupy the cavity of the original spore, whose wall was broken during development of the gametophyte.

Among phanerogams (seed plants), only two orders of gymnosperms, the Cycadales and the Ginkgoales, have ciliated motile sperm cells; all others (higher gymnosperms and angiosperms) have nonmotile sperm cells or sperm nuclei. The evolutionary change which made the fertilization process independent of a liquid milieu was the germination of the megaspores and the development of the megagametophyte *in situ*; that is, on the maternal plant (sporophyte), which, by nourishing the megagametophyte, allows fertilization and the development of an embryo within a storage and propagation organ, the seed. In gymnosperms and angiosperms, the homologue of the megasporangium of heterosporous pteridophytes is the ovule, made of vegetative tissue (nucellus) containing one or more megasporogenous cells and surrounded by one or two integuments which leave a free passage, the micropyle. The ovules are borne on the carpels, the homologues of the megasporophylls, either at their axil (naked ovules: gymnosperms) or within a cavity (pistil) made of one or more carpels (angiosperms). The homologue of the microsporophylls are the stamens in which the anther contains the microsporogenous tissue. Pistils and/or stamens, sometimes accompanied by other types of transformed leaves (e.g. calix, corolla) form the flower, the typical reproductive organ of phanerogams.

In gymnosperms, the megaspore mother cells (2n) in the ovules produce by meiosis a tetrad of megaspores (n), one of which forms a female gametophyte consisting of a haploid tissue in which one or more archegonia each with an egg cell are differentiated. Tetrads of microspores (n) are produced meiotically from microspore mother cells in the anther. The nucleus of the microspore divides to form, within the spore wall, the mature pollen grain; that is, the male gametophyte consisting of some vegetative cells and two male gametes. The pollen grains are shed from the anther and, transported by wind or some other agent, may fall on the micropyle of an ovule. Here they germinate by producing a pollen tube which elongates to transport the male gametes into the proximity of the archegonia. After fertilization, the embryo (the new sporophyte) becomes embedded in a nutritive tissue (the vegetative portion of the female gametophyte) and is protected by the seed coat derived from the ovular integument.

In angiosperms, the pollen grains germinate on the stigma of the pistil and the pollen tubes grow in the stylar tissue to reach the female gametophyte (embryo sac) in the ovule. For a review on fertilization in higher plants, reference is made to Linskens (1969). In the ovule, the nucellus bears one or, less frequently, two or more megaspore mother cells, which undergo meiosis. Following meiosis, a variety of patterns in the development of the female gametophyte can be observed. Since an extensive literature has accumulated on the nuclear cytology of the female gametophyte in angiosperms (Maheshwari, 1950; Battaglia, 1951*a*; Johri, 1963) it will be considered in some detail.

Depending upon the number of megaspore nuclei which participate in the formation of the female gametophyte, three types are distinguished:

(i) monosporic, when one of the four megaspores divides to form the gametophyte;

(ii) bisporic, when the second meiotic division is not followed by cell wall formation. Of the two dyad cells (coenocytes with two haploid nuclei each) thus formed, one develops into a gametophyte;

(iii) tetrasporic, when both meiotic divisions are not followed by cell wall formation and the four nuclei are involved in the development of the gametophyte.

The most common types of embryo sac are shown schematically in Figs. 1.5 and 1.6, which are reproduced from Battaglia (1951*a*). In Battaglia's interpretation, the reduction of the female gametophyte from gymnosperms to angiosperms is characterized by the disappearance of the archegonium and a strong reduction of the somatogenic phase; that is, the number of mitotic divisions which follow sporogenesis (tetrad formation). In angiosperm embryo sacs, the maximum number of somatogenic divisions is three (Normal or *Polygonum* type, monosporic) and the minimum is one as in the tetrasporic types *Plumbagella*, *Adoxa* and *Plumbago*. In the development of the *Polygonum* type, which is found in the majority of angiosperms, one of the four megaspores of the tetrad enlarges (vacuolization) and then undergoes the first somatogenic division to form a binucleate cell in which the two haploid nuclei are brought to occupy the cell poles (polarization). Following the second and third somatogenic divisions, a coenocyte with eight haploid nuclei is produced which undergoes cellularization (wall formation). At maturity, when the egg cell has obtained its typical shape and size (oomorphogenesis), the female gametophyte consists of an egg cell, two synergids,

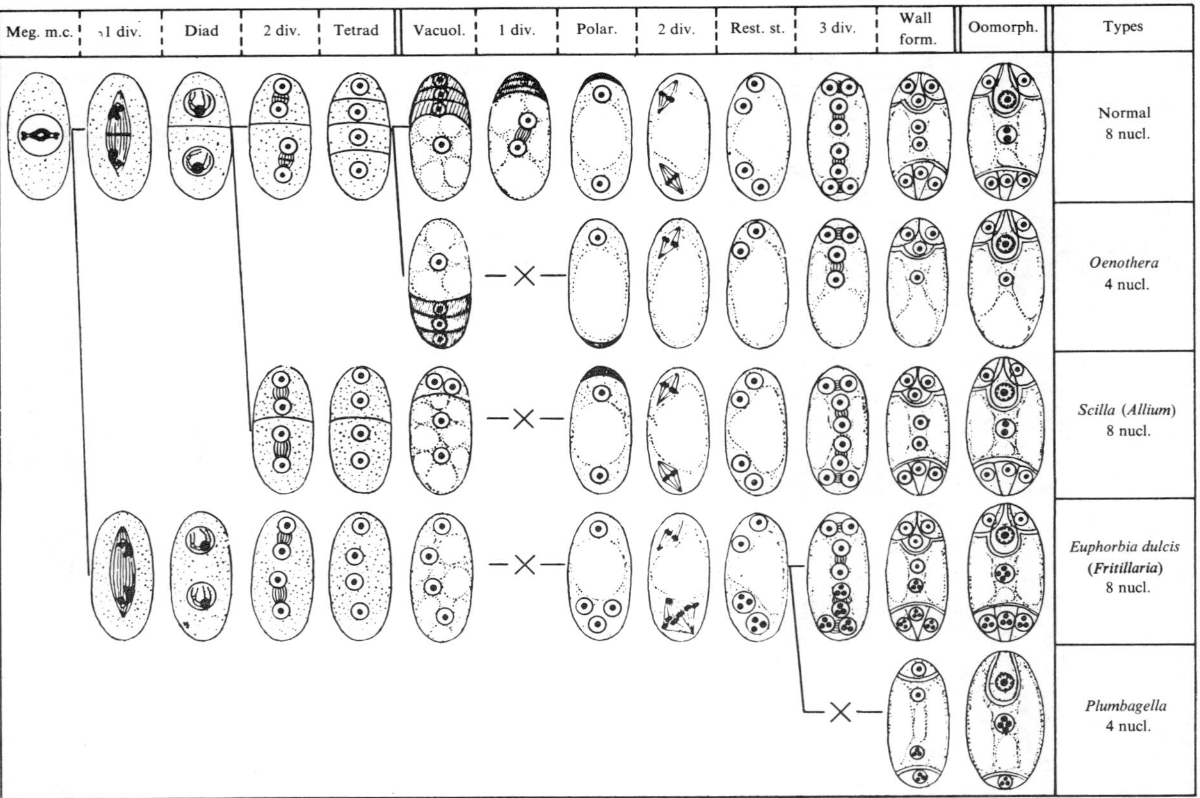

Fig. 1.5. Examples of well-known types of embryo sac development in angiosperms following the interpretation of Battaglia. The ploidy of nuclei is given by the number of nucleoli, from one (haploid) to four (tetraploid). The nucleus of the megaspore mother cell is figured with one bivalent. Vacuol: vacuolization; polar.: polarization; oomorph.: oomorphogenesis (after Battaglia, 1951 *a*).

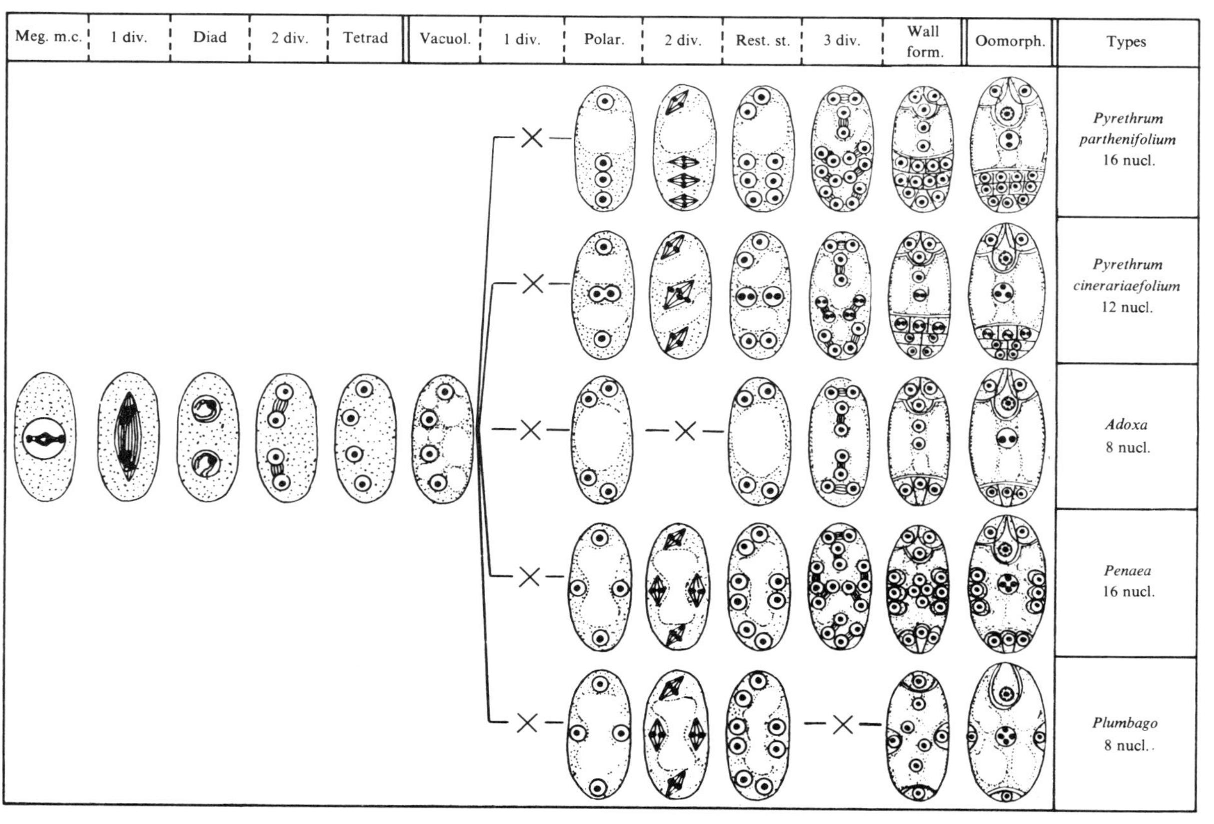

Fig. 1.6. Other examples of well-known types of embryo sac development in angiosperms following the interpretation of Battaglia. For details, see legend to Fig. 1.5 (after Battaglia 1951 *a*).

a central cell (proendospermatic cell of Chiarugi, 1927) with two polar nuclei which fuse to form a diploid secondary nucleus and three antipodals in a chalazal position. In all other embryo sac types shown in Figs. 1.5 and 1.6, the first somatogenic division does not occur; in a few tetrasporic types either the second or the third somatogenic division is lacking. An interesting aspect of the development of some tetrasporic embryo sacs is fusion of interphase nuclei or spindles. In *Euphorbia dulcis*, following a 1+3 polarization of the four megaspore nuclei, the three chalazal nuclei fuse to form a triploid nucleus; in *Fritillaria*, when the three chalazal nuclei divide they fuse their spindles into a bipolar triploid spindle. In both cases, an eight-nucleate embryo sac is formed which contains four haploid micropylar nuclei and four triploid chalazal nuclei. Nuclear fusion is also found in *Pyrethrum cinerariaefolium*, which is characterized by a 1+2+1 polarization. As pointed out by Chiarugi (1950), the fusion of the polar nuclei to produce a diploid or polyploid secondary nucleus is one of the most outstanding features in the construction of the angiosperm embryo sac. The ploidy level of the secondary nucleus is further increased by its fusion with one of the two sperm nuclei during the process of double fertilization which leads to the formation of a diploid zygote and a polyploid proendospermatic cell, which is the initial starting point of the development of a nutritive tissue, the endosperm (see below). According to Chiarugi (1950), the central polyploidization in the angiosperm embryo sac is an adaptation to a dehydration gradient starting from the micropylar and chalazal poles, which may be regarded as the zones of nutrient supply to the gametophyte; by this means, the haploid gametophytic generation is thought to correct its physiological inferiority towards the diploid sporophytic tissues in which it is immersed.

As to the male gametophyte of angiosperms, it originates from a microspore which develops mitotically into a mature pollen grain containing a vegetative cell and two sperm cells and is surrounded by a spore coat (Steffen, 1963). The microspores are produced in tetrads in each of the many microspore mother cells which form the sporogenous tissue in an anther loculus. The first microspore division ends with a typical cell wall which is unpenetrated by plasmodesmata or wider protoplasmic interconnections (Angold, 1968; Heslop-Harrison, 1968): a vegetative and a generative cell are formed. Thereafter, the division of the generative cell produces two separate sperm cells. Pollen grains are shed from the anther in a binucleate or trinucleate state. In the first case, the division of the generative cell to form the two sperm cells occurs in the pollen tube which is growing to reach the egg cell. Binucleate pollen grains characterize roughly 2/3 of angiosperms, namely 170 families; in fifty-four families, trinucleate pollen grains occur. In twenty-four other families, genera with binucleate or trinucleate pollen grains have been recorded, whereas genera in which bi- and trinucleate pollen species have been found are rare (Brewbaker, 1959; Brewbaker & Emery, 1962).

An interesting aspect of the development of the pollen grain, which has been recently clarified concerns the timing of DNA synthesis. In the classical *Trades-*

cantia material, studies on the type of radiation-induced chromosome damage at pollen tube mitosis supported the conclusion that DNA synthesis is completed and the chromosomes effectively double in the generative cell prior to pollen shedding. This was later confirmed by a cytophotometric analysis which showed that the generative nucleus of *Tradescantia paludosa* pollen was in the DNA-doubled state (2C); that is, twice the DNA content (1C) of a haploid nucleus before DNA duplication (Rasch & Woodard, 1959). In species with trinucleate pollen, two sets of observations supported the suggestion that sperm nuclei are introduced to the embryo sac in the DNA-doubled state (Brewbaker & Emery, 1962): (i) the immediacy of endosperm division in many species following fertilization and (ii) after irradiation of trinucleate maize pollen, the recovery of 'fractional' aberrations affecting only part and averaging about 50 per cent of the endosperm: a result suggesting that chromosomes in maize sperm are effectively double in the mature pollen. The above suggestions have found confirmation in a comparative DNA cytophotometric analysis on the development of the pollen grains in tobacco, a binucleate pollen species, and barley, a trinucleate pollen species (D'Amato, Devreux & Scarascia-Mugnozza, 1965): its results are schematically reported in Fig. 1.7. In tobacco, the mature pollen grain bears a vegetative and a generative nucleus both of which are in the DNA-doubled condition or 2C; in the mature barley pollen all three nuclei, the vegetative and the two sperm nuclei, are 2C. A situation identical to that of barley occurs in *Lolium*, another genus of Graminaceae, in which all seven species investigated have been shown to shed trinucleate pollen with 2C DNA content in the vegetative and the two sperm nuclei (Manthriratna & Hayward, 1973). Other recent data on DNA replication in pollen refer to four mutants of *Petunia hybrida*: in the generative nucleus, DNA synthesis was not completed at pollen shedding and only came to an end in the pollen tube; in one mutant, the vegetative nucleus of some pollen grains showed an increase over 1C of the DNA content (Hesemann, 1971). More recently, it was found, in the varieties Amsel and Wisa and the two intervarietal F_1 hybrids of barley, that both the vegetative and the two sperm nuclei in mature pollen grains had a DNA content greater than 1C but lower than 2C (Hesemann, 1973).

These observations on DNA replication in the pollen grain are important for two reasons:

(i) DNA replication in the vegetative nucleus, a nucleus which does not undergo mitosis, must be taken as evidence of functional activity, in contrast with the view of some authors (e.g. Brewbaker & Emery, 1962) that the vegetative nucleus is degenerative and with no important function in the later growth of the pollen grain. In *Tradescantia paludosa*, the vegetative nucleus has been shown to be very active in RNA synthesis, although it has not been determined what fraction of it is mRNA, rRNA or tRNA (Mascarenhas, 1966). Duplication of the genome (2C DNA content) and/or selective replication of some portions of the genome (DNA amplification) in the vegetative cell nucleus are both mechanisms which increase the availability of templates which might be needed for an increased synthesis of

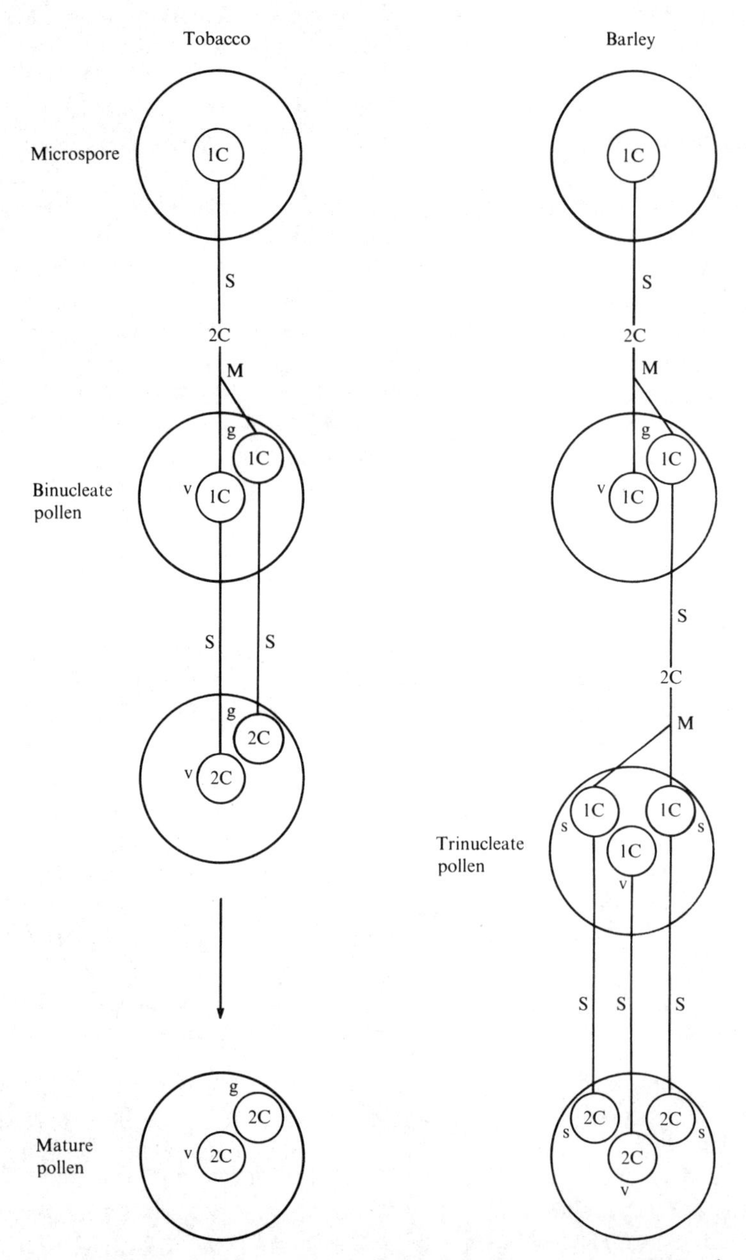

Fig. 1.7. Schematic representation of the development of mature pollen grains from microspores in tobacco and barley. 1C: nuclear DNA content before replication; S: DNA synthesis; 2C: nuclear DNA content at the end of S; M: mitosis; v, g, s: vegetative, generative and sperm nuclei respectively. Based on the observations of D'Amato *et al.* (1965).

some types of RNA necessary for the development of the pollen grain and/or the pollen tube. Investigations on this aspect using the presently available techniques of molecular biology would be highly desirable;

(ii) in barley and *Lolium* the sperm nuclei participate in fertilization in the DNA-doubled (2C) state. Since in these species, as well as in other trinucleate pollen species, the interval between fertilization and division of the endosperm nucleus and of the zygote is rather short, it seems probable that both the egg nucleus and the secondary nucleus of the embryo sac have undergone DNA duplication before being fertilized. However, this assumption needs confirmation by direct DNA measurement.

It seems probable that the participation in fertilization of gametes which have doubled their DNA before fusing may be a more general phenomenon than generally appreciated. Evidence, or indications, that gametes or sexually competent nuclei have doubled nuclear DNA before karyogamy is available for the male and female pronuclei of some mammals (Austin & Amoroso, 1959; Oprescu & Thibault, 1965), of the sea urchin (Hinegardner, Rao & Feldman, 1964) and of the grasshopper *Melanoplus differentialis* (Swift & Kleinfeld, 1953), in the egg nucleus of the fern *Pteridium aquilinum* (Bell, 1960), in the sperm of the liverwort *Sphaerocarpos donnellii* (Reitberger, 1964) and in the prefusion nuclei in the crozier of the ascomycetes *Neottiella rutilans* and *Neurospora* (Rossen & Westergaard 1966; Westergaard & Wettstein, 1968, 1972).

Haplo-dikaryotic organisms

In the higher fungi, the Euascomycetes and the Basidiomycetes, an integral part of the life cycle is a stable association of paired haploid nuclei (dikaryophase), which is propagated vegetatively. The dikaryophase is the substitute for a vegetative diplophase and must be regarded as the genetic and physiological equivalent of the diploid (Raper & Flexer, 1970). In the haplodikaryotic life cycle, there occurs an alternation of haploid and dikaryotic phases of varying durations; the dikaryophase ends with the fusion of two haploid nuclei into a single diploid cell which undergoes meiosis to form spores. Although basically similar, the life cycles of the Euascomycetes and Basidiomycetes show clear differences. In the Euascomycetes, the vegetative phase which originates from the spore (haplophase) dominates in the life cycle; the dikaryotic condition is initiated by fusion of haploid cells produced by differentiated sexual organs. During fertilization within the female organ (ascogonium) the nuclei from both mates do not fuse but become associated in pairs in a common cytoplasm (plasmogamy). These dikarions propagate through a process of synchronous division of the paired nuclei and their movement into dikaryotic filaments (ascogenous hyphae) which grow out of the ascogonium; thus the whole of the ascogenous hyphae (dikaryophase) is an obligate parasite of the maternal haploid mycelium. Following fusion of nuclei (karyogamy) in one of the dikaryotic cells of the ascogenous hypha, the diploid

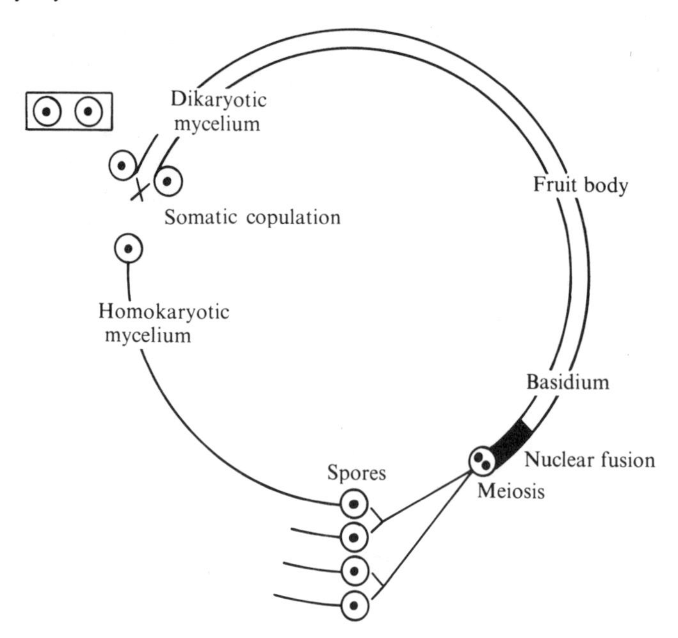

Fig. 1.8. Diagram of a haplo-dikaryotic life cycle (higher Basidiomycetes). In species with self-sterile homokaryotic mycelia, two mycelia of compatible mating types are needed to produce the dikaryon; in species with self-fertile mycelia, one homokaryotic mycelium produces the dikaryon. In this diagram the dikaryophase is represented by two parallel thin lines and the outline of a cell with two haploid nuclei. For the other schematic representations, see legend to Fig. 1.4.

nucleus undergoes meiosis immediately. As a rule, each sporangium (ascus) contains eight spores (ascospores), because meiosis is followed by one mitosis.

In the Basidiomycetes, the dikaryophase, which can live independently of the original haploid mycelium, is the dominating phase in the life cycle. Since this class of fungi produces no specialized sex organs, cells of ordinary vegetative hyphae have assumed the function of plasmogamy. In the life cycle, each spore (basidiospore) germinates to produce a haploid vegetative mycelium made of uninucleate cells with identical genotype (homokaryon). Homokaryotic mycelia give rise to dikaryotic mycelia. In about 10% of the species studied so far, dikaryotic mycelia originate from a single homokaryotic mycelium (self-fertility); in the majority of species, two homokaryotic mycelia of compatible mating types are needed for dikaryotization, which is achieved by nuclear migration from one cell of a mycelium (donor) to one cell of the other mycelium (acceptor). The dikaryotic mycelium consists of binucleate cells with a characteristic appendage, the clamp connection, at each septum. In the dikaryotic mycelium, the final act of the sexual cycle, karyogamy, produces a diploid cell which immediately undergoes meiosis to form a sporangium (basidium) with four spores. The life cycle of higher Basidiomycetes is shown in Fig. 1.8.

TABLE 1.1. *Distribution of life cycles in eukaryotic organisms* (from Raper & Flexer, 1970)

Organisms	Haploid	Haplo-diploid	Haplo dikaryotic	Diploid
Protozoa	×	×		×
Higher animals				×
Myxomycetes		×		
Algae				
Green	×	×		
Red	×	×		
Brown		×		×
Diatoms				×
Golden	×			
Fungi				
Uniflagellate water molds	×	×		× (rare)
Biflagellate water molds	?			×
Zygomycetes	×			
Hemiascomycetes	×	×		
Euascomycetes			×	
Basidiomycetes			×	
Embryophyta				
Mosses and liverworts		×		
Ferns, etc.		×		
Seed plants		×		(effectively → ×)

Diploidy versus haploidy in life cycles

From a survey of the distribution of life cycles in eukaryotes (Table 1.1) it is apparent that the universal presence of diploidy in the more highly advanced forms argues strongly that the diploid condition confers significant biological advantages (Raper & Flexer, 1970). As pointed out by these authors, the difference in genetic endowment between diploids and haploids has several consequences, some of which confer obvious advantages, others are of questionable significance. At the population level, two obvious advantages of diploidy, with its dual set of genes, are a wider range of accommodative responses to unfavorable conditions and a tremendous increase in the plasticity and potential variability of the organism.

In contrast with animals, in which diploidy became fixed early in evolution, plants have approached effective diploidy in a long evolutionary trend which marked the progressive reduction of the haploid (gametophytic) phase in a diplo-haploid life cycle. The essential features of this evolution, which established the primacy of the land plants through the progressive dominance of the diplophase, were the anchorage to the soil of the diplophyte and achievement (by means of apical meristems) of a system of unlimited growth which tremendously increased the contact of the plant with both water and atmosphere. In the pteridophytes dominance of the diplophase is clearly expressed in the differentiation of a variety of tissues and in the habit of unlimited growth. Because of the difficulty – sometimes the impossibility – of finding gametophytes (prothalli) of ferns, lycopods etc. in nature and the rhizomatous habit of the diploid plants (sporophytes), these species today depend more upon vegetative than upon sexual reproduction (Wetmore *et al.*, 1963). Although the development of a gametophytic generation in the life cycle of a pteridophyte would seem not to be essential for the survival of the species, it serves the scope of exploiting the variation produced by genetic segregation and recombination at spore formation. Most probably, the independent life of the prothallus, by greatly increasing the probability of death of individuals, strongly reduces the exploitation of genetic variation. This situation radically changed in evolution when in a new type of plant, the seed plant, the female gamethophytes were made to develop on the diplophyte, the maternal plant; this allowed full exploitation of genetic variation. For this reason, diploidy in the diplo-haploid life cycle of a seed plant has the same genetic value as diploidy in the diploid life cycle of an animal.

Alternation of diploid and haploid generation in the diplo-haploid life cycle of plants raises the problem of the relations of the morphological and physiological characteristics of a generation to its genomic constitution. In life cycles with alternation of homomorphic generations (some algae), the diploid nucleus must be conditioned to react like a haploid nucleus; in vascular plants, the two alternating generations are strikingly different, but these differences are by no means related to genomatic constitution. In this context, the experimental work done on ferns is highly instructive. When excised young leaves of *Todea barbara* were cultured *in vitro* in a medium with a low sucrose concentration, they differentiated prothalli, diploid like the original explant, which produced upon fertilization tetraploid plants. Repetition of the process allowed differentiation *in vitro* of tetraploid prothalli which gave upon fertilization octoploid embryos. Thus, in *Todea barbara* a series of haploid, diploid and tetraploid prothalli and of diploid, tetraploid and octoploid young sporophytes was made available; prothalli of different ploidy were phenotypically indistinguishable as were the young sporophytes at different ploidy levels (Wetmore *et al.*, 1963). These results are not restricted to *Todea*, but can be repeated in other fern species (Bristow, 1962; Wetmore *et al.*, 1963). Both in *Todea* (Wetmore *et al.*, 1964) and in *Pteris cretica* (Bristow, 1962), control over the type of differentiation is exerted by sucrose: at low concentration, or in the absence, of sucrose, gametophytes differentiate; at higher sucrose levels, sporophytes are formed. If we now turn to the situation in nature, we see that in ferns the zygote, or the diploid egg in the case of apomictic species, develops within the archegonial jacket into a massive three-dimensional structure, the embryo; by contrast, the unenclosed spores, which develop on a surface, give rise to a two-dimensional structure, the prothallus. Obviously, the determination of the prothallial or the sporophytic pattern is not a commitment because of DNA differences in spore and zygote, but is strictly dependent on the milieu in which spore and zygote develop (Wetmore *et al.*, 1963).

Questions of a similar nature are raised by an analysis of the alternation of morphologically and physiologically distinct forms in the diploid life cycle of some animals as exemplified by pedogenesis in the sheep liver fluke and metamorphosis in insects. Actually, in the liver fluke the three larval stages (miracidium, redia and cercaria) and the adult possess the same set of genes; which means that a single genome is able to express several patterns of genetic potentialities in response to external influences, selection pressure, physiological adaptation, etc. Obviously, control of gene action in a genome is the basis for the expression of genetic potentialities of a cell. In view of its importance in development and differentiation, this question will be explored more fully in another chapter.

Unusual and anomalous forms of fertilization

Time of sperm entry into egg cytoplasm

In many groups of plants and animals, natural selection has favoured the establishment of different modes of development of new individuals in the absence of gamete fusion (amphimixis) which are described under the collective terms of apomixis and parthenogenesis. Since apomixis and parthenogenesis are very frequently bound to modifications or suppression of the meiotic process, they will be treated in detail, because of their importance in development, after analysis of the course of meiosis and of its genotypic control (Chapter 3). In this chapter, it seems pertinent to consider briefly some unusual and anomalous forms of fertilization which may occur sporadically in plants and animals. In contrast with flowering plants, in which the sperm enters the egg cytoplasm when the egg is mature (Linskens, 1969), sperm penetration occurs at different stages of egg maturation in different animals (Table 1.2). In sea urchins and coelenterates, the sperm penetrates into the egg at the end of the meiotic divisions (maturation divisions) which have produced an egg with a haploid pronucleus and two or three haploid polar bodies (the two mitotic derivatives of the first polar body and the

TABLE 1.2. *Stage of egg maturation at which sperm penetration occurs in different animals* (from Austin, 1965)

Young primary oocyte	Fully grown primary oocyte	First metaphase	Second metaphase	Female pronucleus
Brachycoelium	*Ascaris*	*Aphryotrocha*	*Amphioxus*	Coelenterates
Dinophilus	*Dicyema*	*Cerebratulus*	Most mammals	Echinoids
Histriobdella	Dog and fox	*Chaetopterus*	*Siredon*	
Otomesostoma	*Grantia*	*Dentalium*		
Peripatopsis	*Myzostoma*	Many insects		
Saccocirrus	*Nereis*	*Pectinaria*		
	Spisula			
	Thalassema			

second polar body). In an extreme case, that of some annelids and plathyhelminths, the spermatozoon enters the young primary oocyte, in whose cytoplasm it remains unchanged until the oocyte completes maturation. In general, however, the time of sperm entry is closely associated with that of maturation, whose completion seems to be conditioned by the presence of a sperm in the egg cytoplasm. Thus meiosis, for example, proceeds spontaneously to either the first or the second metaphase and is held at these points pending sperm penetration (Austin, 1965). The need for sperm entry for egg maturation has also been demonstrated in species or biotypes of animals developing by pseudogamic parthenogenesis; for example, in the planarians (Benazzi-Lentati, 1970).

Polyspermy and polyandry

In animals with large yolky eggs (birds, reptiles and elasmobranchs) or with moderately yolkladen eggs (urodeles, insects) several spermatozoa enter an egg, but only one male gamete fuses with the female pronucleus. To this physiological polyspermy a pathological polyspermy may be contrasted; namely, the case in which the female pronucleus fuses with two or more male gametes (polyandry). In mammals, as in most animal groups, the zona pellucida of the egg is generally penetrable to only one spermatozoon, but in aged eggs the block to polyspermy does not operate fully. Thus, in rats, pigs and rabbits, mating late in oestrus can increase the rate of polyspermy 5 to 16 times over the spontaneous value (Austin, 1960, 1965; Hunter, 1967). Polyandry may lead to the production of triploid and tetraploid embryos in many animal species. These embryos, like the polyploid embryos resulting from polygyny (see below), generally die before completing development. However, experimentally produced triploid toads, frogs and newts developed through metamorphosis. In the fowl, survival of triploids is sometimes possible. Ohno, Kittrell, Christian, Stenius & Witt (1963) described a fully grown hen with cock's comb and hackles with a 3A+ZZ chromosome complement. More recently, Donner, Chyle & Sanerova (1969) have described four triploid individuals (three embryos and a chicken), all males (ZZZ). This finding strongly supports the origin of these individuals by polyandry (note that the homogametic sex in the fowl is the male).

Polyspermy, the discharge of more than one pollen tube into an embryo sac, has been reported in some angiosperms and fusion of the egg nucleus with several male gametes has been observed (Linskens, 1969). But polyandrous development does not seem to have ever been ascertained in plants. Polyspermy in an embryo sac generally leads to consequences other than polyandry; namely, additional embryo(s) due to fusion of sperm with a synergid and/or an antipodal cell (polyembryony).

Polygyny

Fertilization involving one male pronucleus and two female pronuclei to produce triploid embryos is much rarer than polyandry under natural conditions. The rate of polygyny is, however, much increased, as in the case of polyandry, by insemination of aged eggs (Austin, 1960). Polygyny arises through suppression of the first or second polar body: at anaphase I or at anaphase II the two groups of chromosomes are retained within the egg.

Aneugamy

Aneugamy (Austin, 1960) occurs when fertilization involves a diploid (unreduced) gamete: namely, male 2n gamete+female n pronucleus or vice versa. Aneugamy is very rare. Consequently, most of the triploid embryos in animals are due to polyandry or polygyny (Austin, 1969).

Androgenesis

Androgenesis, the development of a haploid embryo from a male gamete which divides in the cytoplasm of an egg whose pronucleus has degenerated (or for some reason has not fused with the sperm), is an exceptional event in nature (for a review on experimental androgenesis in animals, see Beetschen, 1973). In the axolotl, using recessive markers in the male parent, Humphrey & Fankauser (1957) found that 0.074% of the eggs developed into haploid androgenetic larvae which, like the haploid parthenogenetic larvae, died before metamorphosis. In crosses between two species of silkworm, *Bombyx mandarina* and *B. mori* a very few eggs hatched and gave fertile diploid androgenetic males. These were due to the division in the egg cytoplasm of the fusion product of two male nuclei in a polyspermic egg (Astaurov & Ostriakowa-Varshaver, 1957). In both axolotl and *Bombyx* hybrids, the frequency of androgenesis was increased by temperature shocks; in *Bombyx* is was also increased by X-irradiation of fertilized eggs.

Androgenetic haploids in plants have been obtained following interspecific hybridization or by pollinating irradiated plants with unirradiated pollen. As in animals, spontaneous androgenetic haploids may occur in the progeny of some plant species (Kimber & Riley, 1963). True androgenesis, that is, development from male germ cells has been achieved in plants through culture of anthers or pollen grains *in vitro* (Chapter 9).

Partial fertilization

In his classical work on fertilization in sea urchins, Boveri (1888–9) found that sometimes the sperm does not immediately fuse with the egg pronucleus. Instead, the latter divides and of the resulting two blastomeres one contains one of the division products of the egg pronucleus only whereas the other contains the other division product and the sperm nucleus. Following fusion of this nucleus with the egg-derived nucleus in its blastomere, the egg becomes a mosaic of unfertilized (haploid) and fertilized (diploid) nuclei. This type of fertilization, which Boveri (1888) called partial fertilization, may be occasionally responsible for the formation of haploid–diploid gynanders in bees (Chapter 4).

2

The cell cycle

The cell nucleus in eukaryotes

The cell nucleus is the carrier of heredity in the form of DNA molecules and is the production site of the various types of RNA needed for the synthesis of all proteins.

In eukaryotic cells, the nuclear constituents are enclosed in an envelope that consists of two distinct parallel membranes which are regularly interrupted by pores about 400–1000 Å in diameter. The two membranes are 70–80 Å thick and have the unit membrane structure typical of the lipoprotein membrane systems of the cytoplasm known as endoplasmic reticulum. Indeed, in some materials, it has been shown that the nuclear envelope is formed at telophase by elements of the endoplasmic reticulum and at interphase the outer nuclear membrane is in direct connection with the endoplasmic reticulum; this allows a spatial distribution of functional activities in the cytoplasm (Wischnitzer, 1973). This condition strikingly contrasts with that of the cell in protokaryotes (bacteria and blue-green algae) which lack both a typical membrane-bounded nucleus and a demonstrable endoplasmic reticulum (Porter, 1963).

From isolated eukaryotic nuclei, the substance constituting the chromosomes (chromatin) can be isolated; it is a complex of DNA with histone* and nonhistone proteins as well as a particular type of RNA (chromosomal RNA) (Bonner, Dahmus, Fambrough, Huang, Marushige & Tuan, 1968). It is generally accepted that in most, if not all, eukaryotes the amount of nuclear DNA is disproportional to the number of informational genes that the organism can carry; this implies the presence of some excess DNA sequences with other, possibly structural and/or regulatory, functions. This genome complexity, now known for a number of animal and plant species, was first demonstrated by an analysis of the kinetics of the renaturation reaction carried out on isolated DNA following denaturation (Britten & Kohne, 1968). Take as an example the extensively studied genome of the mouse: about 70% is made of unique (mostly 1-copy) DNA, about 20% of related but not identical DNA sequences with 10^2 to 10^5 copies each, and about 10% of very similar sequences of about 130 nucleotide pairs of repeat units 7 or 8 base pairs long and are present in several 10^6 copies. This mouse DNA is a light, adenine–thymine (A–T) rich (34% guanine–cytosine, G–C, content) satellite DNA

* Histone proteins do not occur in the chromosomes of fungi (Leighton *et al.*, 1971) and dinoflagellates (Soyer, 1971).

which can be separated from the main DNA (42% G–C content) by ultracentrifugation (Southern, 1960, 1975; Walker, 1971). The localization of the satellite DNA in the mouse genome has been clearly defined, at the level of both interphase nuclei and metaphase chromosomes, by means of autoradiography following two different techniques: (i) molecular hybridization of DNA of cytological preparations with radioactive RNA transcribed *in vitro* from the satellite DNA, and (ii) annealing of DNA of cytological preparations with radioactive single-stranded satellite DNA. The mouse satellite is mainly located at the centromere regions of all chromosomes, the Y excluded: all these regions represent constitutive heterochromatin (Jones, 1970; Pardue & Gall, 1970; Rae & Franke, 1972). In other animal and plant species, in addition to the unique DNA sequences which are generally regarded as informational sequences, DNA sequences with an intermediate and high degree of repetition and redundancy are present (Walker, 1971; Ingle *et al.*, 1975*b*). All highly repetitive DNAs renature very rapidly following denaturation. In species where highly repetitive DNAs have been analyzed, e.g. mouse, guinea pig (Southern, 1970), *Drosophila virilis* (Gall, 1973), species of crabs (Beattie & Skinner, 1972), they have been found to be composed of repeating units a few nucleotide pairs in length. Highly repetitive DNAs have been localized by hybridization *in situ* in centromeric heterochromatin, and in other regions of constitutive heterochromatin in different species, e.g. *Drosophila* (Jones & Robertson, 1970; Rae, 1970; Botchan, Kram, Schmid & Hearst, 1971; Gall, Cohen & Polan, 1971), *Rhynchosciara* (Eckhardt & Gall, 1971), *Triturus* (Barsacchi & Gall, 1972) and *Microtus agrestis* (Arrighi, Hsu, Saunders & Saunders, 1970). On the contrary, the A–T rich satellite DNA of *Xenopus mulleri* shows no homology to centromeric heterochromatin: it hybridizes to discrete bands in most of the chromosomes (Pardue, 1974).

As to moderately repetitive DNA sequences, there is now a substantial amount of evidence, based on biochemical and electron microscope analyses, that unique and moderately repetitive sequences alternate over much of the genome in different species (review in Swift, 1974).

Both in heterochromatin and euchromatin, DNA is complexed with histones. Histones are small (102 to about 220 component amino acids) basic proteins of which there are only five major fractions (Table 2.1) with striking similarities in composition between plants and animals. Each of the five fractions possess some microheterogeneity resulting from small differences in size and charge (Johnson, Douvas & Bonner, 1974). Histones combine strongly with the DNA phosphates through their cationic residues. In different tissues the mass ratio of total histone to DNA ranges from 0.70 to 1.30, which is below the stoichiometric ratio (1.35) of one basic amino acid residue to one DNA phosphate (Bonner *et al.*, 1968). Hence, not all DNA phosphate is bound to histone. Histones are regarded, for the most part at least, as general repressors of gene activity, masking DNA for RNA transcription unspecifically. There is evidence that the interaction of histones with DNA does not employ a particular group of DNA sequences, all DNA

TABLE 2.1. *Characteristics and nomenclature of the five principal fractions of histone* (data and references in Johnson *et al.* 1974)

Class	Nomenclature	Lysine/ arginine	Total residues	Molecular weight
Lysine-rich	I or f 1	20	218	21000
Slightly lysine-rich	IIb1 or f 2a2	1.25	129	14000
	IIb2 or f 2b	2.5	125	13774
Arginine-rich	III or f 3	0.67	135	15324
	IV or f 2a1	0.79	102	11282

sequences being equally suitable for binding of histones (Zimmerman & Levin, 1975).

Chromosomes also contain a complex of acidic (nonhistone) proteins of which 50 to 100 species have been detected by electrophoresis; they range in molecular weight from under 10000 to over 150000. Since acidic proteins are difficult to remove from chromatin, they are also known as residual proteins. The nuclear content of acidic proteins varies from tissue to tissue: in some tissues the acidic protein content varies inversely with histone content. The nonhistone proteins are currently regarded as regulators of gene expression in eukaryotic cells (Johnson *et al.*, 1974).

In addition to chromosomal proteins, another class of molecules, chromosomal RNA, has been postulated to act as derepressors of DNA (Frenster, 1969). Chromosomal RNA is a special class of RNA molecules which are on the one hand firmly bound to chromosomal proteins and on the other hand bound in native chromatin in RNase resistant form. The DNA sequences to which chromosomal RNA binds are middle repetitive ones. The average chain length of chromosomal RNA is 40–50 nucleotides (3S). (Bonner *et al.*, 1968; Patel & Holoubek, 1973; Bonner, 1975). Although it has been proposed that chromosomal RNA may be an artifact resulting from degradation of tRNA (Heyden & Zachau, 1971) or that it may not exist at all (Artman & Roth, 1971), there is evidence that it is a special class of RNA (Holmes, Mayfield, Sander & Bonner, 1972; Patel & Holoubek, 1973).

In addition to DNA, proteins and chromosomal RNA, isolated chromosomes have been found to contain probably lipids and certainly several ions, mostly of calcium and magnesium, which are considered to play an important role in the structural integrity of chromosomes (Steffensen, 1959).

Chemical analysis of cells and isolated nuclei also reveals several types of RNA transcribed on different nucleotide sequences of DNA (Sirlin, 1972). The better known among these are the following:

(i) labile RNA, also known as heterogeneous or polydisperse RNA, shows heterogeneity in size, a DNA-like base ratio and an apparent lack of secondary structure. Because these features are also found in messenger RNA (mRNA), some labile RNA is believed to be mRNA;

(ii) ribosomal RNA (rRNA) constitutes up to 90% of the total cellular RNA and is mostly found in cytoplasmic ribosomes; it has considerable secondary structure. The two major rRNA species originate from separate sequences in a common precursor; they have a sedimentation value (in Svedbergs) of 25S and 28S the large rRNA species (in plants and mammals respectively) and of 18S the small rRNA species (both plants and animals). The approximate number of nucleotides in the two molecules is calculated to be 5000 and 2000 nucleotides respectively. The small rRNA species (5S) contains about 120 nucleotides and is intimately associated with the large ribosomal subunit. 5S RNA occurs in equimolar amounts with the large rRNAs;

(iii) transfer RNA (tRNA) mediates transfer of amino acids to the nascent peptide, contains about 75 ∓ 5 nucleotides and has a secondary structure with a cloverleaf arrangement, which allows 60% base pairing:

(iv) mRNA has the unique role of conveying the genetic information in DNA to the protein synthesizing apparatus of the cell. Messenger RNA, which is the RNA most sensitive to endogenous nucleases, seems to be associated with non-ribosomal protein from its initial formation in chromatin to its final destination in polysomes. Association of mRNA with proteins in particles known as informosomes probably serves the scope of stabilizing mRNA against chemical instability and enzymatic degradation.

Patterns of the cell cycle

In multicellular organisms, development starts from a single cell (e.g. a zygote, a spore, a conidium) which undergoes a series of mitotic divisions. In highly differentiated forms, cells are integrated at supracellular level in tissues and organs making a functional whole, the organism. During development and growth, four categories of proliferating cell populations may be distinguished (John & Lewis, 1969):

(i) cleaving populations in early embryos and persistent meristems in plants;

(ii) differentiating populations in embryonic tissues;

(iii) renewal populations such as those in skin, intestinal epithelia and the cork cambium of plants;

(iv) germ line populations.

Before exploring in some detail the dynamics of proliferating cell populations, we need to consider the essential features of the cell cycle (for an extensive review on the subject, see Mitchison, 1971). Knowledge in the field has much advanced in the last two decades as a result of the application of three important techniques in cell biological research. One was the development of methods for the induction of synchronous divisions in cell cultures, which opened the way to a biochemical analysis of the different phases of the cell cycle. The second technique was the use of microspectrophotometry in the quantitative determination, in individual

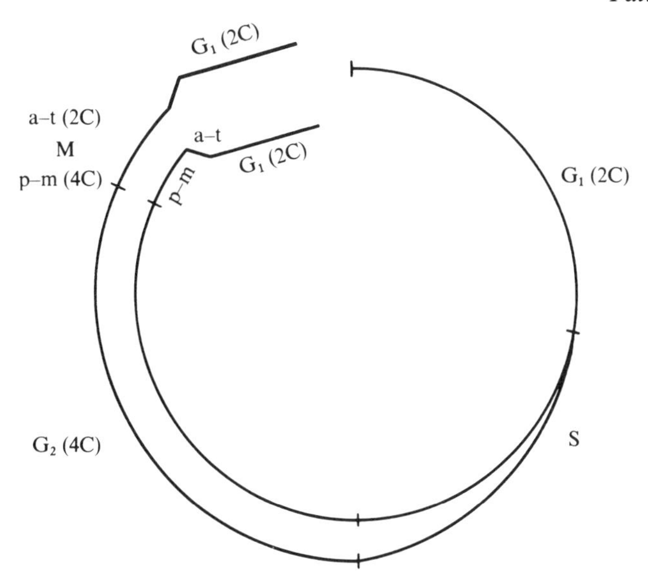

Fig. 2.1. Diagram of the diploid cell cycle illustrating the case where both daughter nuclei re-enter the cycle (i.e. behave as G_1). G_1: pre-DNA synthesis phase; S: DNA synthesis; G_2: post-DNA synthesis phase; M: mitosis; p, m, a, t: prophase, metaphase, anaphase and telophase respectively. Nuclear DNA content is reported in parenthesis.

cells, of different cell constituents subjected to specific staining reactions. The third technique was the introduction of autoradiography following feeding of cells with compounds containing radioactive isotopes, the nuclides used being first radiophosphorus (^{32}P) and later tritium (^{3}H). By the early fifties these techniques had rapidly expanded our knowledge on the cell cycle; especially on interphase, whose analysis had previously been of necessity restricted to determinations of cellular and nuclear volume changes as indicators of interphasic growth (Jacobj, 1925; Schreiber, 1949–50). From a study on ^{32}P incorporation into nuclear DNA, Howard & Pelc (1953) concluded, in agreement with previous investigators who had used UV-absorption and Feulgen-cytophotometry (Walker & Yates, 1952; Pätau & Swift, 1953), that in higher cells the DNA doubles during a restricted period of interphase. They also introduced a nomeclature of the cell cycle which is now universally accepted: the phase of DNA synthesis (S) is preceded by a gap (G_1, the first gap) in which no DNA synthesis occurs and is followed by a second gap (G_2) in which there is also no DNA synthesis. After G_2, the nucleus enters mitosis (M) to produce two daughter nuclei each of which, if dividing again, represents the new G_1 (Fig. 2.1).

Leaving aside exceptions, some of which will become apparent when dealing with proliferating cell populations, the sequence G_1–S–G_2–M can be taken as the typical phase sequence in the mitotic cycle of animals and plants.

G_1 phase

The G_1 phase is the most variable phase in the cell cycle: the faster the cell division rate the shorter the phase. In some cell types, G_1 may be lacking: in other cell types, with a very low mitotic rate, it lasts for most of the interphase duration (references in Fossati-Tallard, 1967; Mitchison, 1971). From the start of G_1 to the end of interphase all major RNA species are simultaneously and continuously synthesised (Bello, 1968; Pfeiffer, 1968). As a consequence of the initial burst of rRNA synthesis at early G_1, the nucleolus – which has reformed at the telophase of the previous division – starts to increase in size. Since at G_1 the decision is taken for the nucleus either to divide again or to differentiate, the G_1 cell is synthetically very active. In case the G_1 nucleus proceeds in the cell division cycle and enters DNA replication, proteins are required for the initiation of replication. This synthesis of proteins *de novo*, which occurs in G_1, provides the necessary enzymes mediating DNA biosynthesis, some possible 'initiator protein' and other proteins as well (references in Barlow, 1972*a*). Since the major quantity of protein synthesized during G_1 accumulates in the cytoplasm (Zetterberg, 1966) it seems probable, as indicated by the experimental evidence, that DNA synthesis is not initiated until the nucleus is surrounded by a certain amount of cytoplasmic protein. In addition to cytoplasmic factors, the physicochemical properties of the deoxyribonucleoprotein complex plays an important role in the initiation of DNA replication (Zetterberg, 1970; Mitchison, 1971; Sirlin, 1972).

S phase

It has been suggested by Comings & Kakefuda (1968) that DNA replication is initiated at the nuclear membrane. Using electron microscope autoradiography after a short-duration treatment (pulse) with ^{3}H-thymidine of synchronized human cells in culture, they observed that in cells at the start of S the radioactivity was restricted to the inner nuclear membrane and the periphery of the nucleolus. More recently, experiments by others have shown the association of the newly synthesized DNA with the nuclear membrane, but there is controversy as to whether this association occurs throughout S, or at the beginning or the end of S (references in Barlow, 1972*a*). Some authors even disagree that for replication to occur DNA or chromatin needs to be in association with the nuclear envelope (Fakan, Turner, Pagano & Hancock, 1972; Oppenheim & Wahrman, 1973). Once initiated, DNA synthesis can be studied at the chromosome level by analyzing autoradiographs of metaphase chromosomes collected at different times after a pulse treatment of dividing cells with ^{3}H-thymidine: the sequence of chromosome labeling reflects the sequence of DNA replication as it occurred in the preceding S phase. An analysis of this type was first made on root tip chromosomes of *Crepis capillaris* by Taylor (1958) who found that towards the end of the S phase, more DNA was synthesized around the centromeres than at the distal ends of the

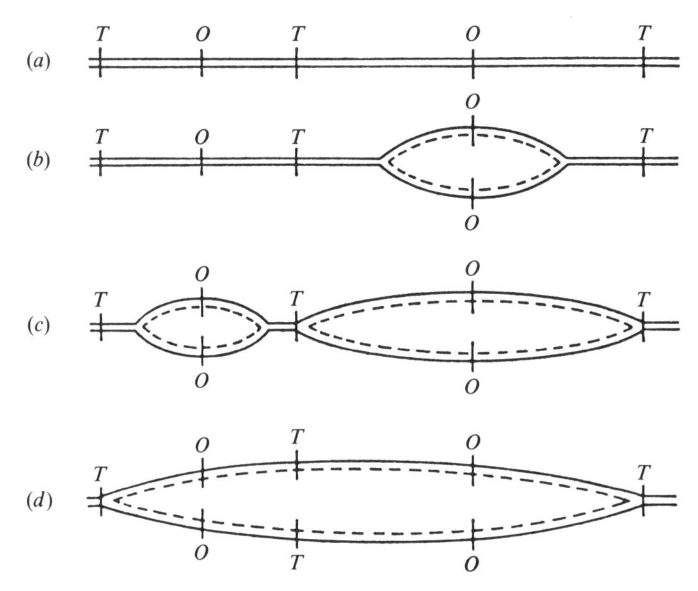

Fig. 2.2. Bidirectional model for DNA replication. Each pair of horizontal lines represent a double helical DNA molecule (——, parental chain; - - - -, newly synthesised chain). Position of origins (*O*) and termini (*T*) are shown by vertical lines. (*a*) stage prior to replication; (*b*) replication started in right-hand unit; (*c*) replication started in left-hand unit and completed at termini of right-hand unit; (*d*) replication completed in both units (after Huberman & Riggs, 1968).

chromosomes. Taylor (1958) assumed that DNA replication started at the chromosome ends and proceeded zipper-like towards the centromeres. Three years later, Wimber (1961) described in *Tradescantia paludosa* a reverse pattern to that in *Crepis*: the chromosome regions most heavily labeled at the end of S were the distal ends of the chromosomes. More recent investigations have demonstrated that chromosomal DNA consists of many units of DNA replication, or replicons, in which replication is asynchronous and may occur in opposite direction (Fig. 2.2) (Huberman & Riggs, 1968; Lark, Consigli & Toliver, 1971; Kriegstein & Hogness, 1974). This mode of DNA replication in eukaryotes raises the question whether each chromosome contains more than one DNA molecule or is made by only one DNA molecule, in which replication starts at many points (review in Du Praw, 1970). Evidence that, in four species of *Drosophila* including *D. melanogaster*, the DNA molecules must run the length of the chromosome and cannot be discontinuous at the centromere has been presented by Kavenoff & Zimm (1973). By using a new technique, the viscoelastic technique, they demonstrated the presence of molecules of a size commensurable with that of the chromosomes (molecular weights from 20×10^9 to 80×10^9).

Asynchrony in initiation and completion of DNA synthesis is now well docu-

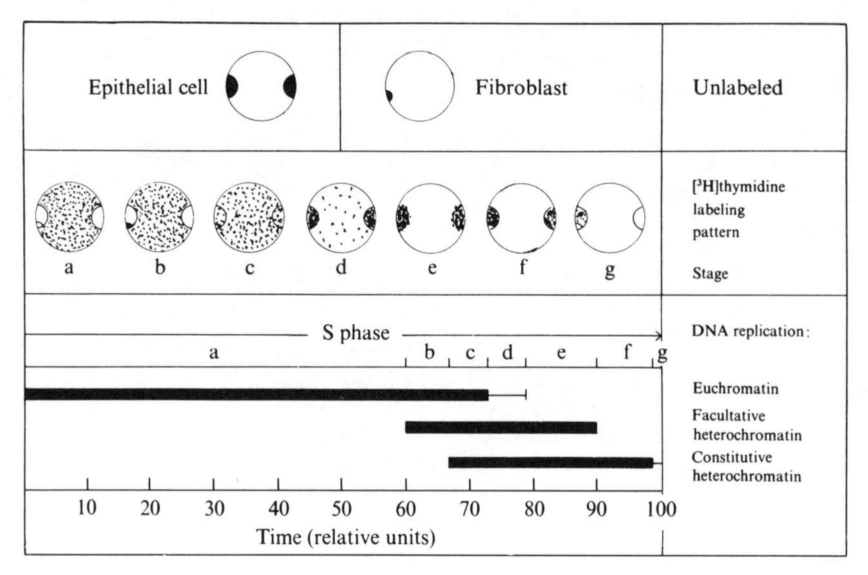

Fig. 2.3. Schematic representation of DNA synthesis in female *Microtus agrestis*. Above: nucleus of ephithelium cell with two large chromocenters (constitutive heterochromatin) and nucleus of fibroblast with sex chromatin (facultative heterochromatin), unlabeled. Middle: pattern of labeling by ^{3}H-thymidine of epithelium cells and fibroblasts in different stages of the S phase. Below: duration (relative units) of DNA synthesis of euchromatin, facultative heterochromatin and constitutive heterochromatin (after Pera, 1970).

mented in a variety of animal and plant species following the initial discovery by Lima-de-Faria (1959) that the heterochromatic X chromosome in spermatocytes of the male grasshopper *Melanoplus differentialis* and the heterochromatic chromosome segments of rye replicated in the later part of the S phase. Late replication of heterochromatin does not necessarily imply that heterochromatin finishes replication after euchromatin; this situation occurs in some materials, whilst in others heterochromatin seems to finish replication at much the same time as euchromatin (references in Lima-de-Faria, 1969, and in Barlow, 1972*a*). At any rate, late DNA replication is shown not only by constitutive but also by facultative heterochromatin (that is, euchromatin in a condensed, transcriptionally inactive state) as exemplified by the haploid heterochromatic set in male mealy bugs (Brown & Nur, 1964), the inactivated single X chromosome in male locusts, and the inactivated X chromosome (sex chromatin body) in the female of mammals including man (review in Lima-de-Faria, 1969). Fig. 2.3 shows the sequence of DNA replication in euchromatin, facultative and constitutive heterochromatin in cultured cells of the femae field-vole (*Microtus agrestis*). Facultative heterochromatin in female mammals, in which the randomly inactivated X replicates late and the other X replicates early, clearly shows that the time of replication is independent of the replicon base composition (obviously, the two Xs are identical in base composition) but depends in some way upon the condensed state of chromatin (Lima-de-Faria, 1969). Thus in *Melanoplus* males the X chromosome no longer shows a differential replication pattern in the early spermatogonial cell generations in which it is not in a condensed (heteropyknotic) state (Nicklas & Jaqua, 1965). That the time of replication during the S phase is, however, a reflection of the base composition of DNA has been found in a number of organisms by means of buoyant density determinations in CsCl gradients and/or direct determinations of the base composition of DNAs. By an analysis of the main band DNA (mDNA) it has been found that the DNA to be synthesized first is slightly more dense than the DNA which replicates later; this implies – in terms of base composition – a shift during the S phase from a relatively G–C rich to a relatively A–T rich DNA (Tobia, Schildkraut & Maio, 1970; Bostock & Prescott, 1971; Comings, 1972). However, Comings (1972) has also found that the DNA synthesized in the first 30 minutes of synthesis in Chinese hamster cells is slightly A–T rich. As to satellite DNAs, it has been reported that the light A–T rich mouse satellite is synthesized either predominantly at the end of S (Tobia *et al.*, 1970; Flamm, Bernheim & Brubacker, 1971) or preferentially during the third quarter of S (Bostock & Prescott, 1971). In the kangaroo rat, *Dipodomys ordii*, the two heavy satellites with densities of 1.707 and 1.714 g/cm^3 (density of mDNA: 1.699 and 1.702 g/cm^3) – which comprise about 60% of the total nuclear DNA, predominantly occupy the shorter arms and kinetochore regions of chromosomes, and are largely or entirely restricted to the heterochromatic regions of chromosomes – replicate predominantly in the second half of S (Bostock, Prescott & Hatch, 1972). The recent introduction in cytological research of the quinacrine fluorescence, pioneered by Casperson's group (1968), and of the Giemsa staining after denaturation–reannealing of chromosomal DNA (Pardue & Gall, 1970; Arrighi & Hsu, 1971) are rapidly becoming new important tools for the cytological localization of repetitive DNA sequences and, in many cases, of A–T rich DNA (review in Schnedl, 1974). The binding of ^{3}H-actinomycin D to the DNA of nuclei and chromosomes in fixed preparations also appears promising (Avanzi, 1972; Cionini & Avanzi, 1972; Jaworska, Avanzi & Lima-de-Faria, 1973).

DNA in eukaryotic chromosomes replicates by the semiconservative mechanism. Direct evidence for semiconservative replication was first obtained in mammals by means of density labeling of replicating DNA with 5-bromodeoxyuridine (Chun & Littlefield, 1963) and later demonstrated in tobacco by ^{15}N density labeling of DNA (Filner, 1965). When DNA is labeled with ^{3}H-thymidine and the chromosomes are analyzed in the first, second and third division after precursor application – as in the now classical work of Taylor, Woods & Hughes (1957) – semiconservative segregation of DNA is observed. This semiconservative segregation of DNA would be direct evidence of semiconservative replication if the chromosome were found to be unineme, that is, to consist of only one DNA double helix in its transversal dimension; but the actual structure of eukaryotic chromosomes is unresolved as yet (reviews in Wolff, 1969; Prescott, 1970; Ris & Kubai, 1970).

Because of the asynchrony of DNA replications, RNA synthesis still continues during S (it is generally assumed that replicating DNA segments do not transcribe). Active translation also occurs during S; it is required for the synthesis both of the enzymes needed for DNA replication itself and of the chromosomal proteins. As shown by extensive cytophotometric, autoradiographic and biochemical analyses in a number of plant and animal materials, histone synthesis occurs concurrently with DNA replication, so that the histone content doubles with the doubling of DNA (Alfert, 1958; Bloch, Macquigg, Brack & Wu, 1967; Hardin, Einem & Lindsay, 1967; Das & Alfert, 1968; Butler & Mueller, 1974). Histone is synthesized in the cytoplasm, from which it migrates to the nucleus. Evidence for cytoplasmic synthesis of histones is provided by the demonstration that in HeLa cells a histone mRNA (9S) appears in small cytoplasmic polysomes and is translated during the S period (Gallwitz & Mueller, 1969; Butler & Mueller, 1973). Like other nuclear proteins, the bulk of acidic chromosomal proteins is made in the cytoplasm; their synthesis is generally not in coordination with DNA synthesis, although some particular nonhistone proteins are clearly increased during S (Stein & Baserga, 1970; Gerner & Humphrey, 1973).

G_2 phase

Protein synthesis continues during G_2 but the protein spectrum which obtains in this phase is quite different from the protein spectrum of the preceding phases (Kolodny & Gross, 1969). In addition to the proteins responsible for chromosome coiling (condensation) at mitosis (see below), the protein components of the microtubules of the mitotic spindle – the dimeric 6S proteins (molecular weight of about 120000) or tubulins – are probably formed in G_2 and begin to organize into microtubules at the G_2-early prophase transition (Margulis, 1973).

Mitosis

When the nucleus enters mitosis, chromosomes begin to condense. Since the histone to DNA ratio remains practically unchanged from the S through the M phases whilst acidic chromosomal proteins relatively increase by a factor of about 2.5, some acidic proteins must be implicated in chromosome condensation (Sirlin, 1972); also some histone might be involved in the process (Johnson *et al.*, 1974). That during mitosis factors prevail which induce chromosome condensation is clearly demonstrated by the condensation of chromosomes in interphase nuclei when a non-mitotic and a mitotic HeLa cell are fused (Rao & Johnson, 1970). Due to chromosome condensation, transcription is extremely reduced and does not occur from late prophase to late anaphase–early telophase (Fig. 2.4); translation is reduced by 60–70% during mitosis; cytoplasmic polysomes being degraded for the most part (references in Sirlin, 1972).

Chromosome condensation, which becomes clear at prophase when the chromosomes show their typical two-chromatid structure, reaches a maximum at meta-

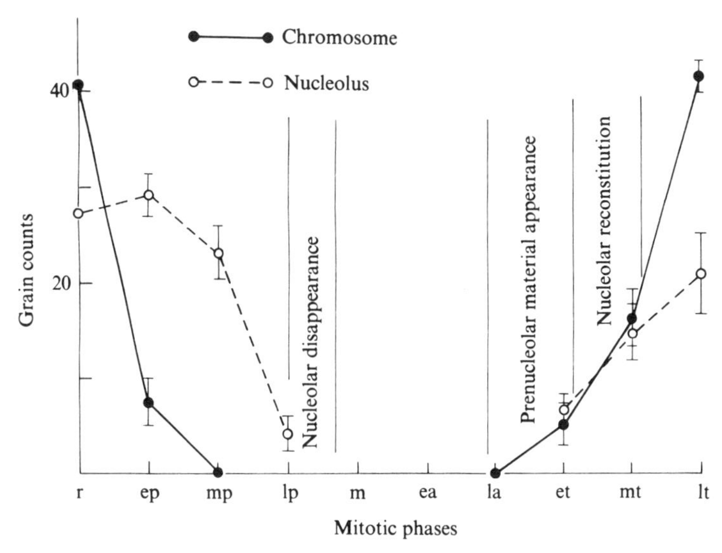

Fig. 2.4. Average grain counts on chromosomes and nucleolus of *Luzula purpurea* at different stages of the cell cycle. Material: root tips treated with ^{3}H-uridine for 10 minutes (after Kusanagi, 1964).

phase. Later in prophase, in animals and higher plants, the nuclear envelope breaks down (see, however, Wada, 1976) and the nucleolus or nucleoli dissolve. In the transition from prophase to metaphase (prometaphase) the movement of chromosomes starts in connection with the organization of the mitotic spindle with its continuous fibers (bundles of microtubules) running from pole to pole, and kinetochore fibers running from sister kinetochores (i.e. chromatid kinetochores) to the poles. This movement culminates at metaphase in the orientation, at the spindle equator, of the chromosomes so that each of the sister kinetochores lies toward each of the spindle poles. Anaphase begins when sister kinetochores separate from each other and move toward opposite poles. The reconstitution of the nuclear envelope around the two anaphase chromosome groups, the uncoiling (decondensation) of chromosomes to return to the interphase condition, and the formation of one or more new nucleoli are the main nuclear changes in telophase.

Nuclear division is usually followed by the division of the original cell into two daughter cells (cytokinesis). The mechanism of cytokinesis differs in animal and plant cells. The former generally divide by a cleavage furrow: the cytoplasm constricts progressively until division is complete; formation of the new plasmalemmas takes place in the stalk at the base of the furrow. In plant cells, typically enclosed in a pecto-cellulosic wall, new walls and new plasmalemmas are both formed at cytokinesis. By relaxation of the continuous spindle fibers at telophase the spindle becomes a barrel-shaped structure (phragmoplast), in the equator of which a great many Golgi vesicles containing pectic substances fuse to form the middle lamella flanked on both sides by a plasmalemma. Against the middle lamella more wall material (primary walls) is deposited. As we shall see in other sections of this book, the presence of a middle lamella between cells has profound implications in histogenesis and organogenesis in plants. For an extensive discussion on nuclear and cell division, reference is made to the recent reviews by Forer (1969), Nicklas (1971) and Margulis (1973).

Duration of the cell cycle and its phases

Knowledge on the duration of the cell cycle and its component phases is essential for an adequate analysis of proliferating cell systems. To this end, the technique of labeled mitoses, inaugurated by Howard & Pelc (1953) and fully developed by Quastler & Sherman (1959), has proved to be most useful. The technique, which is based on the scoring of labeled mitoses in materials fixed at various intervals after a pulse label with ^{3}H-thymidine, gives the total cycle time (T) and the duration of the individual cell cycle phases. Another method is the determination of the minimum mitotic cycle time as the period of time between a short colchicine treatment and the initial appearance of tetraploid cells in division (Van't Hof & Sparrow, 1963). Although, in one and the same material, discrepancies in mitotic cycle time may be found when applying the two methods (Van't Hof, 1965*a*) the colchicine technique deserves more attention than it has so far received (Bertalanffy, 1964). In proliferating complex systems with concurrent cell differentiation (e.g., apical meristem), the colchicine technique combined with the labeled mitoses technique permits a clear discrimination between DNA synthesis for mitosis (cycling cells) and DNA synthesis for differentiation (noncycling cells undergoing chromosome endoreduplication and/or extra DNA synthesis, i.e. amplification).

In some diploid plant species a simple relation has been found between the minimum mitotic cycle time, the interphase nuclear volume and the nuclear DNA content. These relationships are independent of chromosome number and the DNA amount per chromosome (Table 2.2). In many species of animals and plants a relationship seems to exist between the duration of the S phase and the DNA content (Van't Hof, 1965*b*; Evans, Rees, Snell & Sun, 1972; Bennett, 1972): observations made on seven species of angiosperms are reported in Table 2.3. Such a relationship does not hold when cells with different chromosome number in the same species are analyzed. In diploid and haploid embryos of *Xenopus laevis*, the endoderm cells have an S phase of the same duration (Graham, 1966). In plants, the duration of the S phase has been found to be similar in the diploid and in its autotetraploid form in each of the four species investigated (Troy & Wimber, 1968). Diploid and tetraploid cells also have a cell cycle time and S phase of similar duration when analyzed within the same tissue, as in the case of the mixoploid (diplo-tetraploid) root apices obtained by colchicine treatment in *Pisum sativum* (Van't Hof, 1966) and *Vicia faba* (Friedberg & Davidson, 1970). These results

TABLE 2.2. *Nuclear characteristics of meristematic cells from various plant species* (from Van't Hof & Sparrow, 1963)

Species	Minimum cycle time (h)	Interphase nuclear volume (μm^3)	DNA per cell (pg)	2n	DNA per chromosome (pg)
Trillium erectum	29	1175	120	10	12
Tulipa Kaufmanniana	23	800	93.7	24	3.91
Tradescantia paludosa	18	640	59.4	12	4.95
Vicia faba	13	377	39.4	12	3.2
Pisum sativum	10	200	11.67	14	0.83
Helianthus annuus	9	195	9.85	34	0.29

TABLE 2.3. *Relationships among DNA content, duration of cycle time and duration of the S phase* (condensed from Van't Hof, 1965*b*, Table 1)

Species	DNA/cell (pg)	Duration of the mitotic cycle (h)	Duration of S (h)
Crepis capillaris	3.82	10.75	3.25
Impatiens balsamina	5.14	8.8	3.9
Lycopersicon esculentum	8.44	10.6	4.3
Allium fistulosum	41.0	18.8	10.3
Allium cepa	54.3	17.4	10.9
Tradescantia paludosa	59.4	20.0	10.8
Allium tuberosum	66.3	20.6	11.8

clearly show that the specific rate of DNA synthesis in diploid and autotetraploid cells of a species is the same: doubling of chromosome number doubles the absolute rate of DNA synthesis.

An important factor influencing the duration of the cell cycle is heterochromatin (Ayonoadu & Rees, 1968; Barlow & Vosa, 1969; Barlow, 1972*b*; Nagl, 1974).

Dynamics of proliferating cell populations

Natural synchrony

In the life history of plants and animals synchronous nuclear and cell divisions may occur at particular stages of development. Erickson (1964) and Agrell (1964) have reviewed the many cases of natural synchrony which have been described in the plant and animal kingdom respectively. In plants, no synchrony occurs in proliferating somatic tissues (meristems); striking examples of synchrony are found in reproductive tissues and cell groups such as female gametophytes of angiosperms and gymnosperms, endosperm, gametangia, the antheridia and sporangia of bryophytes and pteridophytes, microsporogenous tissue and microsporocyte masses in the anther of *Trillium*, *Lilium* etc. Synchrony of nuclear division also characterizes the early stages of embryo development in gymnosperms (in angiosperms, *Paeonia* excluded, the embryo is cellular since the first division of the zygote and the cell divisions are asynchronous). When, as is commonly the case, the synchronous mitoses are not followed by cell division, there is formation of an apocyte from an initial uninucleate cell (e.g. a proendospermatic cell, a zygote in a gymnosperm, an initial cell of a sporogenous tissue). Taking as an example the Cycad *Dioon*, a megaspore produces through seven synchronous mitoses an apocytial female gametophyte with 128 nuclei; after further mitoses, the apocyte undergoes cellularization (wall formation) when it contains about 1000 free nuclei.

Use has been made of the natural synchrony of meiotic division in meiocytes and mitotic divisions in the microspores of some plant species in studies on chromosome breakage, cell permeability, respiration, nucleic acid and protein synthesis. Some important results on the physiology of meiosis, obtained by Stern and his colleagues with the meiocytes of *Lilium*, will be considered in some detail in another section of this book. Obviously, division synchrony in apocytia and syncytia can be traced to the presence of a common cytoplasm. In species with synchronous meiosis in all meiocytes of each anther loculus, complete cytoplasmic continuity is established by a system of cellular interconnections. When at metaphase II the links between contiguous meiocytes are broken, synchrony is progressively lost. In some massulate orchids, which produce pollen in aggregates of hundreds of grains, the nuclei of all cells in each massula are synchronized from meiosis throughout pollen mitosis; here again, cytoplasmic connections between cells are present (Heslop-Harrison, 1966, 1968).

In animals (Agrell, 1964), synchronous divisions are common in early embryogenesis; there are, however, very many animal species, including mammals, in which divisions are already asynchronous at the two-blastomere stage. Synchrony in early embryos may last for a number of cell cycles: three in the sea urchin, nine in the holothurian *Synapta digitata* and twelve to fifteen in amphibians. Because of the lack of cell growth between cell divisions, the very large egg cell splits into an increasing number of smaller and smaller cells. In the amphibians, whose egg increases in volume by more than 100000-fold during oogenesis, the 20000 cells formed at the end of the division synchrony period are still considerably larger than adult cells. The synchronous division cycle in early embryos is very rapid: there is no G_1, G_2 is very short and S takes most of interphase. In the sea urchin, DNA synthesis starts at telophase (Lison & Pasteels, 1951; Hinegardner *et al.*, 1964); after the period of division synchrony, the cell cycle duration progressively increases mostly by an extension of the interphase duration, while duration of mitosis shortens for a while, probably due to the increased nucleocytoplasmic ratio

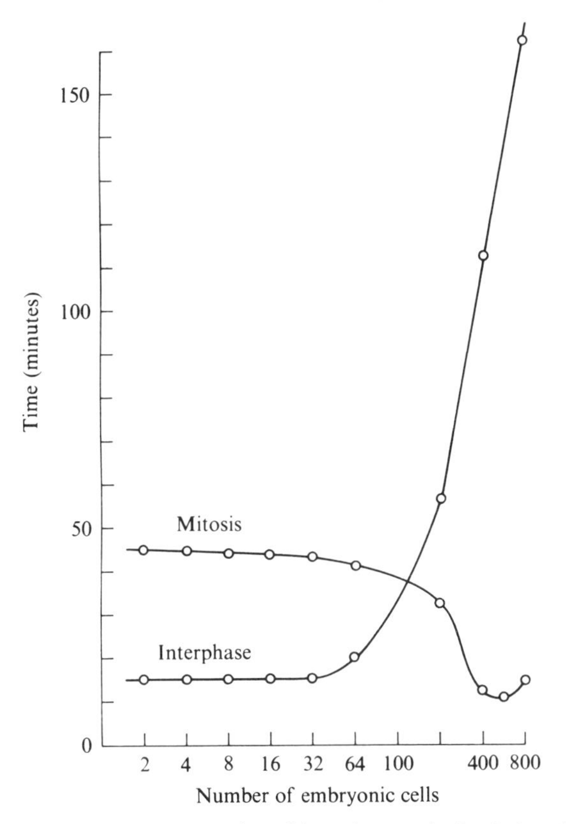

Fig. 2.5. Changes in the average duration of interphase and mitosis (prophase to telophase) during the early development of the sea urchin embryo *Paracentrotus lividus* (after Agrell, 1964).

(Agrell, 1957) (Fig. 2.5). Thus, during embryogenesis in animals of the division-synchrony group, the cell cycle changes from a cycle with no G_1 to a cycle similar to that of adult dividing cells: it becomes regulated at the G_1 period.

Apical meristems

An unique feature in development and growth of vascular plants descends from the localization of embryonic tissues at the ends of the axial organs, roots and shoots: throughout their life, these apical meristems form new tissues and organs which are added to those formed during embryogenesis. Because of this plants have been defined as organisms with continued embryogenesis (Bower, 1930) or recurrent ontogenesis (Chiarugi, 1952). Apical meristems have been made the subject of extensive anatomical and physiological investigation. In the last two

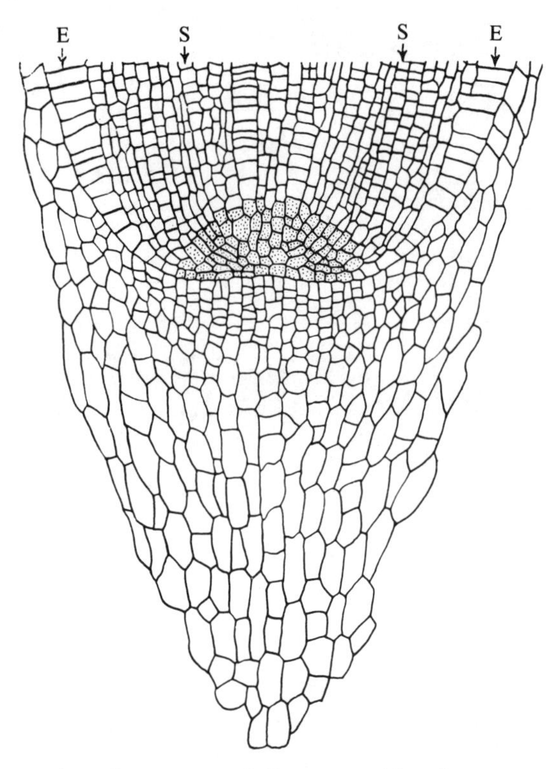

Fig. 2.6. Median section of root apex of *Zea mays* with quiescent center shaded. E: epidermis; S: outer layer of stele (after Clowes, 1959*a*).

decades, much attention has been devoted to the dynamics of meristem cell populations with special emphasis on root apices (mostly angiosperms), which provide a much easier and more appropriate experimental material than shoot apices (Clowes, 1961; Torrey, 1965; Nougarède, 1967). It is now clear that the discovery of the quiescent center (Clowes, 1956), by offering a fresh interpretation of the structure and mode of functioning of the angiosperm root apex, has had an important part in accelerating our knowledge on the dynamics of root apical meristems. Up to that time, the old histogen theory of Hanstein (1868), further specified by Janczewski (1874), was still accepted. The theory postulated that the angiosperm root apex is made of three histogens – plerome, periblem and dermatogen – which are the meristematic parts of the stele, cortex and epidermis respectively; a fourth histogen, the calyptrogen, was assumed to give rise in some species to the root cap. Acceptance of this theory implied that each histogen had an initial or a small group of initials in active division. Because of the radial symmetry of the apex it was natural to assume that the initials lay in the axis of the apex. From an analysis of mitotic frequencies in longitudinal radial sections of onion and wheat roots, Buvat & Genèves (1951) and Buvat & Liard (1951)

TABLE 2.4. *Durations in hours of the mitotic cycle in four parts of the root apices of four species, calculated with the colchicine method* (from Clowes, 1967*b*)

Species	Cap initials	Quiescent center (Qc)	Stele just above Qc	Stele 200 μm above Qc
Maize	12	174	28	29
Broad bean	44	292	37	26
White mustard	35	520	32	25
Garlic	33	173	35	33

concluded that the cells previously believed to be initials divided very rarely or not at all. A few years earlier, D'Amato & Avanzi (1948) reported that in onion roots treated for a prolonged period with colchicine the cells near to the cap were diploid or, at the most, tetraploid whereas the other cells of the meristem reached the 16-ploidy level; they did not, however, realize the significance of this observation. In 1956, Clowes compared rates of DNA synthesis in apical meristems of *Zea mays* roots fed with ^{32}P-phosphate and ^{14}C-adenine; he demonstrated that a portion of the root apex, which consisted of the poles of the stele and the cortex-epidermis complex forming a rough hemisphere, was not meristematic. This region, which contains the initials of classical anatomy but does not include the cap initials, was called the quiescent center (Fig. 2.6). Among the primary roots of the best known species, *Zea* has 600 cells in its quiescent center compared with 125000 actively dividing cells and *Vicia faba* has 1000 out of 250000. Of the meristematic cells all around the quiescent center, those on its surface can be usefully regarded as the initials of the meristem. The cells of the quiescent center are carried forward passively by the growth and division of the surrounding meristem (Clowes, 1959*a*, *b*, 1964, 1965). The presence of the quiescent center has been clearly demonstrated in a variety of apices, both in intact roots (Clowes, 1959, 1967*a* and *b*, 1972*b*; Thompson & Clowes, 1968; Byrne & Heimsch, 1970; Avanzi, Bruni & Tagliasacchi, 1974) and excised roots grown in well-defined media (Thomas, 1967; Phillips & Torrey, 1971; Torrey, 1972; Webster & Langenauer, 1973).

Clowes and his associates were the first to measure the duration of the mitotic cycle and of its component phases in different regions of the root apex in some plant species; the main results of their observations are reported in Tables 2.4 and 2.5. In the four species listed in Table 2.4 the duration of the mitotic cycle in the quiescent center is from 5 to 17 times longer than in the meristematic cells of the stele. In maize and in garlic the duration $T-G_1$ in the quiescent center is not much different from T in the stele (Table 2.5): this implies that once a quiescent center cell has entered the S phase it will proceed in the mitotic cycle with a speed comparable to that of an ordinary meristematic cell. Of special interest is the very rapid mitotic cycle of cap initials in *Zea* resulting in a negative G_1 (-1 hour). As

TABLE 2.5. *Duration in hours of the mitotic cycle (T) and its component phases G_1, S, G_2 and M, determined by pulse labelling with ^{3}H-thymidine* (from Clowes, 1967*b*)

	Cap initials	Quiescent center (Qc)	Stele just above Qc	Stele 200 μm above Qc
		Maize		
T	14	174	22	23
G_1	−1	151	2	4
S	8	9	11	9
G_2	5	11	7	6
M	2	3	2	4
		Garlic		
T	27	173	26	26
G_1	3	142	4	4
S	13	17	12	12
G_2	6	8	4	4
M	5	6	6	6

shown by autoradiographic analysis, DNA synthesis in these cells begins during telophase, a situation which parallels that found in the very rapidly dividing cells of early embryos of sea urchins.

That the majority of cells in the quiescent center are in the G_1 (2C) condition has been shown by both the biochemical (Jensen, 1958) and the cytophotometric analysis (Clowes, 1968) of the nuclear DNA content per cell. In the quiescent center of maize, a discrepancy has been found between the relative percentages of 2C (G_1), 2–4C (S) and 4C (G_2) DNA contents as measured by the cytophotometer and those of the three phases of the mitotic cycle as calculated by pulse labeling: 53, 43 and 4% and 87, 5 and 8% respectively. According to Clowes (1968, 1972*b*), a possible explanation of the discrepancy is that cells in the quiescent center may have been stopped during S by whatever induces quiescence in the originally meristematic tissue. It seems possible, however, that in the quiescent center cells, or in some of them, DNA synthesis takes two distinct forms: synthesis for mitosis (S phase) and extra synthesis (DNA amplification) in non-cycling cells. This possibility is strongly suggested by the labeling patterns of the cells of the quiescent center and of the cap initials in root apices of *Calystegia soldanella* fed with tritiated thymidine and uridine. Within the quiescent center four types of cells were resolved, each characterized by some morphological differentiation and/or by a typical pattern of thymidine and uridine incorporation. Based on the labeling pattern, the suggestion was advanced that some cells of the quiescent center may undergo selective replication of particular DNA base sequences which are active in transcription (Avanzi *et al.*, 1974).

DNA amplification, if occurring in the quiescent center cells, might reflect the demand for an increased amount of DNA templates needed for the transcription of messengers involved in some specific syntheses. This would fit the idea advanced by Torrey (1963, 1972) that the quiescent center may be a site of the specialized biosynthesis of a hormone essential to cell division, whose supra-optimal concentration in the quiescent center would be responsible for cell quiescence. Although Torrey's idea seems to be contradicted by the experimental evidence discussed below and there is no indication that the synthesis of a hormone occurs in the quiescent center, this idea and the interpretation of Avanzi *et al.* (1974) are interesting because they try to characterize in positive terms the quiescent center, which had so far been defined on 'negative' features: the lowest average DNA content (due to the high proportion of G_1 cells), the lowest RNA content per cell, the lowest rates of DNA, RNA and protein synthesis, the fewest ribosomes per unit volume of cytoplasm, the fewest mitochondria, the smallest Golgi bodies and the relatively thick cell walls (Clowes, 1968; Clowes & Juniper, 1964; Pilet & Lance-Nougarède, 1965; Barlow, 1970).

Since mitoses, rare as they may be, do occur in the quiescent center, the question may be raised whether there is a uniform degree of quiescence in the cells of this region. In cultured roots of *Convolvulus arvensis*, indications have been obtained that in the quiescent center there may be a gradient in the length of cell cycles such that progressively longer ones occur toward its central region (Phillips & Torrey, 1971; see also Webster & Langenauer (1973) for the excised roots of *Zea mays*). Although one can visualize this gradient related to a hormone gradient (Torrey, 1972), it seems more likely that it reflects the developmental history of the quiescent center. In *Sinapis alba*, all cells in the embryonic root apices are meristematic: the quiescent center appears after seed germination (when the radicles are about 5 mm long, the quiescent center consists of between four and forty cells) and increases with root growth until it includes 500 or more cells in the full-grown roots (Clowes, 1958). In *Malva silvestris* the quiescent center is established during embryogeny and the number of its constituent cells rapidly increases in an advanced stage of root development (Byrne & Heimsch, 1970). Independently of the time of initial formation, in *Sinapis*, *Malva* and other species, the quiescent center grows centrifugally from a focal point (which includes the cells which first entered quiescence) by the continuous addition of previously meristematic cells which enter quiescence in a clear time progression.

The long G_1 period confers on the quiescent center cells obvious advantages over meristematic cells in the root apex. When roots are X-irradiated, the meristematic cells – most of which are in S, G_2 and mitosis – are more damaged than the quiescent center cells which are mostly in G_1. Loss of chromosome material in the first mitosis post-treatment strongly reduces or abolishes mitotic activity in the meristem; in these conditions, the cells of the quiescent center enter mitosis and, in case of strong radiation damage, can regenerate the meristem. The same result is obtained by internal irradiation by beta particles, as it occurs in chromosomes of meristematic cells subjected to a long treatment with ^{3}H-thymidine (Clowes, 1959*b*, 1964, 1965; Clowes & Hall, 1962). A lowering of the reproductive integrity of meristematic cells can be induced by factors responsible for dormancy, e.g. cold. In the root tip of *Zea mays*, quite moderate cold (+5 °C) induces mitotic disturbances and chromosome breakage which strongly reduces the mitotic rate of the meristem surrounding the quiescent center: this stimulates the quiescent center cells into mitosis when favourable tempertures return (Clowes & Stewart, 1967). Another aspect of the control of the activity of the quiescent center has been found in excised roots of *Zea.* Upon addition of sucrose to roots which have been deprived of carbohydrate, both the cells of the quiescent center and those of the surrounding meristem undergo DNA synthesis. Following the onset of proliferative activity in the meristem, DNA synthesis in the quiescent center cells is again arrested (Webster & Langenauer, 1973). All these observations suggest that the mitotically active cells surrounding the quiescent center may be the source of some factors responsible for maintaining low rates of division in the quiescent center (Webster & Langenauer, 1973). Other, probably complex, factors operate in the regulation of mitotic rates in the various regions of an apical meristem, as shown by the removal of the distal half of the cap from roots of *Zea mays* (Clowes, 1972*a*).

From a developmental point of view, the quiescent center is a geometrical necessity in the organization of the root apex (Clowes, 1961, 1968). This concept has received further support from the recent investigations on root apices with a single apical cell in pteridophytes. In these plants, the root originates from a single initial which, after enlarging radially, cuts off cells on four sides, so that a tetrahedral cell is left in the middle. From this time on, the apical cell, with its typical tetrahedral form, is found in roots of any age in a central position just behind the root cap. For nearly one century, it has been stated that the apical cell is a meristematic cell, which divides by cutting off cells on all of its four faces (e.g. *Marsilea*) or on the three basiscopic faces (e.g. *Azolla*). In 1960, Gifford found that the apical cell of the roots in the water fern *Ceratopteris thalictroides* invariably incorporated ^{3}H-thymidine. On the assumption that DNA synthesis in a nucleus can generally be accepted as evidence for subsequent mitosis, Gifford concluded that apical cells actually divide, perhaps at a higher rate than envisaged by previous workers. Since the autoradiographic method cannot discriminate between DNA synthesis in preparation for mitosis and DNA synthesis responsible for chromosome endoreduplication and/or DNA amplification, correct information on the structure and mode of functioning of root meristems with a single apical cell can only be obtained from autoradiographic techniques supplemented by cytochemical and cytophotometric procedures. Investigations combining these techniques in five species of leptosporangiate ferns (D'Amato & Avanzi, 1965; Avanzi & D'Amato, 1967, 1970) have demonstrated (Table 2.6) that the apical cells, which divide very rarely in established apices, are mostly held in G_2 with a 4C nuclear DNA content (58 out of the 78 apical cells which fell within the diploid

TABLE 2.6. *DNA content in the apical cells of root apices and root primordia of five species of leptosporangiate ferns (in brackets, the number of apical cells in mitosis, either diploid, 4C, or tetraploid, 8C)* (from Avanzi & D'Amato 1967).

Species	Material	Number of apical cells measured	Number of apical cells with DNA content*					
			2C	2–4C	4C	4–8C	8C	8–16C
Marsilea strigosa	Root primordia	62	0	7	27 (4)	17	9 (1)	2
	Root apices	13	0	1	7 (1)	1	3 (1)	1
Blechnum brasiliense	Root apices	26	1	2	8	6	9	0
Blechnum gibbum	Root apices	30	0	4	8	15	2	1
Polypodium aureum	Root apices	25	0	3	5	13	4	0
Ceratopteris thalictroides	Root apices	10	0	2	3	1	2	2
	Total	166	1	19	58 (5)	53	29 (2)	6

* 2–4C, 4–8C and 8–16C: DNA contents intermediate between 2C and 4C, 4C and 8C, 8C and 16C.

TABLE 2.7. *DNA content of apical cell, and cells adjoining the apical cell in the root apices of three species of leptosporangiate ferns* (from Avanzi & D'Amato, 1970)

Species	Number of apical cells with DNA content						Number of adjoining cells with DNA content					
	2C	2–4C	4C	4–8C	8C	8–16C	2C	2–4C	4C	4–8C	8C	8–16C
Blechnum brasiliense	0	2	8	3	9	0	3	18	21	13	2	0
Blechnum gibbum	0	4	8	15	2	1	8	25	26	13	0	0
Polypodium aureum	0	3	5	13	4	0	7	34	13	12	0	0
Total	0	9	21	31	15	1	18	77	60	38	2	0

* 2–4C, 4–8C and 8–16C: DNA contents intermediate between 2C and 4C, 4C and 8C, 8C and 16C.

limits, i.e. from 2C to 4C) and a great proportion of them show a DNA content greater than 4C, with intermediate values of 8C–16C in extreme cases. It has been found that cells adjoining the apical cell in established root apices also accumulate in G_2 (4C), but many of them increase their DNA over the 4C value (Table 2.7). These data and other observations reported in the original papers have led to a new interpretation of the apical structure in roots of cryptogams with a 'single apical cell'. It proposes that the apex contains a multicellular quiescent center whose size is related to the developmental stage of the root. The quiescent center

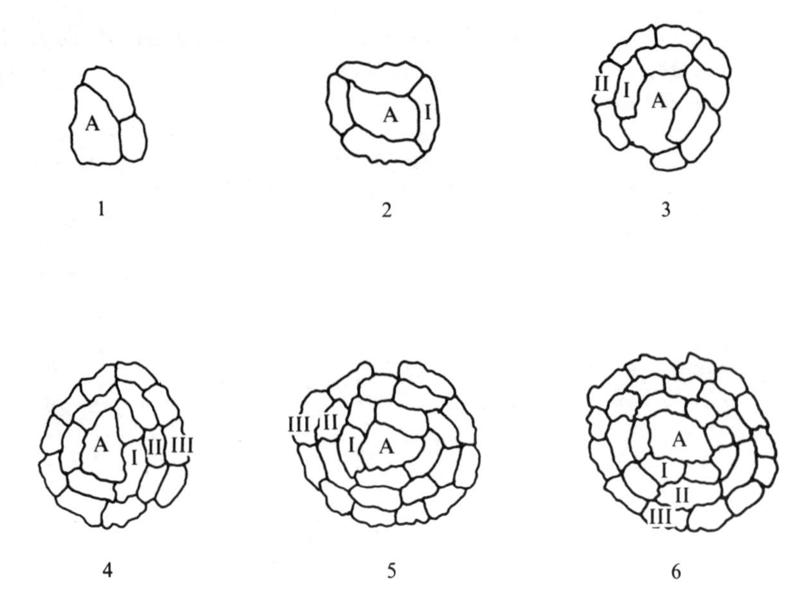

Fig. 2.7. Transverse section of lateral root primordia of *Marsilea strigosa* in different stages of development, numbered as in Tables 2.8 and 2.9. A: apical cell; I, II and III: first, second and third ring of cells surrounding the apical (after Cremonin, 1974).

starts to be formed when the initial cell, which has divided to form a primordium, stops dividing and becomes a specialized cell. With further development cells adjoining the apical acquire the physiological characteristics of the apical cell, so that the quiescent center grows by centrifugal apposition of cells around the apical cell. It is also proposed that the quiescent center cells in cryptogams are engaged, probably concomitant with the DNA amplification processes, in the macromolecular synthesis necessary for the attainment and maintenance of the apical structure (Avanzi & D'Amato, 1970).

That mitotic quiescence spreads centrifugally from the apical cell outwards during root development is documented by the observations of Cremonini (1974). In longitudinal sections of roots of the water fern *Marsilea strigosa* many initials of lateral roots and lateral root primordia of different ages can be found. Because of the radial symmetry of the cells around the apical cell (Fig. 2.7) it is easy to localise mitoses in the apical cell itself and in the cells arranged in rings around it. Such an analysis (Tables 2.8 and 2.9) demonstrates beyond doubt that the initial (apical) cell is the first to become quiescent, followed in a regular sequence by cells in the first and successive rings.

The block in G_2 (4C) and the ability to proceed eventually in further DNA synthesis (see also Michaux, 1970) distinguishes the quiescent center cells of cryptogamic roots from those of angiosperm roots. In angiosperms, the block in G_1 confers on the quiescent center the typical feature of a reserve meristem

TABLE 2.8. *Localization of mitoses in lateral root primordia of* Marsilea strigosa *at different developmental stages* (from Cremonini, 1974)

Developmental stage of primordia*	Number of primordia observed	Number of mitoses in the primordia	Initial (apical) cells in mitosis		Other cells in mitosis					
					I ring		II ring		III ring	
			No.	%	No.	%	No.	%	No.	%
1	159	26	26	100.0	—	—	—	—	—	—
2	222	38	22	57.9	16	42.1	—	—	—	—
3	161	47	12	25.5	19	40.4	16	34.1	—	—
4	113	45	1	2.2	8	17.8	33	73.3	3	6.7
5	62	13	—	—	—	—	2	15.4	11	84.6
6	46	13	—	—	—	—	1	7.7	12	92.3

* See Fig. 2.7.

TABLE 2.9. *Mitotic index (percentage of mitoses on total cells observed) in lateral root primordia of* Marsilea strigosa *at different developmental stages* (from Cremonini, 1974)

Developmental stage of primordia*	Number of primordia observed	Mitotic index in different cells			
		Initials (apical cells)	I ring	II ring	III ring
1	159	59.1			
2	222	15.5	17.6		
3	161	8.6	8.8	7.8	
4	113	6.2	5.7	6.1	5.0
5	62	0.0	0.0	4.3	4.5
6	46	0.0	0.0	4.0	3.5

* See Fig. 2.7.

eventually available for root apex regeneration; in cryptogams, the quiescent center cells possess many features typical of specialized cells (Sossountzov, 1969; Avanzi & D'Amato, 1970), whose regenerative abilities may be reduced, although no evidence for this exists at present. Perhaps some functional similarity between the two types of quiescent center might be established in the future if it can be shown that, as indicated by Avanzi *et al.* (1974), quiescent center cells of angiosperm root apices can undergo extra DNA synthesis (amplification).

This section would be incomplete if no mention were made of the shoot apex. In the horsetail, *Equisetum arvense*, the shoot apex bears a conspicuous apical cell whose nucleus has been found to have 4C (G_2) or higher (up to 8–16C) DNA contents; the apical cell is flanked by the so-called segment cells, most of which show DNA values of 4C or higher. Thus, the most distal portion of the shoot apex is made of differentiated cells (apical and segment cells) whilst the meristematic region of the apex is the subapical region (D'Amato & Avanzi, 1968). From evidence of a different type this situation also occurs in the shoot apex of other cryptogams 'with single apical cells' (Sossountzov, 1965, 1969) leading to the supposition that in cryptogams the apical zone of the shoot apex is the equivalent of the root quiescent center (Avanzi & D'Amato, 1970). For a review on the nuclear cytology of apical meristems with single apical cells, reference is made to D'Amato (1975*b*).

Despite extensive anatomical studies (reviews in Clowes, 1961; Nougarède, 1965, 1967), knowledge on mitotic rates in angiosperm shoot apices is still very scarce. There has been much discussion on the relative importance of mitosis in the summit and flanks of shoot apices. In *Datura* the flank cells divide about 2.1 times as fast as the summit cells; in *Trifolium* the ratio is about 1.4 and in *Solanum* it is 1.6 (Leshem & Clowes, 1972). Although being less mitotically active than the flanks, the summit of the shoot apex represents a condition much different from the quiescent center of root meristems whose cells divide some 10 times more slowly than those of adjacent regions (Leshem & Clowes, 1972).

Imaginal disks

As we have seen in the preceding chapter, the imaginal disks start to be formed around the time of blastoderm formation during the embryogenesis of holometabolous insects. By means of different analyses, the most useful being those on genetic mosaics, it has been possible to estimate roughly the number of primitive cells that form the initial population of a disk (imaginal cells): this ranges from two for the eye anlage to about forty for the dorsal mesothoracic (wing) disk (Nöthiger, 1972). In *Drosophila*, the growth dynamics of imaginal disks has been investigated and data are available for all major disks (Fig. 2.8). All imaginal cells remain stationary during embryogenesis and enter the proliferative state during larval development. An exception is represented by the abdominal histoblasts which remain stationary throughout larval development and begin to increase after puparium formation at a rapid rate (cell cycle time of 2–3 hours). For the other disks, the average length of the cell cycle varies between 6 and 12 hours, these disks reach maturity (mitotic quiescence) at the end of the larval period with the possible exception of the wing disk (curves 1 and 2 in Fig. 2.8) (references in Nöthiger, 1972). A detailed analysis of the behaviour of cells in a disk shows that differences in mitotic rates between groups of cells occur; at some time in disk development, mitotic activity may become restricted to one or some parts only of a disk (references in Nöthiger, 1972).

Differential rates of cell division in imaginal disks are most probably a necessity

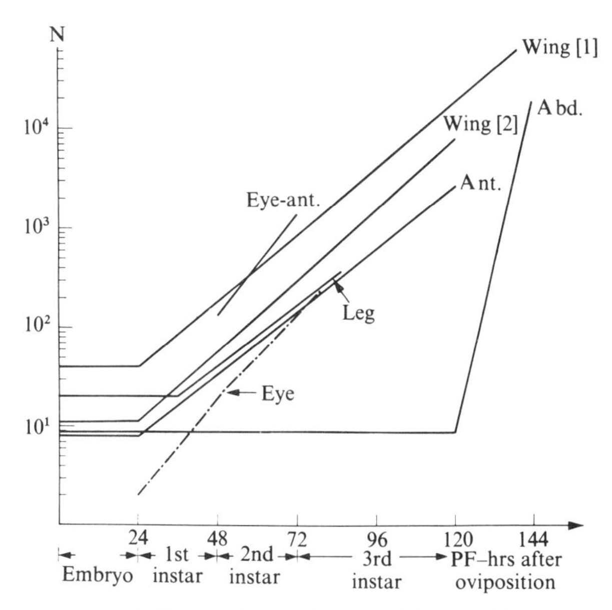

Fig. 2.8. Growth curves of different imaginal disks of *Drosophila*. N: number of cells estimated from genetic mosaics except for eye-antenna, which is based on cell counts; abd.: abdominal histoblasts; ant.: antennal disk; eye: presumptive cells of the eye region; eye-ant.: eye-antennal disk; leg: leg disk; wing: wing disk; PF: puparium formation. Based on the data of different laboratories (after Nöthiger, 1972).

for the development and maintenance of organization; the same applies to the root apical meristems which we have already discussed, and to other proliferating cell systems to be considered later.

Renewing cell populations in animals

Among the many renewal cell populations in animals, we intend to discuss two distinct types of tissue: namely, one with a slow acting renewal rate (mouse epidermis and chicken duodenal epithelium) and the other with a fast acting renewal rate (esophageal epithelium in chickens).

In mouse epidermis, the basal cell layer (stratum germinativum), which is a single layer, has an average cell generation time of about 24 days (Sherman, Quaestler & Wimber, 1961). In spite of its rather long generation time, from 3 to 6 weeks in different samples of skin, epidermis is still regarded as a renewing cell population because its rate of cell production exceeds that of body growth (references in Gelfant, 1962). In intact epidermis, only 1.7% of the cells are in S and less than 0.1% are in mitosis, so that at any given time about 98% of the cells in the basal layer of epidermis are not moving through the cell cycle. In the unstimulated control epidermis under continuous exposure to ^{3}H-thymidine the few mitoses collected are all labeled; they demonstrate the small population of normally cycling cells, characterized by a long G_1 (2–5 days), a short G_2 (4 hours), an S period of 10 hours and a mitosis lasting about 3 hours. Wounding or hair plucking stimulates division in part of the noncycling cells (Gelfant, 1962, 1963, 1966). About 49–54 hours after wounding or hair plucking, mitoses begin to appear both in ear and in body skin epidermis. When the experiments are carried out in the presence of ^{3}H-thymidine, two categories of mitoses are found: labeled and unlabeled. Labeled mitoses, are cells which were in G_1 when the experiment began and moved through the cell cycle in the usual manner ($G_1 \rightarrow S \rightarrow G_2 \rightarrow M$); unlabeled mitoses are cells which were in G_2 at the time of stimulation and entered mitosis without DNA synthesis. The fraction of cells held in G_2 in the basal layer of the mouse epidermis is estimated to be 5–10%. Studies *in vitro* have shown that wounding alone is not able to induce G_2 nuclei to enter mitosis in the ear epidermis: some additional physiological factor is required. This requirement is usually satisfied by adding glucose to the standard medium; but this additional physiological requirement can also be satisfied by altering the sodium or potassium concentration in the standard medium alone. These results have led Gelfant (1966) to postulate the existence of G_2 physiological subpopulations which have specific and different requirements for mitosis.

Gelfant (1966) has suggested that the existence of G_2 population cells may be related only to tissues having a slow acting renewal rate. A comparison of the behaviour of two different epithelial tissues of the chicken may illustrate this point. In chickens, the single-layered epithelium of the duodenal crypts manifests rapid cell proliferation, whilst the stratified epithelium of the esophagus is characterized by a low cell division rate. In newly hatched chickens, cellular proliferation can be depressed by starvation; when the animals are fed, cell division is stimulated. As shown by an analysis of the mitotic index and of the DNA labeling index, the duodenal epithelium consists of only one population of cells held in G_1; on the contrary, the esophageal epithelium contains a group of cells held in G_1 and another far smaller group of cells held in G_2 (Cameron & Cleffmann, 1964). These observations, and others of a similar nature to be found in the literature, raise the question of the mechanisms possibly involved in the positioning of proliferating cells in one or another phase of the cell cycle. In the last 10 years or so information on this and related aspects has been obtained from studies on natural quiescence in apical meristems.

Natural quiescence of apical meristems

A great event in the evolutionary history of vascular plants was the establishment, with the phanerogams, of a propagation and storage organ, the seed, which contains in a quiescent state the plant embryo. When the seed germinates, the apical shoot and root meristems of the embryo resume active metabolism and continue their development and growth in the seedling and the adult plant. Since in the seed plants the meristem cell line of the shoot is destined to become a germ line (it will form the sporogenous cells and tissues in the flower), the seed is the most appropriate device for the natural long-term conservation, in the form of a somatic meristem line, of the genetic information of any given species of phanerogams (D'Amato, 1975*a*).

In an embryo which develops within a seed, cell division in the apical meristems can proceed until a stage of maturation when the seed dries and becomes quiescent. An important question raised by seed maturation refers to the phase(s) of the cell cycle in which embryo meristem cells are blocked. For a long time it has been assumed that meristem cells in dry seeds are held in G_1. This assumption was substantiated by the general, but not exclusive, occurrence of chromosome-type aberrations in the first mitotic cycle in meristems of seedlings raised from seeds irradiated in the dry state. In 1963, however, a new picture arose from an X-irradiation experiment on dry seeds (karyopses) of *Triticum durum*, which also included DNA cytophotometric and autoradiographic analyses. It was shown that root meristems in the dry seed contained both 2C and 4C nuclei and that in irradiated seeds germinated in the presence of ^{3}H-thymidine practically all mitoses with chromosome aberrations were labeled (G_1 cells in the dry seed) whilst of the mitoses with chromatid aberrations part were labeled (G_1 cells in dry seed) and part were not (G_2 cells in dry seed). This experiment provided the first evidence that most, if not all, 2C and 4C nuclei in the resting seed meristem were actually nuclei in the G_1 and G_2 phases of the nuclear cycle and that G_1 cells in a metabolically inactive state could respond to radiation with either chromosome or chromatid aberrations (Avanzi, Brunori, D'Amato, Nuti-Ronchi, Scarascia-Mugnozza, 1963). This double radiation response of G_1 cells is also known in proliferating animal and plant cells (Hsu, Dewey & Humphrey, 1962; Evans & Savage, 1963; Wolff, & Luippold, 1964); in association with cytochemical tests (Ancora, Brunori & Martini, 1972) it helps in discriminating between early or mid G_1 (chromosome-type response) and late G_1 (chromatid-type response). Clear results come from an irradiation experiment on dry seeds of *Vicia faba* in which the radicle meristem contains both 2C (G_1) and 4C (G_2) cells. Since the radiation exposure used did not alter the sequential release of cells from quiescence into mitosis, it was possible to demonstrate the following progression in mitotic activation: G_2 (unlabeled cells, chromatid aberrations), late G_1 (labeled cells, chromatid aberrations), early (or mid) G_1 (labeled cells, chromosome aberrations) (Fig. 2.9).

Since the early observations in 1963, the use of DNA labeling and/or DNA cytophotometry, sometimes combined with irradiation, has greatly extended our knowledge on the nuclear conditions in apical meristem cells in quiescent seeds. The situation can be summarized as follows:

(i) all cells in G_1: root apex of *Pinus pinea* (Brunori & D'Amato, 1967), *Lactuca sativa* (Feinbrun & Klein, 1962; Brunori & D'Amato, 1967), *Helianthus annuus*

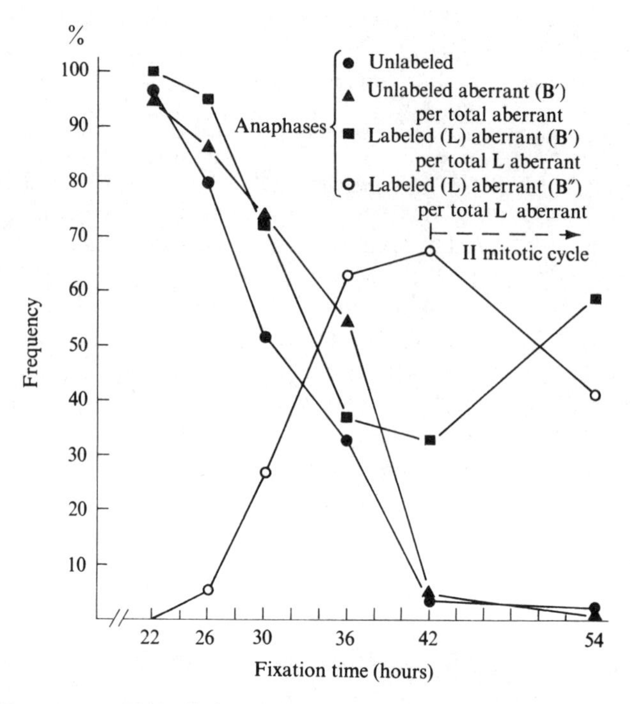

Fig. 2.9. Response to X-irradiation (1500 R) of G_1 and G_2 cells of the radicle meristem of resting seeds of *Vicia faba*. Based on the data of Brunori *et al.* (1966) (after D'Amato, 1972*a*).

(Brunori, Georgieva & D'Amato, 1970), *Haplopappus gracilis* (Röbbelen & Nirula, 1965), *Allium cepa* (Brunori & Ancora, 1968; Bryant, 1969); shoot apex of *Triticum durum* (Avanzi, Brunori & D'Amato, 1969), *Pinus pinea* (Brunori & D'Amato, 1967) and *Oryza sativa* (Matsubayashi & Yamaguchi, 1971);

(ii) cells in G_1 and in G_2, the proportion of G_2 cells varying from 1.6% (*Triticum dicoccum*) to 39% (*Triticum durum*): root and shoot apex of *Zea mays* (Stein & Quastler, 1963), root apex of *Vicia faba* (Brunori, Avanzi & D'Amato, 1966; Davidson, 1966; Jakob & Bovey, 1969), *Triticum durum* (Avanzi *et al.*, 1969), *Malva silvestris* (Byrne & Heimsch, 1970) and *Pisum sativum*. In this species, the interesting observation was made that the radicle apex in resting seeds also contained cells with 2C–4C intermediate DNA values (S phase cells), whose frequency was very different in the two harvest years investigated (Bogdanov & Jordansky, 1964; Bogdanov, Liapunova & Sherudilo, 1967).

The course of events leading to cell quiescence in embryonic meristems at seed maturation has been investigated in *Vicia faba* and *Triticum durum*. In *Vicia faba* (Fig. 2.10) it has been observed that, at a given time in embryogenesis, the radicle meristem shows a mitotic index comparable to that of a seedling root meristem

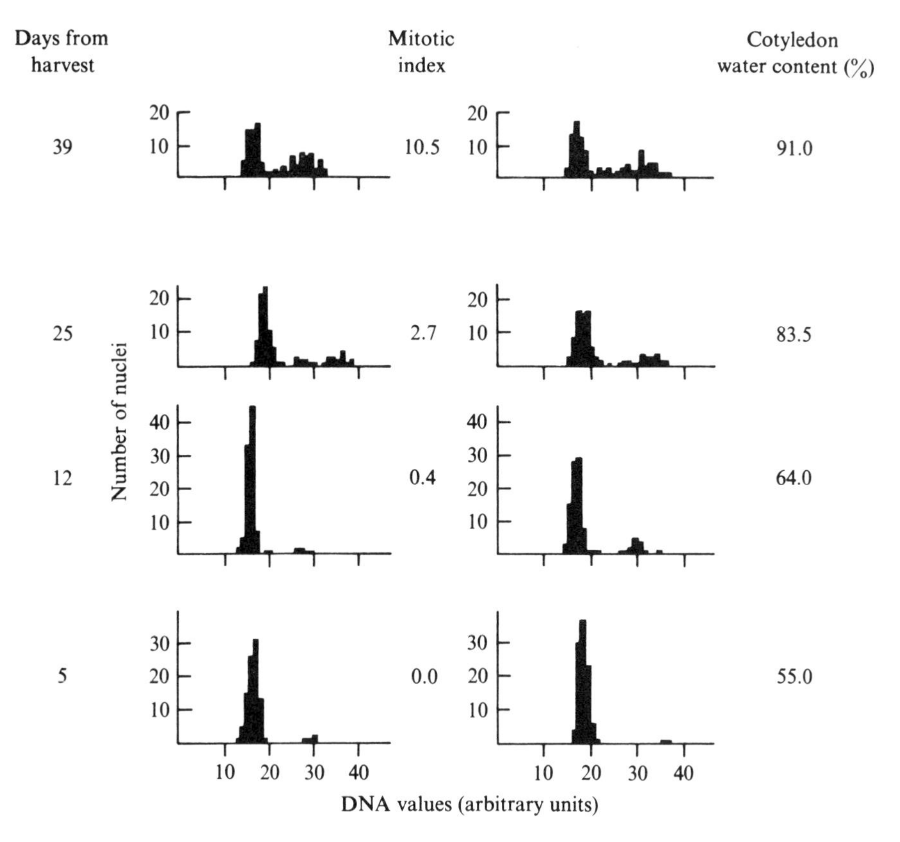

Fig. 2.10. Distribution of DNA contents in interphase nuclei of root meristems of embryos of *Vicia faba* at different times from seed harvest. Based on the data of Brunori (1967) (after D'Amato, 1972*a*).

(about 10%) and comprises cells with the expected interphasic DNA classes: 2C (G_1), 2C–4C (S) and 4C (G_2). At later development, the mitotic index decreases and DNA synthesis is clearly reduced until it is stopped (only two DNA classes: 2C and 4C). Because some residual mitotic activity is left, the proportion of 4C (G_2) to 2C (G_1) cells is reduced in the week before maturation. Since the primary root meristem of *Vicia* is left as a mixed population of G_1 and G_2 cells in the dry seed, it must be assumed that the interval between inhibition of DNA synthesis and inhibition of mitosis is not long enough to allow the passage into mitoses of all G_2 cells: in other words, there occurs an incomplete 'phase of depletion of G_2 cells' (Brunori & D'Amato, 1967). The importance of the depletion phase in the development of embryonic meristems in seeds is demonstrated by a comparative analysis of different meristems of the *Triticum durum* embryo, which have been shown to follow a different developmental history, whose final expression is the

TABLE 2.10. *Frequency of nuclei with 2C and 4C DNA contents in meristems of five embryos in dry (resting) seeds of* Triticum durum (based on data from Avanzi *et al.* 1969)

	Embryos									
	No. 1		No. 2		No. 3		No. 4		No. 5	
Meristems	2C	4C	2C	4C	2C	4C	2C	4C	2C	4C
Primary root	77	23	86	14	61	39	72	28	80	20
Root of I pair	90	10	91	9	95	5	94	6	92	8
Root of II pair	100	0	100	0	92	8	100	0	100	0
I leaf	100	0	100	0	100	0	100	0	100	0
II leaf	100	0	100	0	100	0	100	0	100	0
III leaf	50	0	50	0	50	0	50	0	50	0
Shoot	30	0	30	0	30	0	30	0	30	0

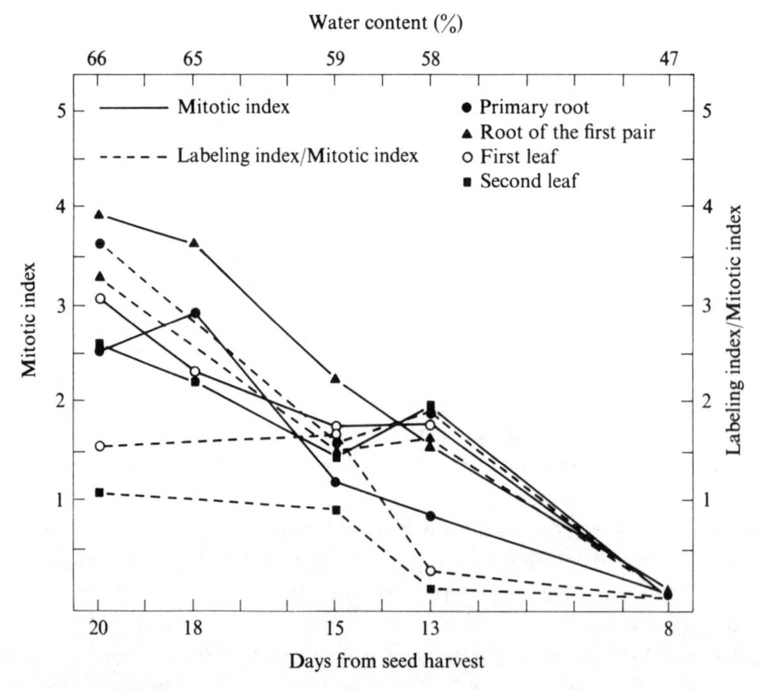

Fig. 2.11. Mitotic index and labeling index: mitotic index ratio in different meristems of embryos of *Triticum durum* at different times from seed harvest. Based on the data of Avanzi *et al.* (1969) (after D'Amato, 1972*a*).

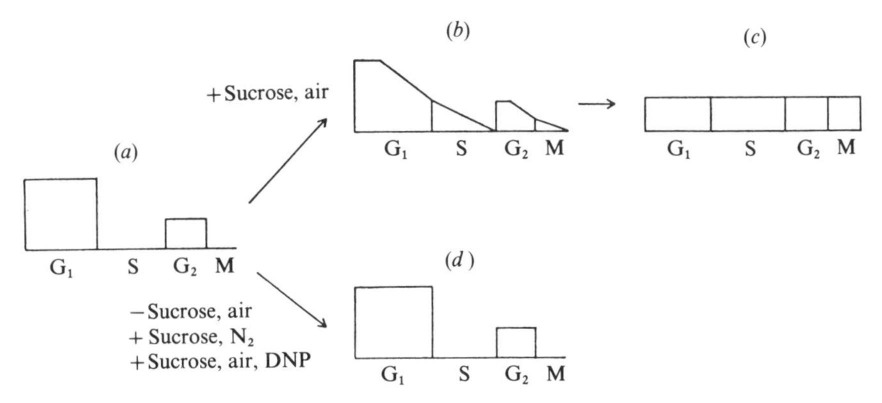

Fig. 2.12. Diagrammatic representation of cell distributions in apical meristems of excised roots at *Pisum sativum* grown *in vitro*. (*a*) stationary phase roots; (*b*) following transfer to sucrose in air; (*c*) proliferative phase roots; (*d*) stationary phase roots in absence of sucrose, or in presence of sucrose, under nitrogen, or in presence of sucrose, air and 2,4-dinitrophenol (after Webster & Van't Hof, 1969).

different cellular composition in the quiescent seed (Table 2.10). The shoot apical meristem ends its development rather early during embryogenesis, at a seed water content of 66%, when the other meristems are still growing by mitosis. But the relative durations of S and M in the meristems, as given by the labeling index/mitotic index ratio, are different, with a very clear contrast beween meristems destined to be a G_1 cell population at seed maturity (I and II leaf) and meristems to be a mixed G_1 and G_2 cell population (primary root and I-pair roots) (Fig. 2.11). Clearly in the leaf meristems the 100% efficiency of the depletion phase is achieved by inhibition of DNA synthesis at an earlier time than inhibition of mitosis (for more details, see Avanzi *et al.*, 1969; D'Amato, 1972*a*).

Although water availability may be a limiting factor in the development of embryonic meristems in a ripening seed, an energy requirement might be involved. In excised pea root tips in culture (Van't Hof, 1968, 1974; Webster & Van't Hof, 1969, 1970) a stationary phase in the meristems (cells in G_1 and G_2) can be established by carbohydrate starvation. Upon the provision of both sucrose and oxygen, the cell distribution changes due to the initiation of DNA synthesis and mitosis. The stationary phase can also be maintained in a sucrose-containing medium provided that the roots are grown either in a nitrogen atmosphere or in air with 2,4-dinitrophenol added (Fig. 2.12). In stationary phase meristems, cells can be positioned in the S phase by transfer to a medium with sucrose, and retained in S by hypoxia; these S cells resume DNA synthesis and enter mitosis when hypoxia terminates. Since in the excised pea root system cells collect preferentially in G_1 'the relative probabilities that a cell would remain in a given period would be $G_1 > G_2 >$ (S or M)' (Van't Hof, 1968; Van't Hof & Kovacs,

1972). This situation finds a striking parallel in the nuclear conditions established in embryonic meristems when the seed ripens.

Natural quiescence of shoot meristems occurs during normal growth of plants. In *Tradescantia paludosa*, the suppression of growth of axillary buds by apical dominance results in a clear zonation of the apex; whereas mitosis occurs, albeit infrequently, throughout the rest of the meristem, it is absent within a group of cells located in the summit of the apex. These cells have large nuclei with a much lower histone content than that in the subjacent cells of the meristem and most of them contain 2C DNA (Dwidevi & Naylor, 1968); it is not known whether the few 2C–4C cells found in the apex summit are S cells or cells engaged in extra DNA synthesis (amplification). Following shoot decapitation (removal of the terminal bud), the axillary buds are released from inhibition: the cells of the summit zone now enter DNA synthesis and mitosis, and a DNA/histone ratio similar to that of a proliferating shoot meristem is re-established (Dwidevi & Naylor, 1968). A shoot apex zonation identical to that occurring in axillary buds inhibited by apical dominance can be induced in the terminal bud by limiting either the level of nitrogen or the intensity of light available to the plant (Yun & Naylor, 1973).

3

Meiosis: its course, modification and suppression

The course of meiosis

As we have seen in Chapter 1, meiosis is the nuclear process, typical of sexually reproducing eukaryotes, which precedes the formation of gametes or spores. Normal meiosis, which consists of two successive nuclear divisions in the meiocyte (the cell undergoing meiosis), serves the following main functions:

(i) it provides for the random assortment of paired chromosomes and the halving (reduction) of the somatic chromosome number which was established at zygote formation;
(ii) it determines the segregation of alleles as well as the random recombination of independent (i.e. unlinked) genes (interchromosomal genetic recombination) and the nonrandom recombination of linked genes (intrachromosomal genetic recombination).

In organisms having chromosomes with localized centromeres, meiosis is prereductional; that is, the homologous chromosomes of each chromosome pair (bivalent), except for the cross-over segments, are segregated to opposite poles at the anaphase of the first meiotic division. In organisms with diffuse centromeres meiosis is postreductional (see later).

According to Westergaard (1964) and Rossen & Westergaard (1966), three types of meiosis can be recognized:

(i) The *Lilium* type represents meiosis in diplontic and haplo–diplontic species in which karyogamy (sexual nuclear fusion) and meiosis are separated by diploid mitoses. In the *Lilium* type, meiotic chromosome replication takes place at the last premeiotic interphase; moreover, there occurs a typical leptotene stage and the chromosomes are in a much extended state when undergoing homologous pairing;
(ii) The *Neurospora* type is found in haplontic fungi in which there is no time interval between karyogamy and the onset of meiosis. The *Neurospora* type is distinguished by two important features: there is no leptotene stage, and homologous pairing occurs between highly contracted chromosomes;
(iii) The *Clamydomonas* type is found in haplontic species (e.g. many species of algae) in which the zygote enters a resting stage of different durations (hours, weeks or months) and meiosis is initiated when the zygote germinates.

In the following pages the course of meiosis of the *Lilium* type will be described, but some interesting aspects of the *Neurospora* type meiosis which have emerged in the last few years will be considered. Recent descriptions of meiosis are found in Taylor (1967), Barry (1969), Walters (1970) and Kezer & Macgregor (1971).

Premeiotic interphase

Autoradiographic and cytophotometic analyses on several species of plants and animals have shown that DNA replicates in the interphase nucleus and, in some species at least, DNA duplication (4C nuclear DNA content) is accomplished before meiotic prophase begins (Taylor, 1967; Bogdanov, Liapunova, Sherudilo & Antropova, 1968; Antropova & Bogdanov, 1970; Bogdanov & Antropova, 1971; Liapunova & Babadjanian, 1973). That the chromosomes are effectively double before meiotic prophase is also shown by the irradiation response (chromatid-type aberrations) of nuclei at the end of premeiotic interphase (Sauerland, 1956; Mitra, 1958). Meiosis also starts from a DNA-doubled diploid (4C) nucleus in two ascomycetes, *Neottiella* and *Neurospora*, with a *Neurospora*-type meiosis: the initial nucleus of the ascus enters meiosis immediately after its formaton by fusion, in the crozier, of the haploid nuclei in a chromatin condensed, DNA doubled (2C, G_2) state (Rossen & Westergaard, 1966; Westergaard & Wettstein, 1968, 1972).

In the *Lilium*-type meiosis, histone synthesis occurs during the premeiotic interphase. This has been shown by biochemical analyses taking advantage of the division synchrony in the pollen mother cells of *Lilium longiflorum* and *L. candidum* (Sheridan & Stern, 1967; Bogdanov, Strokova & Reznickova, 1973: Strokov, Bogdanov & Reznickova, 1973), and by cytophotometric and interference microscopic analyses. Cytophotometry of DNA (Feulgen) and histones (Fast green at pH 8.2) has shown an uncoupling of DNA and histone synthesis in the premeiotic interphase in spermatocytes of *Acheta domesticus* (Bogdanov, Liapunova, Sherudilo & Antropova, 1968), *Pyrrhocoris apterus* (Antropova & Bogdanov, 1970) and *Triturus vulgaris* (Bogdanov & Antropova, 1971). In these cases, when the spermatocyte nucleus has attained the 4C DNA content, the nuclear histone content amounts to about 3C: the doubling of histones is completed in early meiotic prophase. Histone synthesis at meiotic prophase has been confirmed by means of interference microscopy in the spermatocytes of *Acheta* (Liapunova & Babadjanian, 1973) and by biochemical analyses in the meiocytes of *L. candidum* (Bogdanov *et al.*, 1973; Strokov *et al.*, 1973). Histones are not the only molecules to be synthesized at meiotic prophase; there also occurs synthesis of DNA and proteins other than histone. We will return to these aspects of the biochemistry of meiosis later on.

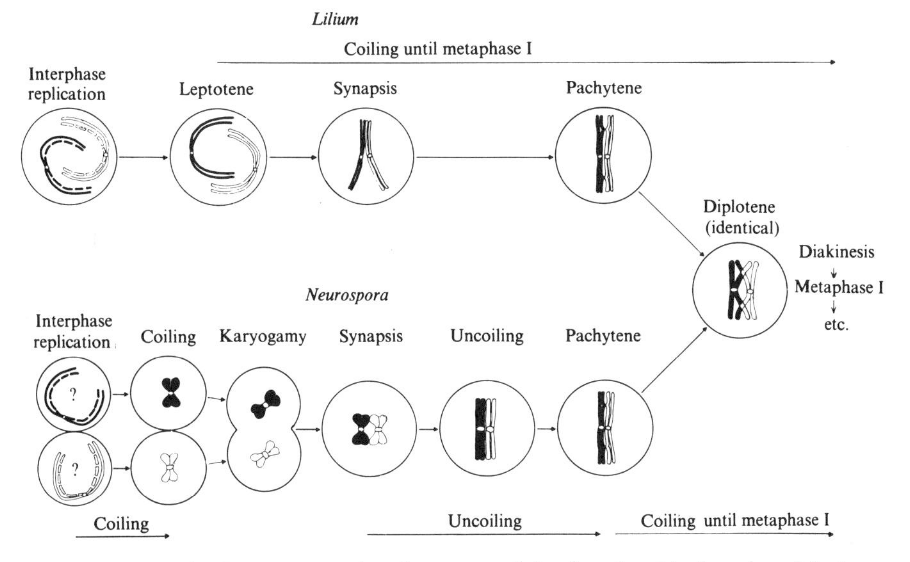

Fig. 3.1. Schematic diagram comparing chromosome behaviour (one bivalent shown) in the *Lilium* and *Neurospora* types of meiosis. In the *Neurospora* type the diploid nucleus entering meiosis results from the fusion of two haploid cells in the DNA-doubled (2C) state. The question marks indicate that the precise time and actual mechanism of chromosome replication in fungi is not known (after Rossen & Westergaard, 1966).

Meiotic prophase: cytological aspects

The first generally accepted stage of meiosis is leptotene, in which the chromosomes are seen as very thin strands with chromomeric structure and are unpaired and randomly distributed in the nucleus. In many, especially animal, meiocytes the chromosomes are polarized in the nucleus with one of their ends attached to a small area of the nuclear envelope from which the chromosomes extend across the nucleus: this arrangement (bouquet stage) will persist until pachytene (Moens, 1969; Kezer & Macgregor, 1971). Leptotene terminates when each chromosome of the complement pairs (synapses) with its homologue (zygotene). Although some contraction has already occurred, the chromosomes paired in zygotene are still much despiralized. In contrast with this situation, which is typical of the *Lilium*-type meiosis, synapsis occurs between strongly contracted chromosomes in the species characterized by *Neurospora*-type meiosis (Fig. 3.1). During pachytene the paired chromosomes (bivalents) undergo further contraction becoming several times shorter than at leptotene. However, in the *Neurospora*-type meiosis, there occurs a gradual uncoiling of the chromosome pairs, so that the degree of coiling of bivalents at pachytene in the two types of meiosis is comparable. At pachytene, in favourable materials, each of the paired chromosomes is seen to consist of two chromatids, and there are one or several reciprocal exchanges of segments between homologous nonsister chromatids recognizable as cross configurations (chiasmata). In many animal and plant species, of both the *Lilium* and *Neurospora* type, pachytene is followed by a diffuse stage in which the DNA that was folded into the chromomeres of the pachytene bivalent becomes extended into Feulgen-positive lateral loops (Moens, 1964; Barry, 1969; Rossen & Westergaard, 1966; Kezer & Macgregor, 1971). This situation is most clearly seen in the lampbrush condition of the bivalents in young amphibian oocytes (Gall, 1954; Callan & Lloyd, 1960). From the diffuse stage the nucleus passes into diplotene by a progressive chromosome shortening, which also involves a complete or incomplete folding of the lateral loops into the axes of the bivalents. At diplotene, the homologous chromosomes of each bivalent repel one another at the level of the centromeres and adjacent segments. Separation of the centromeres is regularly reductional, i.e. sister chromatids remain together at the centromere during diplotene (prereductional meiosis). That segregation of centromeres is regularly reductional can be seen in cytologically favorable materials, but it is convincingly demonstrated by autoradiographic analyses of the spermatocytes of grasshoppers (Taylor, 1965), and by extensive analyses of chromosome segregation at meiosis in plants and in animals heterozygous for a heteromorphic chromosome pair or an interchange difference (references in John & Lewis, 1965, Table 72). Genetic proof of the regularly reductional chromosome separation at the centromere is provided by the genetic analysis of the spores of individual tetrads (tetrad analysis) in many fungi (Esser & Künen, 1967), in the liverwort *Sphaerocarpos donnellii* (Abel, 1967) and in a few angiosperms which shed pollen in tetrads, e.g. *Epilobium* (Michaelis, 1966). Tetrad analysis also demonstrates that the reciprocal exchanges of chromosome segments, which occur at the point of the chiasmata and lead to recombination of linked genes (crossing-over), take place in a four-chromatid bivalent (tetrad of some authors), each crossing-over involving two non-sister chromatids in the bivalent. For the time and place of meiotic crossing-over, see Henderson (1970).

Meiotic prophase ends with diakinesis, a stage in which chromosome contraction is near maximum and the nucleolus or nucleoli usually disappear.

Electron micrographs of chromosomes paired at meiotic prophase has revealed, in a great number of eukaryotes from protists to man, a tripartite structure which was first observed in the crayfish *Procambarus clarkii* and named synaptinemal complex (Moses, 1956). Review articles on the mechanisms of synapsis and the synaptinemal complexes have been written by Moses (1968), Wettstein & Sotelo (1971), Westergaard & Wettstein (1972) and Moens (1973). In longitudinal section the synaptinemal complex joining the two paired homologous chromosomes of a bivalent consists, in Westergaard and Wettstein's (1972) terminology, of two lateral components and one central region which is composed of a central component surrounded by a less electron-dense space. The lateral components are held in register at a distance of about 1000 Å by the central region, whose width is an extremely uniform dimension. The three-dimensional structure of the lateral component is a flattened rod with a diameter of about 500 Å. Uniformity in dimension is also found in the central component. For details, see Table 1 in Westergaard & Wettstein (1972) which gives the structure and dimensions of the synaptinemal complexes (61 genera of eukaryotes) reported since the review by Moses (1968). In *Neottiella rutilans* (*Neurospora*-type meiosis), as well as in other ascomycetes, the lateral components have a banded structure, while the central component is amorphous (no regularly repeating substructure). Central components are generally amorphous, but in insects they have a lattice structure. If the ascomycetes, two species of grasshopper and one species of Pentatomida are excluded, the remaining species listed by Westergaard & Wettstein (1972) show amorphous lateral components. The synaptinemal complex joins the homologs over their whole length, it joins the homologous centromeres and proceeds without interruption from euchromatin to heterochromatin. Cross-sections of bivalents show that they form a tube of chromatin with the synaptinemal complex in the central lumen and that the complex is associated throughout its lateral components with only a small portion of each homolog. Both the central and the lateral components of the synaptinemal complex consist of proteins and RNA. When homologous chromosomes repel at diplotene, the synaptinemal complex is shed from the bivalents except where the chromosomes are held together by chiasmata.

It is generally accepted that the synaptinemal complex is needed for crossing-over and chiasma formation. There seem to be, however, a few exceptions, such as the presence of the complex in meiocytes which are known to lack genetic exchange and/or chiasmata. There is also the converse situation, in which high frequencies of genetic exchanges, as determined by genetic tests, are not accompanied by a detectable synaptinemal complex. This condition has been found in the males of some strains of *Drosophila ananassae* which undergo meiotic exchange. It raises questions concerning the involvement of the complex in crossing-over (Grell, Bank & Gassner, 1972). In this connection, it is to be recalled that genetic exchange is accomplished in protokaryotes and in somatic higher cells which lack the synaptinemal complex.

Meiotic prophase: biochemical aspects

If we now turn to the biochemistry of meiotic prophase, we may consider some syntheses which might be in some way related to the process of chromosome pairing and/or crossing-over.

In the meiocytes of *Lilium* species and hybrids and of *Trillium erectum*, which progress synchronously through the meiotic cycle, it has been shown that most of the nuclear DNA replication occurs during the premeiotic S phase, but a small amount of DNA synthesis occurs at zygotene and pachytene (Hotta, Ito & Stern, 1966; Hotta & Stern, 1971*a*; Stern & Hotta, 1973). The DNA synthesized during zygotene (ZDNA), approximately 0.3% of the total nuclear DNA, has a higher density (48–50% G–C content) (Table 3.1). ZDNA is not rDNA, which in *Lilium*

TABLE 3.1. *Composition of bulk and zygotene DNA in liliaceous plants* (from Hotta & Stern, 1971*a*)

Plant	Bulk DNA density (g/ml)	Zygotene DNA	
		Density (g/ml)	% G+C
Trillium erectum	1.705	1.713	—
Lilium longiflorum	1.702	1.712	49.8
speciosum	1.700	1.713	49.5
tigrinum	1.701	1.710	—
Cinnabar	1.701	1.712	48.9
Bright star	1.701	1.711	47.8
Limelight	1.702	1.713	—

has a still higher density (55% G–C content); ZDNA is present in the genome of the species and its synthesis during zygotene represents a delayed replication (in *Lilium* it occurs 3 to 5 days after the completion of the premeiotic S phase). Autoradiographic data and analyses of the effect of DNA synthesis inhibitors during zygotene (Ito, Hotta & Stern, 1967) may indicate that ZDNA sequences are distributed along the length of each of the chromosomes (Hotta & Stern, 1971*a*). According to these authors, DNA synthesis at zygotene appears to be a necessary, although not sufficient, condition for chromosome pairing. DNA synthesis during pachytene, the stage when genetic recombinaton is assumed to occur, is not of the semiconservative replication type but of the repair replication type. Analyses of DNA–DNA hybridization show that DNA replication at pachytene does not result in a net increase of DNA, which means that the regions undergoing replication must be replacing pre-existing ones. Since, unlike ZDNA, these regions vary in composition and have an average base composition which approximates that of total nuclear DNA, they may indicate repair replication of crossing-overs between homologs (Hotta & Stern, 1971*a*). This possibility is supported by the finding, during zygotene and pachytene in *Lilium*, of enzyme activities involved in DNA repair processes: an endonuclease, a polynucleotide kinase and a ligase (Howell & Stern, 1971). Recent investigations indicate that most of the repair synthesis at pachytene in *Lilium* meiocytes is in moderately repeated DNA sequences (Smyth & Stern, 1973).

Autoradiographic analyses have also demonstrated the incorporation of tritiated thymidine into meiotic chromosomes in newts (Wimber & Prensky, 1963), mice (Mukherjee & Cohen, 1968), man (Lima-de-Faria, German, Ghatnekar, McGovern & Anderson, 1968) and wheat (Riley & Bennett, 1971). In this last species, incorporation of the precursor (indicative of DNA synthesis and/or repair) occurs not only in meiotic prophase but throughout meiosis, even during stages with maximally contracted chromosomes. Biochemical analyses on the meiocytes of

TABLE 3.2. *Histone amount of isolated spermatocyte I nuclei in the cricket,* Acheta domesticus (from Liapunova & Babadjanian, 1973)

	Mean amount of histone per nucleus (pg) and standard error			
Type of histone	A. At the beginning of interphase	B. At the beginning of prophase I	C. At the end of pachytene	C/A
Total histone*	7.30∓0.13	10.48∓0.19	17.40∓0.31	2.4
Fractions groups				
I (F_1)	1.00∓0.04	2.59∓0.11	6.03∓0.26	6.0
II (mainly F_2b and F_2a2)	2.28∓0.12	2.68∓0.15	3.57∓0.19	1.6
III (F_2a1 and F_3)	4.04∓0.09	5.43∓0.11	8.47∓0.18	2.2

* Total histone at the end of interphase in spermatogonia is 13.31∓0.27.

another cereal species, rye, have shown that DNA synthesis occurs not only during zygotene and pachytene–diplotene but also in all stages from metaphase I to telophase II, although the amount of DNA synthesized at meiotic prophase is clearly greater. In contrast to the situation found in *Lilium*, the buoyant density distributions of DNA from zygotene and pachytene–diplotene cells in rye are indistinguishable (Flavell & Walker, 1973). It has been proposed that this difference between *Lilium* and rye might be related to the far shorter duration of meiosis in rye and to the possibility that chromosome pairing in cereals is organized in a premeiotic phase (Flavell & Walkerl, 1973).

During meiotic prophase, several distinctive proteins are synthesized in *Lilium* meiocytes (Hotta, Parchman & Stern, 1968). When the protein synthesis is inhibited with cycloheximide at any stage of meiotic prophase, meiosis, is stopped; meiocytes released from inhibition in late zygotene progress through meiosis but fail to form chiasmata (Parchman & Stern, 1969). Among the proteins synthesized in meiotic prophase, one is similar in many respects to the 'gene 32-protein' which is an essential component of genetic recombination in T_4 bacteriophage (Alberts, 1970). This protein, which is not found in somatic cells and disappears from meiocytes after meiotic prophase, has a very high affinity for single-stranded DNA. In its presence 60% of the denatured lily DNA is converted to a double stranded form at 25 °C in 2 hours. It has been postulated that this protein, which occurs also in rat, bull and human, promotes pairing and crossing-over between homologous chromosomes (Hotta & Stern, 1971*b*, *c*).

We have already seen that in spermatocytes of *Acheta*, *Pyrrhocoris* and *Triturus* histone synthesis in not coupled with DNA synthesis, so that doubling of histone (4C amount) is completed during meiotic prophase. In *Acheta* quantitative interference microscopy has been used to determine the amount of total histone and of the separate groups of histone fractions in isolated nuclei of spermatogonia and spermatocytes. It has been shown (Table 3.2) that the total histone amount per nucleus is about 11 pg (about 3C) at the end of premeiotic interphase, and this increases to 17–18 pg (about 5C) by the end of pachytene. The 3 pg histones in excess of the expected 4C value (= 14–15 pg) were extracted from isolated nuclei at pH 2.2 together with F1 histone (Liapunova & Babadjanian, 1973). Occurrence of an extra amount of histone (14–15% of total histone) has also been demonstrated in the meiocytes of *Lilium candidum* (Strokov *et al.*, 1973). Liapunova & Babadjanian and Strokov *et al.* equate this histone fraction with the meiotic histone first described by Sheridan & Stern (1967) in *Lilium* microsporocytes (it is practically absent in somatic tissues). Strokov *et al.* have suggested that the meiocyte-specific histone fraction they call FM is one of the molecular components of the synaptinemal complex.

From prometaphase I to telophase II

At prometaphase, the nucleolus or nucleoli disappear, the nuclear envelope breaks down, the spindle forms and the bivalents move towards the spindle equator (congression). At metaphase I (by early diplotene in some materials) each member of a chromosome pair (half bivalent) has two sister centromeres, one per chromatid (Brinkley & Stubblefield, 1970; Kezer & Macgregor, 1971). The two sister centromeres of each half bivalent behave, however, as if they were 'functionally single' throughout the first meiotic division. During metaphase I the homologous double centromeres of each bivalent come to lie equidistant from the equator and their respective spindle poles (coorientation). Two sister chromatids (except for cross-overs) are attached to each of the two double centromeres of the bivalent which still shows one chiasma or more. Coorientation is random as to which spindle pole is related to the paternal centromere; a random number of paternal (or maternal) centromeres go to each pole. At anaphase, the homologs separate from one another (half bivalents, each consisting of two chromaids) and move towards their respective poles. The reconstruction at the spindle poles of nuclei with a reduced chromosome number (telophase I) is followed by an interkinesis of variable duration. In some organisms, both animals and plants, telophase I and interkinesis are absent. The first meiotic division may or may not be followed by cell division (cytokinesis).

Meiosis II is mechanically similar to mitosis. After a chromosome stage (prophase II), which does not occur in organisms which omit interkinesis, a metaphase II follows in which the centromeres line up on the equator of the second division spindles. After their movement to the poles, each sister centromere carrying the component chromatid of each chromosome (anaphase II), there occurs reconstitution of interphase nuclei (telophase II). In organisms in which meiosis I is followed by cytokinesis, meiosis II may or may not be followed by cytokinesis. When cytokinesis I does not occur, meiosis II also may not be followed by cytokinesis. In still other cases, simultaneous cytokinesis occurs at telophase II to form four separate cells.

Although it is generally assumed that chromosome pairing occurs at zygotene, it has been reported that in some species of plants and animals the homologous chromosomes may be seen to be paired at the anaphase of the last premeiotic mitosis (Moens, 1964; Feldman, 1966; Brown & Stack, 1968; Stack & Brown, 1969). Premeiotic pairing challenges the classical theory of meiotic chromosome pairing. However, in a cytologically favorable material, *Nicotiana otophora*, it has been shown that the chromosomes do not pair prior to meiosis (Burns, 1972).

Post reductional (inverse) meiosis

The usual sequence of a reductional first and an equational second division in meiosis, as described in the previous section, is inverted in most, but not all, organisms with a diffuse kinetochore (= holokinetic or holocentric chromosomes) (references in White, 1973). The best investigated holokinetic chromosomes are those of Hemiptera (aphids and coccids) (Hughes-Schrader & Ris, 1941; Hughes-Schrader & Schrader, 1961) and of the plant *Luzula purpurea* (Castro, Camara

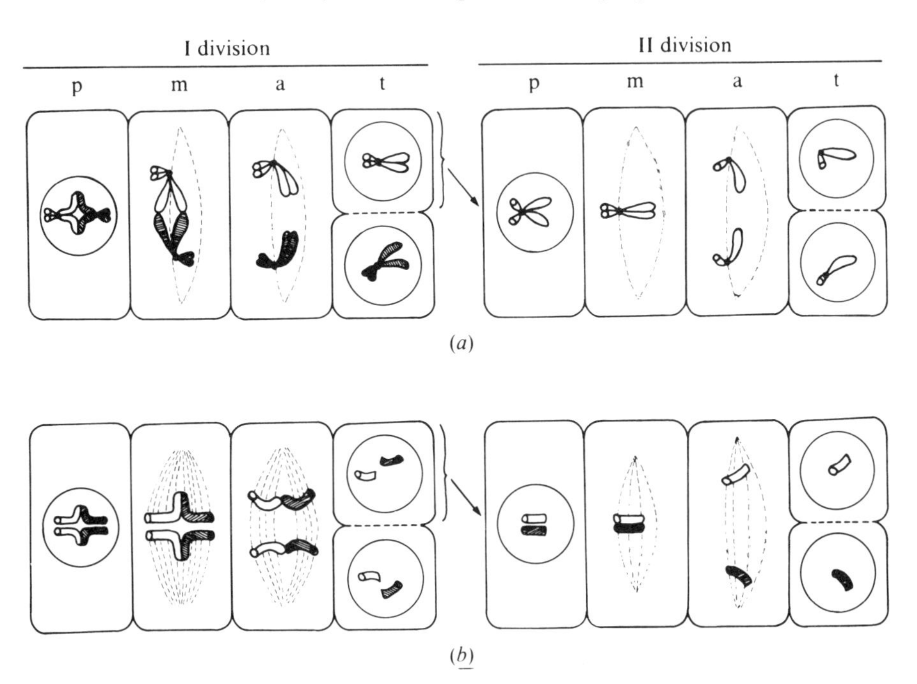

Fig. 3.2. Schematic outline of the pre-reductional (*a*) and post reductional (*b*) types of meiosis. Only one bivalent is shown from prophase I to telophase II. P, m, a, t: prophase, metaphase, anaphase and telophase. Cross-overs are not considered (redrawn from Battaglia & Boyes, 1955).

& Malheiros, 1949; La Cour, 1953; Nordenskiöld, 1962). During mitosis, holokinetic chromosomes show a typical behavior, namely:

(i) at metaphase the chromosome orients with its long axis in the equatorial plane of the spindle;
(ii) chromosomal fibers form throughout the entire length of each chromatid;
(iii) during the anaphase movement the chromatids tend to remain parallel to each other as they separate.

The diffuse nature of the kinetochore is demonstrated by the ability of X-ray induced chromosome fragments to function mitotically, thus being perpetuated through many cell generations.

Postreductional meiosis, the inverted meiotic sequence, has been excellently reviewed by Hughes-Schrader (1948) and Battaglia & Boyes (1955). In this context, suffice it to say that in the organisms with holokinetic chromosomes which undergo inverse meiosis, chiasmata, if present, are completely resolved by metaphase I, and bivalents show sister chromatids (four per bivalent) separated throughout their length. In these conditions, at anaphase I the plane of division coincides with the plane of chromatid separation, which makes evident the equational nature (only for the non-crossover segments of the chromosomes) of the first meiotic division. Thus, each telophase I nucleus receives an unreduced number of chromatids (in prereductional meiosis, however, each telophase nucleus receives a reduced number of chromosomes). At interkinesis the members of each pair of homologous chromatids approach to form pairs which are disjoined in anaphase II; this results in the actual reduction of the chromosome number (Fig. 3.2).

The genotypic control of meiosis

The whole meiotic process is under genetic control. The normal course of meiosis must be viewed as the product of a definite temporal sequence in the action of genes and/or groups of genes. Inactivation or loss of genes involved in the meiotic process may be the result of genotypic changes of a different nature, such as mutations, chromosome structural changes, changes in genetic background and/or balance, etc. In any case, the result is expected to be an abnormality of the meiotic process, more precisely, a block of the meiotic stage controlled by the gene(s) involved. Many examples of a genotypic control of the meiotic process have been reported in the literature; most of them have been described in plants. Compared with the wealth of information on the genetics of some plant and animal species (maize, pea, *Drosophila*, silkworm), our knowledge on the genotypic control of their meiosis is quite inadequate. Table 3.3 includes some typical examples of a genotypic control of the meiotic processes, which will serve as a basis for our present discussion. More data on the genotypic control of meiosis are found in Rees (1961), John & Lewis (1965) and Riley & Law (1965).

TABLE 3.3. *Examples of genotypic control of the meiotic process*

Process (effect)	Species	Genetic control	References
1. Homologous pairing (complete or partial asynapsis)	*Zea mays*	Recessive gene, *as*	Beadle, 1930; Miller, 1963
	Triticum monococcum	Recessive gene, *as*	Smith, 1936
	Gossypium hirsutum × *G. barbadense*	Duplicate recessive gene (15:1 segregation in F2)	Beasley & Brown, 1942
	Hyoscyamus niger	Duplicate recessive gene (15:1 segregation in F2)	Vaarama, 1950
	Triticum durum	Recessive genes	Martini & Bozzini, 1966
2. Homoeologous pairing (prevention)	*Triticum aestivum*	5B, long arm	Sears, 1954, 1958; Sears & Okamoto, 1959; Riley & Chapman, 1958, 1963; Riley & Kempanna, 1963; Riley, 1966, 1967
3. Chiasma formation			
(a) Occurrence (desynapsis)	*Secale cereale*	Recessive gene	Prakken, 1943
	Brassica oleracea	Recessive genes, *ds*	Gottschalk & Konvicka, 1970
	Pisum sativum	Recessive genes, *ds*	Gottschalk & Villalobos-Pietrini, 1965; Gottschalk & Baquar, 1971
	Drosophila subobscura	Recessive gene	Fahmy, 1952
(b) Frequency	*Secale cereale*	Polygenic system	Müntzing & Akdik, 1948; Rees, 1955; Rees & Thompson, 1956
	Zea mays	Polygenic system	Riley & Law, 1965
	Triticum aestivum	Polygenic system	Riley & Law, 1965
(c) Frequency (reduction)	*Triticum aestivum*	Five nullisomics	Riley & Law, 1965
(d) Distribution	*Allium cepa* × *A. fistulosum*	One gene?	Emsweller & Jones, 1945
4. Mitotic spindle (abnormalities)	*Zea mays*	Recessive gene, *dv*	Clark, 1940
	Drosophila simulans	Recessive gene, *dv*	Wald, 1936
5. Chromosome behaviour ('non-disjunction')	*Drosophila melanogaster*	Recessive gene, *c3G*	Gowen, 1933; Lewis, 1948
		Recessive gene, ca^{nd}	Lewis & Gencarella, 1952
		Sex-linked gene	Spieler, 1963

For details, see text.

Asynapsis

A prerequisite for pairing (synapsis) is the homology of chromosomes or chromosome segments. When no homology exists between the chromosomes of the two parents, as in interspecific or intergeneric hybrids, no synapsis occurs at meiosis in the F_1 individual, and the chromosomes remain as univalents (asynapsis). Under particular genetic conditions, asynapsis occurs despite chromosome homology. This shows that homology is a necessary but not sufficient condition for chromosome synapsis. Recessive genes segregating with a typical 3:1 ratio are known to produce asynapsis in many organisms, generally in both the male and the female lines. The first of such genes was discovered in maize and called ' asynaptic' (Beadle, 1930). In plants homozygous for the recessive gene, the number of univalents at diakinesis varies from 0 (= 10 bivalents present) to 20. The possibility of bivalent formation and the demonstration of genetic recombination due to crossing-over in the homozygous recessive plants (Miller, 1963) make the gene better definable as desynaptic rather than asynaptic (for a correct terminology and classification of asynaptic and desynaptic mutants, see Gaul, 1954; Riley & Law, 1965; Martini & Bozzini, 1966). In plants, many cases of genetically controlled complete asynapsis, most commonly due to single recessive genes, have been reported in natural populations and in the progeny of irradiated plants; we may cite as examples *Triticum monococcum* (Smith, 1936), *Gossypium hirsutum* × *G. barbadense* (Beasley & Brown, 1942), *Hyosciamus niger* (Vaarama, 1950) and *Triticum durum* (Martini & Bozzini, 1966). As we will see in the next section of this chapter, asynapsis followed by formation of a restitution nucleus is an important mechanism of apomictic and parthenogenetic development in plants and animals.

Prevention of homoeologous chromosome pairing

More is known about the genetic systems that regulate meiotic chromosome pairing in the genus *Triticum* (Sears, 1954, 1958; Sears & Okamoto, 1959; Riley & Chapman, 1958, 1963; Riley & Kempanna, 1963; Riley, 1966–7) than in any other group of organisms. The bread wheat, *Triticum aestivum*, is an allohexaploid species (2n = 6x = 42) which contains the whole genomes of three diploid progenitor species, each with 2n = 14: *Triticum monococcum* or *T. aegilopoides* (genome A), *Aegilops speltoides* (genome B) and *Aegilops squarrosa* (genome D). Sears (1954, 1958) and Okamoto (1962) have shown that the 42 chromosomes of wheat can be classified into seven groups (now numbered 1–7) of three chromosome pairs each, one belonging to A, the other to B and the third to D (e.g. 1A, 1B, 1D; 2A, 2B, 2D, etc.). In each group, chromosomes of the three genomes are homoeologous; that is, they contain genes with similar functions. Chromosome homoeology between the three genomes is demonstrated a.o. by two sets of data: the absence of a chromosome pair is compensated for by the tetrasomy of a homoeologous pair (e.g. nulli 2A–tetra 2B: 2n = 42) and, in crosses between *Triticum aestivum* and *T. monococcum* (or *Aegilops squarrosa*), the hybrids show homoeologous chromosome pairing. However, in *Triticum aestivum* is found a typically diploid meiosis (21 bivalents due to strict homologous pairing) and consequently a disomic inheritance. Diploidization of the meiotic process in the hexaploid wheat is the result of a mutation in chromosome 5B which prevents homoeologous pairing; this mutation most probably arose after 5B was incorporated in a polyploid *Triticum*, because it does not occur in *Aegilops speltoides*, the donor of the B genome (Riley, 1967). The first evidence for the suppression of homoeologous chromosome pairing by 5B was obtained independently by Sears & Okamoto (1959) and by Riley & Chapman (1958). They demonstrated respectively that when chromosome 5B was absent from crosses between *T. aestivum* and *T. monococcum* there was far more chromosome pairing than normal and that 20-chromosome haploids deficient in chromosome 5B showed numerous bivalents and trivalents (there is little pairing in the euhaploid, n = 21). Further work by Riley's group has demonstrated the full activity of 5B in the normal hexaploid individuals (multivalent formation in 40-chromosome plants nullisomic for chromosome 5B) and has led to the localization of the mutation in the long arm of chromosome 5B.

Desynapsis

In the spermatocytes of *Drosophila melanogaster* and most short-antennae (Brachycera) flies and in the oocytes of *Bombyx mori* (Mittwoch, 1967; White, 1973) chromosomes pair regularly but no chiasmata and no crossing-overs occur (achiasmatic meiosis). Despite the absence of chiasmata, chromosome segregation at anaphase I is normal (prereductional). Achiasmatic meiosis, which has been fixed by evolution in the heterogametic sex of many animal species, is different from desynapsis. In desynaptic meiosis, the chromosomes pair normally at zygotene, but at diplotene they fall apart due to the failure of chiasma formation or to the relaxation of pairing at the points of chiasmata. The result of desynapsis is univalent formation at diakinesis as in asynapsis; indeed, in species in which the pachytene stage is not easily analyzed, difficulties in differentiating desynapsis and asynapsis may arise. Due to the aberrant behaviour of univalents at anaphase I, both asynapsis and desynapsis result in high or total sterility. A case of desynapsis in rye, *Secale cereale*, was found to be controlled by a single recessive gene. The homozygous recessive plants showed in their pollen mother cells seven bivalents until diplotene, but at metaphase I a variable number of univalents (up to 14 in about 2% of meiocytes) was present. Desynapsis also occurred in the female line (embryo sac mother cells) (Prakken, 1943). Genes which are responsible for desynaptic behaviour are known from a variety of plant species (references in Gottschalk & Baquar 1971). The most intensive work on the genetic control of desynapsis has been done on *Pisum sativum;* eighteen desynaptic mutants have

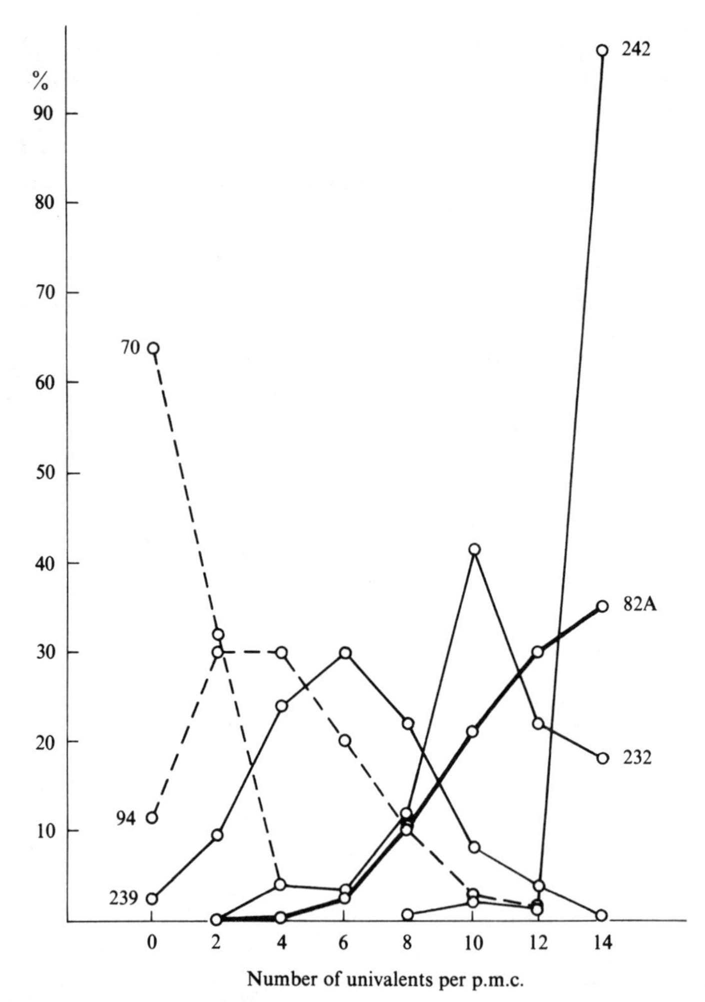

Fig. 3.3. Comparison of the degree of desynapsis (number of univalents per pollen mother cell) in mutants 70 and 94 of *Brassica oleracea* and mutants 239, 232, 242 and 82A of *Pisum sativum* (after Gottschalk & Baquar, 1971).

been analyzed cytogenetically. The genetic analyses have shown that the desynaptic genes in *Pisum* do not belong to a multiple allelic series. This may indicate that a relatively high number of genes exists in the genome of most of the higher plants which cause desynapsis when they are present in a mutated condition (Gottschalk & Baquar, 1971). Desynaptic genes in *Pisum* generally cause a high degree of desynapsis, as can be seen from Fig. 3.3, which also reports data on two desynaptic mutants of *Brassica oleracea* var. *capitata* (Gottschalk & Konvicka, 1970). One case of genetically controlled desynapsis may be cited among the few observed in animals: the 'cross-over suppressor' recessive mutation which is effective only in the homozygous females of *Drosophila subobscura* (Fahmy, 1952). It is probable that desynapsis, or asynapsis, is the primary cause of some of the genetically controlled cases of meiotic chromosome non-disjunction (see below).

Modification of chiasma frequency and distribution

In addition to the major genes responsible for synapsis, which have been discussed in the preceding pages, there are also complex systems of pairing control. Modification of chiasma frequency, a quantitative character, occurs when the established breeding system in a plant species is disrupted. Thus, in rye, the disruption by inbreeding of its normal outbreeding (allogamous) habit clearly reduces chiasma frequency (Müntzing & Akdik, 1948). Rees (1955) and Rees & Thompson (1956) have studied the genetic system responsible for the maintenance of high chiasma frequencies in outbreeding populations of rye. They have shown that high frequencies of chiasmata cannot be simply the reflection of heterozygosity but must derive from the operation and interdependence of many genes (polygenic system). As indicated by the reduction of chiasma frequency in inbred lines of maize, *Zea mays*, another allogamous species, a polygenic system operates in the control of chiasma frequency in this species (Riley & Law, 1965).

An effect analogous to that obtained in rye is found when the normally inbreeding (autogamous) hexaploid wheat, *Triticum aestivum*, is outcrossed: reduction of chiasma frequency and increased univalent frequency occur in F1 individuals of intervarietal crosses. Riley & Law (1965) have discussed the experimental evidence that a polygenic system operates in the maintenance of high chiasma frequency in wheat populations. In wheat, gene interaction in the control of chiasma frequency is also demonstrated by analyses on nullisomics. Nullisomes for five different chromosomes have shown a statistically significant reduction in chiasma frequency as compared to the euploid. The authors suggest that the operation of polygenic systems is probably associated with major gene effects in the regulation of chromosome pairing in many species.

Chiasma distribution also seems to be under genotypic control. In *Allium cepa* (onion), chiasmata are randomly distributed along the chromosomes; in *A. fistulosum*, chiasmata are localized in chromosome segments on both sides of the centromere. In the F1 hybrid between the two species, chiasmata are randomly distributed; in the backcross of F1 to *A. fistulosum*, nearly one half of the individuals had proximal chiasmata whereas the other half had randomly distributed chiasmata (Emsweller & Jones, 1945). This behaviour would indicate a simple mendelian inheritance.

Spindle abnormalities and chromosome loss

In maize, a single recessive gene (*dv*) has been found to produce a divergent spindle at division I in pollen mother cells, but not in embryo sac mother cells. Divergent spindles (spindle fibers diverging instead of converging at the poles) lead to the formation of numerous small nuclei at telophase I (Clark, 1940). This mutation is similar in effect to the *claret* mutation of *Drosophila simulans*, inducing an abnormally wide first meiotic spindle in oocytes which results in the formation of 4–12 nuclei at telophase II (Wald, 1936).

In *Drosophila melanogaster*, several genes have been found which induce chromosome non-disjunction and chromosome loss: for example, the recessive genes *c3G* (Gowen, 1933; Lewis, 1948) and ca^{nd} (*claret* non-disjunction) (Lewis & Gencarella, 1952) located in chromosome III, and a factor located at the proximal end of X-chromosomes which is effective only in females (Spieler, 1963). The genes *c3G* and ca^{nd} have an effect only on homozygous females, the homozygous males showing a normal spermatogenesis. In the case of the sex-linked factor, which induces frequent X-chromosome loss, its action seems to be modified by a complex system of sex-linked factors (Spieler, 1963). Since the genetic analyses are not substantiated by cytological observations, nothing is known of the primary cytological cause of 'non-disjunction' in the *Drosophila* mutants. In some mutants, asynapsis or desynapsis might be at work.

Apomixis in plants and animals

Genetic systems responsible for the abolition of gamete fusion (amphimixis) in the life cycle have been fixed by natural selection in a large number of groups in both the plant and the animal kingdom. Genetically controlled modifications or suppression of the meiotic process leading to the production of unreduced (2n) spores or gametes play an important role in the asexual reproduction of such organisms. In 1908, Winkler included under the term 'apomixis' the various types of reproduction which do not result from the fusion of gametes (amphimixis). Apomixis comprises (i) vegetative propagation, in which a single somatic cell or a group of somatic cells produces a new individual, sometimes through the intermediary of an embryo (adventitious embryony in higher plants) and (ii) parthenogenesis, the formation of an embryo from either a reduced (n) or an unreduced (2n) egg in the absence of fertilization. In a few apomictic angiosperms, the embryo can be formed by division of a cell other than the egg in an unreduced embryo sac (apogamety). Apogamety is a rather common method of apomictic development in pteridophytes (Gustafsson, 1946, 1947; Battaglia, 1963). Among the many reviews on apomixis and parthenogenesis, we may cite those by Gustafsson (1946, 1947), Suomalainen (1950), Battaglia (1951*b*, 1963), Narbel-Hofstetter (1964), Rutishauser (1967) and White (1973).

Apomixis in angiosperms

Our treatment of apomixis in plants will be limited to angiosperms because of the very extensive literature which has accumulated on the cytology and genetics of apomictic development. In sexually reproducing angiosperms, the life cycle comprises the regular alternation of a gametophyte (embryo sac) with a gametic chromosome number (n), and a sporophyte with a somatic chromosome number (2n). In apomictic species or races, owing to failure or suppression of fertilization and meiosis, there is no alternation of nuclear phases in all cases where apomixis occurs regularly generation after generation (obligatory apomixis). In the case of facultative apomixis, apomictic and sexual modes of reproduction coexist. Vegetative reproduction occurs by means of propagules (bulbs, bulbils, runners, adventitious buds, etc.) which generally arise outside the flowering region; but, in extreme cases, they are produced instead of the flowers. In some apomicts, single somatic cells (2n) within the ovule are able to enter an embryogenetic course; an embryo contained in a normally constituted seed is formed (adventitious embryony). Adventitious embryony is one mode of seed formation without fertilization (asexual seed formation or agamospermy). Other methods of agamospermy are haploid parthenogenesis, diploid parthenogenesis (diplospory and apospory) and apogamety (Table 3.4). Since apogamety is an exceptional phenomenon in angiosperms, it will not be dealt with.

Haploid parthenogenesis

By including haploid (reduced) parthenogenesis in the apomictic phenomena in angiosperms, we share Battaglia's (1963) opinion that haploid parthenogenesis is an example of apomixis in reduced (n) embryo sacs. Haploid parthenogenesis in angiosperms is a non-recurrent type of apomixis, as opposed to apomixis in unreduced (2n) embryo sacs which is recurrent (Battaglia, 1963). Naturally occurring haploids have been found in many species of angiosperms (Kimber & Riley, 1963). Development of a haploid embryo from an egg is the result of a retarded or arrested growth of pollen tubes during pollination which, under natural conditions, is generally caused by cold. Cold retards or arrests the elongation of the pollen tubes which have entered the stylar canal but does not abolish their ability to control the synthesis of hormones in the style and ovary (Lund, 1956; Linskens, 1969). This hormonal influence may then stimulate the haploid egg nucleus into parthenogenetic development. The hormonal influence of growing pollen tubes during pollination of ovaries with hundreds or thousands of ovules has also been found to induce occasionally haploid parthenogenetic development in one or a few egg cells.

Stimulation of haploid parthenogenesis by the hormonal influence of pollination is one aspect of pseudogamy; that is, the parthenogenetic development of a female gamete or cell after stimulation, but not fertilization, by a male gamete or gameto-

TABLE 3.4. *Mechanisms of apomictic seed formation* (*agamospermy*) (from Gustafsson, 1946, modified)

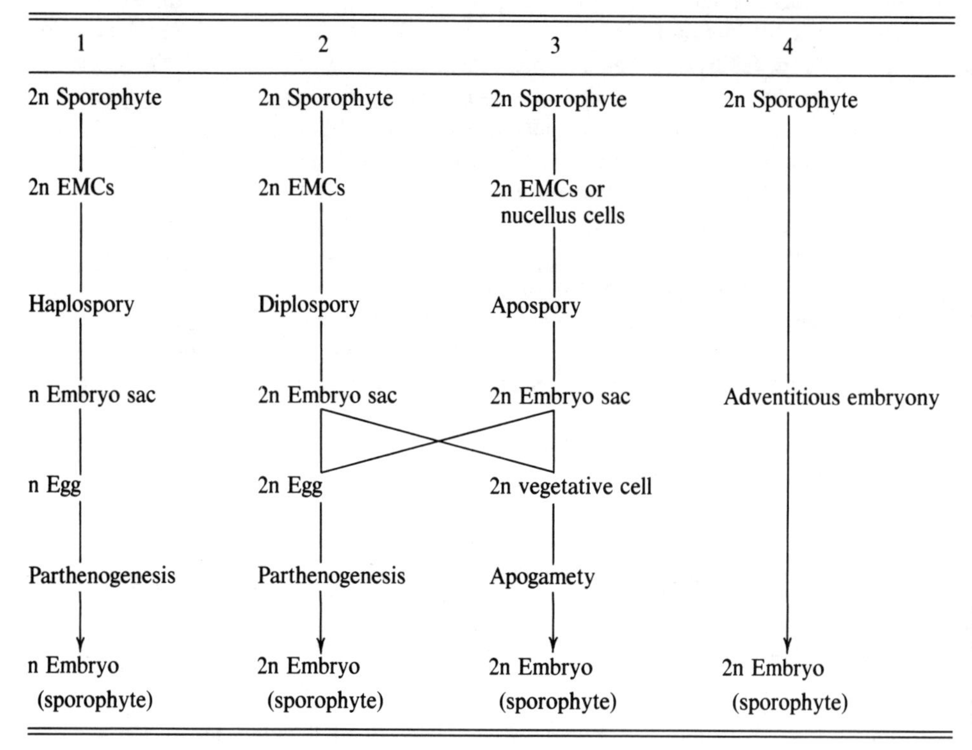

1	2	3	4
2n Sporophyte	2n Sporophyte	2n Sporophyte	2n Sporophyte
2n EMCs	2n EMCs	2n EMCs or nucellus cells	
Haplospory	Diplospory	Apospory	
n Embryo sac	2n Embryo sac	2n Embryo sac	Adventitious embryony
n Egg	2n Egg	2n vegetative cell	
Parthenogenesis	Parthenogenesis	Apogamety	
n Embryo (sporophyte)	2n Embryo (sporophyte)	2n Embryo (sporophyte)	2n Embryo (sporophyte)

n: haploid (reduced) chromosome number; 2n: diploid (unreduced) chromosome number; EMC: embryo sac mother cells.

phyte (Focke, 1881). The most complete cytological studies of natural pseudogamous haploidy in plants are those of Hagerup (1944, 1947) in some orchids. Division of the haploid egg nucleus may occur before the pollen tube opens to discharge the sperm nuclei, before a sperm nucleus enters the egg cell, or after entry of the sperm nuclei(us) into the egg cytoplasm where they (it) degenerate(s). Pseudogamy of this last type is found in many plant and animal species, races or biotypes with unreduced parthenogenesis. Animal biologists frequently speak of gynogenesis (Wilson, 1925) instead of pseudogamy; for reasons of priority, the use of the term gynogenesis should be avoided.

Diploid (*unreduced*) *parthenogenesis*

In most apomictic angiosperms, embryo sacs with the unreduced (2n) chromosome number are formed. Their egg cells produce 2n embryos by parthenogenesis which may be either autonomous (independent of any pollination stimulus) or pseudogamic. Different mechanisms operate in the production of unreduced embryo sacs. The main types of development of unreduced embryo sacs will be described following Battaglia's (1951*a*, *b*, 1963) terminology. For exceptional cases of unreduced embryo sacs due to slight modifications of, or deviations from, the main types, see Gustafsson (1946, 1947), Battaglia (1963), Rutishauser (1967). Of the seven types of development of unreduced embryo sacs, four are due to meiotic disturbances in the embryo sac mother cells leading to unreduced megaspores or megaspore nuclei (aneuspory) and three result from apospory (formation of unreduced embryo sacs by mitotic division of an embryo sac mother cell or a somatic cell of the nucellus.

1. Datura *type* (Fig. 3.4). In the F_2 progeny of a *Datura stramonium* plant pollinated with X-rayed pollen, Satina & Blakeslee (1935) isolated a monogenic recessive mutation responsible for the suppression of meiotic division II in both the female and the male line. The first meiotic division in the embryo sac mother cell was normal. It produced a normal dyad (n = 12) whose cells entered a long interphase stage during which chromosome endoreduplication occurred, at least in the lower chalazal dyad cell which formed the embryo sac. When dividing, this cell showed twelve diplochromosomes (4-chromatid chromosomes) which segregated normally at anaphase (two chromatids of each 'diplo' at each pole) to form two nuclei with 2n (24) chromatids each. After two further divisions an 8-nucleate unreduced embryo sac was formed.

2. Taraxacum *type* (Fig. 3.4). The meiotic prophase ia typically characterized by complete asynapsis or a preponderance of univalents; it is followed by the formation of a restitution nucleus (2n) at metaphase or anaphase I. The second division produces a dyad of cells, the lower of which divides to form an 8-nucleate embryo sac. The *Taraxacum* type occurs in many species of plants, besides *Taraxacum* spp. (list in Battaglia, 1963).

3. Ixeris *type* (Fig. 3.4). As in the *Taraxacum* type, the first meiotic division ends with a restitution (2n) nucleus. The second division is not followed by wall formation and the two nuclei, after polarization, divide to form an 8-nucleate embryo sac. The type, first described in *Ixeris dentata* (Okabe, 1932), has been found in several other species of plants (Battaglia, 1963).

4. Allium nutans *type*. In *Allium nutans* (Håkansson, 1951) and *A. odorum* (Håkansson & Levan, 1957) a premeiotic chromosome doubling occurs in the embryo sac mother cells. In *A. odorum*, a tetraploid species (2n = 4x = 32) which has been studied accurately, chromosome doubling is of the endoreduplication type leading to the formation of 2n diplochromosomes. Each diplochromosome, which has four identical chromatids produced by two successive interphase

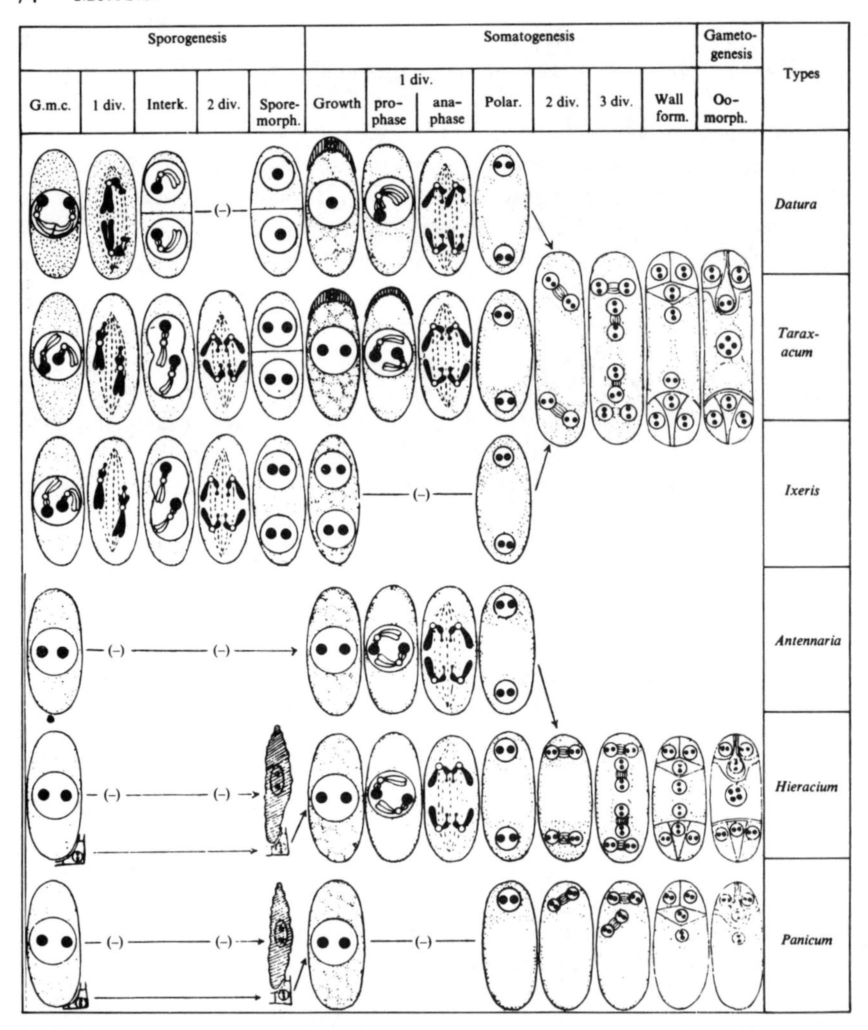

Fig. 3.4. Six types of development of unreduced embryo sac in angiosperms following the interpretation of Battaglia (1951*a*, *b*, 1955*a*). G.M.C.: gynospore mother cell; spore morph.: spore morphogenesis; polar.: polarization; oomorph.: oomorphogenesis. The ploidy level of nuclei is given by the number of nucleoli (one, haploid; two, diploid; four, tetraploid). The nucleus of the megaspore mother cell is figured with one nucleolar bivalent if synapsis occurs or two nucleolar univalents if asynapsis occurs. The upper three embryo sac types are aneusporic. the other three are aposporic (after Battaglia, 1963).

duplications and held in pairs by two centromeres, behaves as a bivalent; at metaphase I, the two centromeres of each diplochromosome are cooriented. This behavior is quite different from a diplochromosome mitosis (Fig. 5.9), in which the two centromeres of each diplochromosome lie in the spindle equator at metaphase (auto-orientation). At the telophase of this peculiar mitosis with diplochromosomes (called 'autobivalents' by Håkansson & Levan, 1957) a dyad of 2n cells is formed, the lower of which gives rise to an 8-nucleate unreduced embryo sac.

5. Antennaria alpina *type* (Fig. 3.4). Meiosis is omitted in the embryo sac mother cell and replaced by a long interphase stage during which the cell grows and vacuolates. Three mitoses lead to the formation of an unreduced 8-nucleate embryo sac. This type of apomictic development, discovered by Juel (1898) in *Antennaria alpina*, is known in many plant species (Battaglia, 1963).

6. Hieracium *type* (Fig. 3.4). This type was discovered by Rosenberg (1906) in some species of *Hieracium* subg. *Pilosella*. An unreduced 8-nucleate embryo sac is produced from a somatic cell of the nucellus whose nucleus undergoes three mitotic divisions. The *Hieracium* type is common among apomictic angiosperms (Battaglia, 1963).

7. Panicum *type* (Fig. 3.4). This type was first described in *Panicum maximum* (Warmke, 1954); it also occurs in other members of the Panicoideae (Battaglia, 1963; Rutishauser, 1967). The unreduced embryo sac is 4-nucleate and consists of two synergids, one egg cell and one polar nucleus.

Apart from cases in which unreduced eggs may be fertilized (references in Battaglia, 1963), unreduced eggs generally develop by parthenogenesis, either autonomous or pseudogamic. An interesting variation in pseudogamy is semigamy, which was discovered and described by Battaglia (1945, 1955*b*) in some apomictic species of *Rudbeckia*. After penetration into the cytoplasm of a unreduced egg, the sperm nucleus divides independently of the egg nucleus, so that variable portions of the embryo possess nuclei of purely male origin. Provided that semigamic embryos reach maturity, semigamy is responsible for the production of plant chimeras consisting of 'maternal' and 'paternal' tissues. Some recent investigations of Turcotte & Feaster (1967) strongly suggest the origin of semigamic plants in crosses between some lines of *Gossypium barbadense* and between *G. barbadense* and *G. hirsutum*.

Adventitious embryony

Another type of agamospermy is the production of an embryo by direct proliferation of a somatic cell of the nucellus. The conditions leading to adventitious embryony are numerous (Gustafsson, 1946, 1947; Rutishauser, 1967); only a few of them will be considered. In *Nigritella nigra* the haploid embryo sac forms normally but degenerates before reaching the adult stage: a nucellus cell then divides to form an adventitious embryo which matures in a seed devoid of endosperm (endosperm-free seeds containing adventitious embryos are also known in other species of orchids). In the orange plant, *Citrus aurantium*, one or more adventitious embryos are developed together with a zygotic embryo within an endosperm containing seed. In *Nothoscordum fragrans*, a diploid species with different chromosomal races which has been investigated by many workers – Levan & Emsweller (1938), D'Amato (1949), Dyer (1967 and here reported references) – a double fertilization leading to a triploid endosperm and a zygote is necessary for the induction of one or more adventitious embryos which replace the zygote. It has been shown by D'Amato (1949) and Dyer (1967) that adventitious embryony is responsible for the constant maintenance in nature of the most interesting race of *N. fragrans* (2n = 19). This race has thirteen long (L) mediocentric and six short (S) subtelocentric chromosomes (13L+6S), two S chromosomes having originated by centromere misdivision (one heterotrivalent LSS at meiosis), and is heterozygous for a reciprocal translocation between two non-homologous L chromosomes (one quadrivalent LLLL at meiosis). Due to the different modes of segregation of the chromosomes of the heterotrivalent and to the deficiency-duplication phenomena bound to structural heterozygosity, gametes with many different chromosome complements are produced: in pollen grains, the chromosome number varies between six and eleven with different proportions of L and S chromosomes and some cases with only L chromosomes (6L, 8L) (Levan & Emsweller, 1938; Dyer, 1967). The ample genetic variability inherent in the meiotic system of the 19-chromosome race of *N. fragrans* is not exploited at all: none of the expected aneuploid or structurally homozygous products of sexual reproduction survive under natural conditions (Dyer, 1967).

Parthenogenesis in animals

In this section, we will discuss first haploid (reduced) parthenogenesis, in which individuals develop parthenogenetically from eggs with a reduced chromosome number, and then diploid (unreduced) parthenogenesis, in which individuals with a somatic (unreduced) chromosome number develop parthenogenetically either from eggs with a reduced (or an unreduced) chromosome number. To distinguish these two types of unreduced parthenogenesis, Suomalainen (1950) introduced two terms which have received ample acceptance in the literature, namely:

(i) automictic parthenogenesis: parthenogenetic development starts from a haploid (reduced) egg and the somatic (zygotic) chromosome number is restored through different mechanisms of chromosome doubling;

(ii) apomictic parthenogenesis: parthenogenetic development from eggs with an unreduced chromosome number, resulting from modification or suppression of the meiotic process in oocytes.

An apomixis resembling the automictic parthenogenesis seems hardly to occur in plants: this condition would corroborate Suomalainen's (1950) view that systems made of the modes of reproduction in plants do not as such apply to animals. Beatty (1967), however, claims that automixis is a type of apomixis. He defines apomixis as development without fusions of reproductive cells of the opposite sex; thus, the term appears to be synonymous with parthenogenesis.

Haploid (reduced) parthenogenesis

Haploid parthenogenesis occurs in four orders of insects (Hymenoptera, Homoptera, Coleoptera, Thysanoptera), in mites (Arachnida) and rotifers (Rotifera) (Suomalainen, 1950; White, 1973). In these groups of animals the males arise from unfertilized eggs and are consequently haploid, while the females are diploid and arise from fertilized eggs. Haploid parthenogenesis is therefore facultative; it may be referred to as arrhenotoky in contrast with thelytoky in which females are produced by diploid parthenogenesis. In species with obligatory thelytoky, males are absent altogether, as in some species of coccids (Nur, 1971). In contrast with haploid arrhenotoky (for the unique type of diploid arrhenotoky in soft scales, see Nur, 1971, 1972), which is restricted to the six animal groups listed above, thelytoky has arisen repeatedly in most of the larger phyla (White, 1973).

Since the frequency of males in the population is determined by the frequency with which unfertilized eggs are laid, the sex ratio in groups with haploid parthenogenesis fluctuates not only between different species, but also within the same species – the sex ratio also depends on environmental factors. In haploid males, the germ line remains haploid; consequently, their spermatogenesis does not involve reduction in chromosome number. In the males of Hymenoptera, which have been studied by many investigators, spermatogenesis consists of one mitotic 'maturation' division, either equal or unequal. In the first case, from a spermatocyte are formed two equivalent spermatids, each of which develops into a functional sperm; in the second case, only the larger of the two cytoplasmically unequal cells (both contain the same chromosome complement) develops into a sperm. For an extensive discussion on haploid parthenogenesis, reference is made to Suomalainen (1950) and White (1973).

Diploid (unreduced) parthenogenesis

Of the many nuclear mechanisms responsible for the parthenogenetic development of embryos with a somatic (2n) chromosome number, only the most important and the most clearly defined will be considered (Table 3.5). More information on the nuclear cytology of diploid parthenogenesis in animals can be found in Suomalainen (1950), Bergerard (1962), Benazzi-Lentati (1970), Nur (1971) and White (1973).

TABLE 3.5. *Summary of the main nuclear mechanisms of parthenogenesis in animals*

Type of parthenogenesis	Premeiotic Chromosome doubling	Meiosis		PBI (and derivative)	PBII	Pro-nucleus	Cleavage I	Cleavage II	Early embryo	Late embryo
		Mitosis I	Mitosis II							
Haploid (reduced)	–	+	+	n	n	n	n	n	n	n
Diploid (unreduced) 1	+	+	+	2n	2n	2n	2n	2n	2n	2n
2	–	Restitution	+	–	2n	2n	2n	2n	2n	2n
3	–	–	+	–	2n	2n	2n	2n	2n	2n
4	–	+	Spindle fusion →	→	→	→	2n	2n	2n	2n
5	–	+	–	2n	–	2n	2n	2n	2n	2n
6	–	+	+	+	n(fusion) →	n →	2n	2n	2n	2n
7	–	+	+	n(fusion) (derivative) →	n →	→	2n	2n	2n	2n
8	–	+	+	n	n	n	n	Spindle fusion	2n	2n
9	–	+	+	n	n	n	n	n	n and 2n (mosaic)	2n

n: haploid (reduced) chromosome number; 2n: diploid (unreduced) chromosome number PB: polar body; +: present; –: absent (for details, see text).

1. *Premeiotic chromosome doubling.* As we have already seen, the chromosome endoreduplication immediately preceding meiosis ensures the constant formation of unreduced embryo sacs (and eggs) in the *Allium nutans* type of apomictic development. Premeiotic chromosome doubling is clearly more frequent among animals with unreduced parthenogenesis. Cases of premeiotic chromosome doubling were first described in parthenogenetic planarians. Lepori (1949, 1950) assumed that an endomitotic chromosome doubling preceding meiosis was responsible for a normal meiosis with 24 bivalents in the oocytes of a triploid ($2n = 3x = 24$) pseudogamic biotype of *Polycelis nigra.* In other triploid pseudogamic biotypes of planarians, chromosome doubling occurs by fusion of the two nuclei produced at the last mitosis of the neoblasts that transform themselves into oogonia (Benazzi-Lentati, 1970). Premeiotic chromosome doubling also occurs in the female line of parthenogenetic Lumbricidae (Muldal, 1952; Omodeo, 1952). In the species studied by Omodeo, chromosome doubling results from the restitution nucleus formation in the last oogonial mitosis. The bivalents formed at mitosis I in oocytes resulting from a premeiotic chromosome doubling are regarded by Håkansson & Levan (1957) to be 'autobivalents' of the type in *Allium odorum.* Autobivalents certainly occur in the oocytes of the parthenogenetic Lumbricidae studied by Muldal (1952), who calls them 'pseudobivalents'. Premeiotic chromosome doubling has also been found in the grasshopper *Moraba virgo*; after chromosome doubling, the oocytes undergo the two meiotic divisions and the egg develops into females (thelytoky) (White, Cheney & Key, 1963).

2. *Restitutional mitosis I.* Univalent formation (asynapsis) followed by a 2n restitution nucleus at mitosis I of meiosis is one of the important nuclear mechanisms operating in the apomictic development of plants (*Taraxacum* and *Ixeris* types). Restitutional mitosis I is not frequently encountered among parthenogenetic animals. In some, but not all, of the oocytes of the parthenogenetic species *Luffia ferchaultella* (Lepidoptera Psychidae), a 2n restitution nucleus is formed at anaphase I which divides equationally to produce a polocyte and a parthenogenetic egg (Narbel-Hofstetter, 1961). Formation of a restitution nucleus at anaphase I had previously been described in the oocytes of another psychid insect, *Solenobia lichenella* (Narbel-Hofstetter, 1950).

3. *Suppression of mitosis I.* Suppression of the I (reductional) division is by far the most important cytological mechanism of the production of unreduced eggs among parthenogenetic animals (Suomalainen, 1950; Narbel-Hofstetter, 1964; Nur, 1971; White, 1973). The oocytes undergo only one 'maturation' division which is typically mitotic (equational); it produces one polar body and one egg, both with 2n chromosomes. When describing this type of single 'maturation' division, many authors have spoken quite improperly of asynapsis and/or univalent division. It is, in fact, well known that true univalents, such as result from either asynapsis or desynapsis, are characterized by aberrant movement, non-disjunction,

centromere misdivision, etc., at mitosis I, the consequence of which is gamete sterility. As abundantly demonstrated in the cytological literature, the only efficient compensatory mechanism for an univalent mitosis I is the formation of a restitution nucleus (2n).

4. *Spindle fusion during mitosis II.* In the parthenogenetic *Apterona helix* (Lepidoptera Psichidae, 2n = 62), the first meiotic division in the oocytes is normal (31 bivalents), but at late anaphase the spindle elongates and cuts itself into two half-sized spindles (one for each group of 31 chromosomes = half bivalents). These two spindles eventually fuse side by side to give the spindle of the second division on which all 62 chromosomes are oriented. This division gives rise to two diploid (2n = 62) nuclei which function as the two first cleavage nuclei. So in *Apterona* no polar body is formed (Narbel, 1946). A parthenogenetic development of the type found in *Apterona* also occurs in a few other animal species (Narbel-Hofstetter, 1964).

5. *Suppression of mitosis II.* In *Rhabditis monohystera* (Nematoda), the oocytes undergo a typical reductional mitosis with the formation of two daughter groups of 10 chromosomes (n) each. At telophase I, the two chromatids of each chromosome (half bivalent) in each nucleus separate from one another; the result is the suppression of mitosis II leading to the formation of one diploid (2n = 20) polar body and one diploid parthenogenetic egg (Belar, 1924; Nigon, 1949). In a few other parthenogenetic species, e.g. *Cognettia glandulosa* (Oligochaeta), the dissociation into chromatids of half bivalents occurs later than in *Rhabditis:* at interkinesis or prophase II. The results is, however, identical: diploid (2n = 108) egg production (Christensen, 1961).

6. *Fusion of egg with polar body II.* In two coccids with facultative parthenogenesis, *Lecanium hesperidum* and *L. hemisphaericum*, the oocyte undergoes a normal meiosis which results in four haploid nuclei, viz. the egg nucleus and the three polar nuclei. In case of parthenogenetic development, the egg nucleus and the nucleus of the polar body II, after fusing at prophase, form the first cleavage spindle with a diploid chromosome number (Thomsen, 1927; Nur, 1971). Fusion of the egg with the polar body II also characterizes the parthenogenetic development of some other species of animals; see Narbel-Hofstetter (1964), Nur (1971), White (1973).

7. *Fusion of polar body II with one derivative of polar body I.* This type of development has been described in great detail by Seiler & Schäffer (1960) in a diploid (2n = 62) parthenogenetic race of *Solenobia triquetrella* (Lepidoptera). In the oocytes, a normal meiosis produces four haploid nuclei in the egg cytoplasm: the two derivatives of polar body I (PBI) the polar body II (PBII) and the female pronucleus. The pronucleus is then displaced to the anterior pole of the egg where

it produces a group of haploid nuclei. Later, spindle fusions at metaphase produce a mixture of haploid, diploid, and polyploid nuclei in the anterior pole of the egg. Meanwhile, one derivative of PBI (the one more peripherally located) degenerates after reaching metaphase. On the other hand, the internal derivative of PBI and PBII fuse at prometaphase to produce a diploid mitotic figure. After this diploid mitosis, which is the first cleavage division, other divisions follow which are synchronous with those of the pronucleus and its derivatives in the anterior pole of the egg. The origin of an embryo with somatic chromosome number from the fusion product of polar nuclei (called 'Richtungskopulationskern', RKK, by Seiler & Schäffer, 1960) has also been demonstrated in the parthenogenetic tetraploid (2n = 4x = 124) race of *Solenobia triquetrella* (Seiler, 1963). Narbel-Hofstetter (1964) has discussed a few cases of parthenogenesis which might be due to the RKK phenomenon.

8. *Nuclear fusion during second cleavage.* In the obligate parthenogenetic coccid *Pulvinaria hydrangeae*, the oocytes produce, by normal meiosis, haploid eggs whose pronucleus divides into two haploid nuclei (first cleavage). These two nuclei fuse at the time of the second division to produce two diploid nuclei which later give rise to the entire embryo (Nur, 1963). Restoration of the diploid chromosome number by fusion of the two haploid products of the female pronucleus during the second cleavage is found in a few other species of coccids (Nur, 1971, 1972).

9. *Chromosome doubling in initially haploid embryos.* In an amphimictic species of phasmids (walking sticks), *Clitumnus extradentatus*, virgin females lay haploid eggs, part of which develop parthenogenetically (facultative parthenogenesis). In *Clitumnus*, strains have been isolated with a very high percentage (up to 95%) of haploid eggs developing parthenogenetically. Since the sex determination mechanism is of the XX–XO type (2n = 38 in the females; 2n = 37 in the males) all haploid eggs have one X. In the parthenogenetic early embryos, which develop slowly, all nuclei are haploid (n = 19): later, a mixture of haploid and diploid (2n = 38) nuclei is found and still later, when the embryo shows a development rate similar to that of the amphimictic (2n = 38 or 37) embryos, only diploid nuclei are found. Restoration of the diploid chromosome number is claimed to result from a metaphase block of haploid mitoses similar to a colchicine induced metaphase block (Bergerard, 1962). The most plausible mechanism of diploidization seems, however, to be endoreduplication leading to diplochromosomes (White, 1973). All diploid parthenogenetic individuals, with 2n = 38 (XX), are female (Bergerard, 1962).

4

Mosaics and chimeras

In 1907, Winkler used the term 'chimera' to designate a plant composed of tissues of two species of *Solanum*, tomato (*S. lycopersicum*, i.e. *Lycopersicum esculentum*) and black nightshade (*S. nigrum*). The plant originated from an adventitious bud which developed from the callus of a graft between the two species. Since that time it has been customary to designate as chimeras individuals with cell populations of more than one karyotype or idiotype (the sum total of hereditary determinants, both chromosomal and extrachromosomal) derived from two independent individuals following such diverse events as double fertilization, chorionic vascular anastomosis, graft or transplantation. Chimeras have thus been distinguished from mosaics, individuals whose cells were all derived from a single fertilization event but in whom subsequent chromosomal or genetic changes had produced more than one genetic cell type. As pointed out by Stern (1968), the distinction between these two types of 'compounds' is not always a clear-cut one, decision on the mode of origin of some compound individuals resting on genetic evidence. Stern therefore uses the term 'mosaic' to designate 'any kind of genetic multiplicity in an individual' and proposes 'to leave to specific inquiry the decision concerning what circumstances led to its origin' (Stern, 1968, p. 86).

In contrast with its wide use in the zoological literature, the term 'mosaicism' in the botanical literature is generally restricted to cases of a rather random distribution of sectors of different cell types within the plant body. But, even in these cases, the terms 'mosaic chimera' and 'sectorial chimera' have been employed. The generalized use of the term 'chimera' in the botanical literature is due to two main reasons (Cramer, 1954):

(i) botanists have called chimera any plant composed of two or more genetic cell types independently of the derivation of the different cell types, either intra-individual or interindividual:

(ii) due to the overall immobility of plant cells (they are surrounded by a relatively rigid wall and are cemented together within the tissues), compound structures with a perfect separation of genetically different cell types are frequently obtained. This situation, which is most striking in shoot apices in which the genetically different cell types are arranged in layers (periclinal chimeras, see further below), can be maintained practically unlimitedly by vegetative propagation. As we shall see, in the aerial organs (stems, leaves, flowers) which develop from such shoot apices, the component tissues (or tissue systems) with different genetic constitutions occupy positions in the organ which are related to the positions in the shoot apex of the corresponding cell types. The regularity and repeatability of the spatial distribution during organogenesis of the different cell lineages in the periclinal compounds are features which seem to be in contrast with the common notion of a mosaic.

The discussion on mosaics and chimeras which follows does not include two types of mosaicism, namely (i) somatic mosaicism due to chromosome endoreduplication, endomitosis or cell fusion and (ii) functional mosaicism due to heterochromatization of particular chromosomes or chromosome sets. The nuclear cytology of somatic tissues in plants and animals and the problem of heterochromatization are dealt with in other chapters.

Mosaics and chimeras in animals

Genetic mosaics in animals and man have been made the subject of a recent very lucid essay by Stern (1968). In this chapter, we will discuss some of the many types of animal and human mosaics, including the mouse chimeras which have been obtained experimentally by fusing two developing eggs during early or mid-cleavage.

Haploid–diploid mosaics

As we have seen in another section of this book, the male bees (drones) develop parthenogenetically from haploid eggs whereas the females are diploid and develop from fertilized eggs. In the honeybee, sex mosaics, known as gynandromorphs or gynanders, were first discovered and described in the early years of the nineteenth century. In 1889, as a result of his classical observations on fertilization in sea urchins, Boveri proposed partial fertlization as a possible mechanism for the origin of gynanders in the honeybee. Partial fertilization would give rise to male parts composed of cells with unfertilized (haploid) nuclei and to female parts composed of cells with fertilized (diploid) nuclei (Fig. 4.1).

In recent times, studies on bee gynanders have been facilitated by the isolation of bee strains predisposed to sex mosaic formation and by the experimental production of gynanders. It has been shown that partial fertilization is only rarely responsible for the origin of gynanders. A far more common mechanism of gynander formation in bees seems to be bound to polyspermy. Among 82 gynanders obtained by chilling the eggs (Drescher & Rothenbuhler, 1963), 80 had male parts of paternal origin and female parts of hybrid phenotype. The origin of these gynanders is best explained by assuming the participation in development of two sperm nuclei, one of which fuses with the egg nucleus and the other divides in parallel with the division of the fusion nucleus (Fig. 4.2). The other two gynanders observed had male parts with maternal phenotype; they were most probably the result of partial fertilization. For a survey of the cytological modes of origin of gynanders in bees and Hymenoptera in general, reference is made to Cooper (1959).

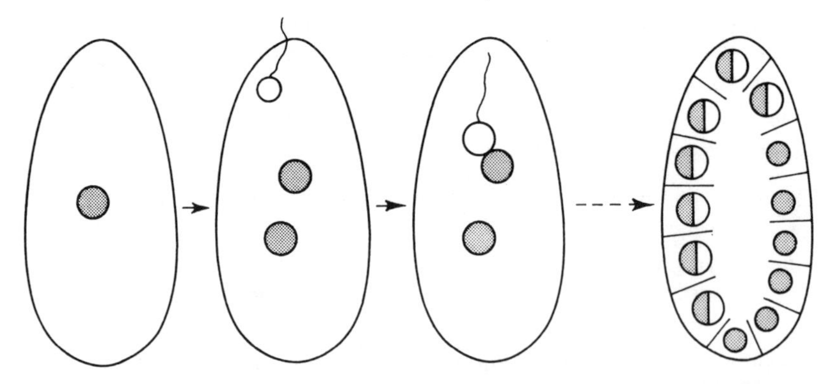

Fig. 4.1. Origin of a gynandric bee through partial fertilization. The egg nucleus divides before fertilization and a sperm nucleus fuses with one of the two nuclei. The gynander is hybrid in its female parts and purely maternal in its male parts (after Stern, 1968).

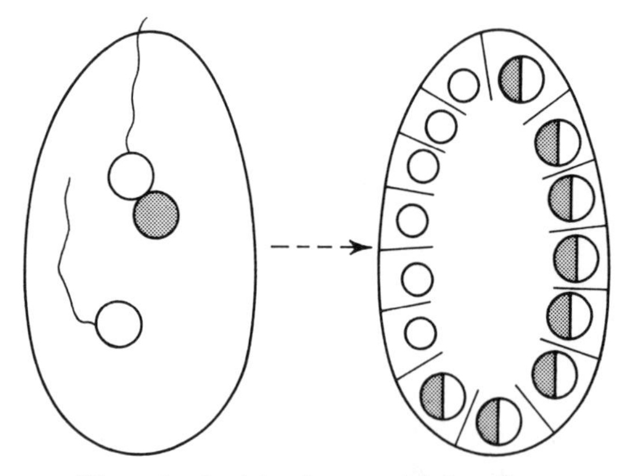

Fig. 4.2. Another possible mode of origin of a gynandric bee. Two sperm nuclei participate in development. One fuses with the egg nucleus, the other is replicated without fusion. The gynander is hybrid in its female parts and purely paternal in its male parts (after Stern, 1968).

In many other species of insects and in species of rotifers and mites, haploidy occurs as a regular phenomenon in the life cycle. The haploid individuals originating from unfertilized eggs are invariably male. This situation is quite different to that found in *Drosophila.* As first proposed by Bridges (1921) and later confirmed by extensive investigations, in *Drosophila* sex determination results from a balance between the number of X-chromosomes and the number of sets of autosomes: diploid (2X2A), triploid (3X3A) and tetraploid (4X4A) flies are all female (in all of them the ratio X:A is 1). In normal diploid males (1X2A) the X:A ratio is 0.5. In case of haploid individuals (1X1A, X:A = 1), the expected sex would be female. No wholly haploid *Drosophila* individuals have ever been found; however, both in

D. melanogaster (Bridges, 1925) and *D. pseudo-obscura* (Crew & Lamy, 1938) haploid–diploid mosaics are not male–female gynanders but all-females mosaics. In *D. pseudo-obscura*, all but one of the forty-five all-female mosaics showed haploid areas carrying the paternal genome (genetic markers). In these same mosaics the diploid parts, which were found to carry the same paternal markers, must have been derived from the same male pronucleus as the haploid area. This indicates that the male pronucleus divided once and then one of its derivatives fused with the female pronucleus or one of its derivatives (Crew & Lamy, 1938).

Sex-chromosome mosaics

In most animals, sex is determined by the presence or absence, or the presence in varying numbers, of sex chromosomes. In these animals, gynanders are diploid mosaics in which male and female parts differ in sex chromosome constitution. Thus in *Drosophila*, with XX–XY sex determination mechanism, the female part of the individual gynander may consist of XX cells and the male part of XY or XO cells (in *Drosophila*, the Y chromosome, though essential for complete spermatogenesis, is not needed for male morphology and behaviour). Due to the practical impossibility of a cytological analysis in adult gynandric flies, X-linked marker genes are very helpful when inquiring the mode of origin of these gynanders. Such an analysis is based on crosses between a *Drosophila* female homozygous for recessive X-linked genes and a male which carries in his single X-chromosome the corresponding dominant alleles. In their classical work on *Drosophila* gynanders, Morgan & Bridges (1919) concluded that the great majority of them start development with two X-chromosomes. Then, generally at the first cleavage, one X-chromosome divides normally, while of the two division products (chromatids) of the other X only one is included in one of the two cleavage nuclei (the other chromatid lags behind on the spindle and is lost). The result is then a bilateral gynander, half female (XX) and half male (XO).

By means of X-linked markers it has been shown that in a few instances *Drosophila* gynanders originate by X-chromosome non-disjunction (both sister chromatids of an X-chromosome are included in the same nucleus). The result is a mosaic composed of three types of cells: XX, the cell progeny which never underwent chromosome non-disjunction, and the two cell progenies of an initial product of chromosome non-disjunction, XXX and XO (Stern, 1960). The use of appropriate X-linked marker genes in *Drosophila* gynanders for a detailed analysis of cell lineages in development has been demonstrated by Sturtevant (1929). In this study, use was made of the X-linked markers yellow and singed or forked, which express themselves in any one of the micro and macrochaetae on the body surface, thus allowing one to assign most of the body areas to the XX or XO condition. It was shown that the majority of gynanders were bilateral, as would be expected if they originated at the first cleavage division, but a good proportion of gynanders were about three-quarters female and one-quarter male or with still

TABLE 4.1. *Sex chromatin incidence (percent) in different tissues from 8 μm sections of an XY/XXY mosaic human fetus and of a normal male and a normal female control fetuses* (from Klinger & Schwarzacher, 1962)

Tissue	Mosaic in sex chromatin positive areas	Fetus overall tissue average	Normal male fetus	Normal female fetus
Epidermis	18.0	10.7	2.8	62.0
Esophagus epithelium	23.0	6.7	1.7	60.5
Liver parenchyma	27.0	10.0	2.0	48.0
Peritoneal epithelium	31.0	15.0	3.0	55.5
Average of all tissues	24.8	9.6	2.3	58.6

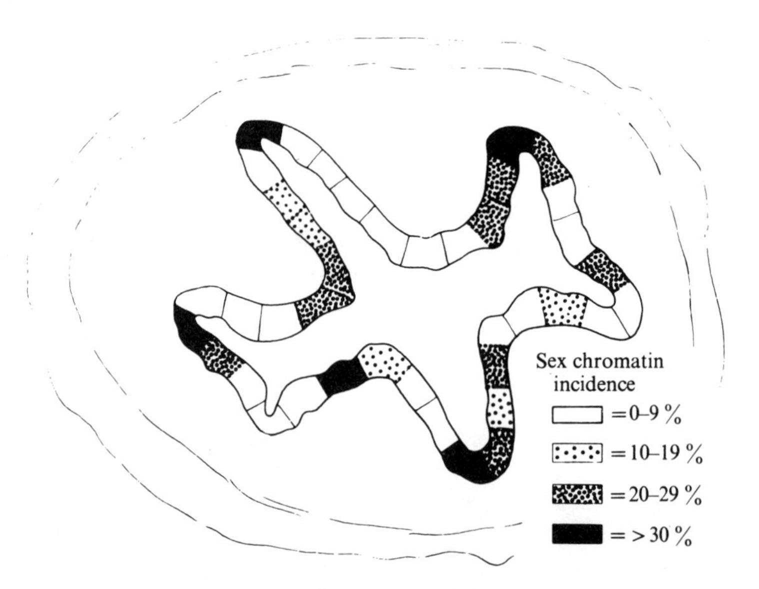

Fig. 4.3. Incidence of sex chromatin positive nuclei in the epithelial layer of the esophagus (transverse section) in an XY/XXY human fetus (after Klinger & Schwarzacher, 1962).

greater female areas. The distribution of the two gynandric components in the adult varied greatly among the flies: more or less bilateral, anterior–posterior etc. This variability in gynandric phenotype is due to the relatively indeterminate embryonic development of insects (details in Stern, 1968).

A species in which much information on sex-chromosome mosaicism has been accumulated in a relatively short time is the human. The cytological analysis of mosaics in man has been greatly facilitated by the possibility of obtaining from a single individual repeated biopses of different types of cells (bone marrow, blood, skin etc.) to be used in short- or long-term cultures. A variety of double and triple sex-chromosome mosaics is at present known in man (reviews by Mittwoch, 1967; John & Lewis, 1968; Stern, 1968). Among these mosaics, the most thoroughly studied has been an XY/XXY fetus which was removed by Caesarean section at the twelfth to thirteenth week of pregnancy (Klinger & Schwarzacher, 1962). In addition to the cell cultures derived from various regions of the body which demonstrated the XY/XXY mosaicism, the authors also studied serial sections of the fetus to determine the incidence of sex chromatin in interphase nuclei (one Barr body in XXY cells due to heterochromatization of one of the X-chromosomes). In the different tissues analyzed, on average about 10% of the nuclei had a Barr body (Table 4.1); the two cell types were frequently distributed in cell clusters (Fig. 4.3). Since the mosaic pattern occurred throughout the entire body, Klinger & Schwarzacher suggested that the mosaic arose through non-disjunction in an XY cell at a very early stage, possibly at the third somatic division, but certainly before tissue differentiation.

Among animals other than man, cases of sex-chromosome mosaicism have been described. We may cite the diploid 60, XX/60, XY mosaicism found in two bovine male pseudohermaphrodites (McFeeley, Hare & Biggers, 1967) and the bovine true hermaphrodite (Dunn, Kenney & Lein, 1968).

Diploid–triploid mosaics

As seen elsewhere in this book, triploidy can be compatible with healthy post-natal life in the fowl. In mammals, triploid embryos, which occasionally occur, are unable to survive, but diploid–triploid embryos may remain viable in post-natal life. A two-year-old boy was found to be a diploid–triploid mosaic: 46,XY/69,XXY (Böök & Santesson, 1960). Diploid–triploid mosaicism has been reported in three more humans ranging in age from 2 to 21 years (Ellis, Marshall, Nomand & Penrose, 1963; Ferrier *et al.*, 1964; Lejeune *et al.*, 1967), in a male tortoiseshell cat (Chu, Thuline & Norby, 1964), in a true hermaphrodite mink (Nes, 1966) and in a 10-month old bovine true hermaphrodite: 60,XX/90, XXY (Dunn, McEntee & Hansel, 1970). A chromosomal analysis of a variety of tissues from the bovine hermaphrodite (Table 4.2) revealed that the triploid cell line was present in the kidneys, right prefemoral lymph node, uterus and gonads. The possible origin of the diploid–triploid cell lines in this case is reported in Table 4.3. Since aneugamy is a very rare event, it seems probable that the triploid condition arose by either polyandry or polygyny. The hypothesis of double fertilization and fusion of the fertilized meiotic products seems the most feasible explanation of this mosaic. It is, however, also possible that the specimen started as a triploid zygote and the diploid cell line originated through the elimination of a whole chromosome set from a triploid nucleus. An interesting aspect brought about by the analyses of Dunn *et al.* is the selective advantage of diploid over

TABLE 4.2. *Incidence of diploid and triploid mitoses in different tissues in a 60,XX/90,XXY bovine true hermaphrodite* (from Dunn *et al.*, 1970)

Tissue	Metaphases classified					Ratio of diploid XX to triploid XXY cells
	Diploid		Triploid			
	XX	XXY	XXY	XXX	Total	
Blood (lymphocytes)	698		1	1	700	99.9:0.1
Bone marrow	221	3	0		224	100:0
Lymph node (prefemoral)	19		9		28	68:32
Skin	200		0		200	100:0
Kidney (left)	193		7		200	96:4
Kidney (right)	169		31		200	84:16
Ovary (left)	178		22		200	89:11
Gonad (right, ovarian part)	150		50		200	75:25
Gonad (right testicular part)	101		99		200	50:50
Uterus	102		98		200	51:49
Total	2031	3	317	1	2352	(87:13)

TABLE 4.3. *Possible origin of the diploid–triploid cell lines in a 60,XX/90,XXY bovine true hermaphrodite* (from Dunn *et al.*, 1970)

Cell line	Paternal gamete(s)	Maternal gamete	Mode of fertilization
60,XX	30,X	30,X	normal
90,XXY	30,X and 30,Y	30,X	polyandry
	30,Y	30,X and 30,X	polygyny
	60,XY	30,X	aneugamy
	30,Y	60,XX	aneugamy

triploid cells in some tissues. This is in agreement with the observations made in the first described diploid–triploid mosaic human individual in whom the frequency of triploid cells decreased from 84% in skin cells at the age of 14 months to 10% at 4.5 years, and from 49% in connective tissue to 7% (Böök, 1964).

For other possible modes of origin of diploid–triploid mosaicism in animals, reference is made to Chu *et al.* (1964) and Nes (1966).

Mutational and somatic crossing-over mosaics

Mosaics due to either spontaneous or induced mutations are studied by analyzing F1 individuals obtained by crossing females homozygous recessive for several marker genes with males homozygous for the corresponding dominant alleles. If a mutation from dominant to recessive or a loss of the dominant occurs in one or more allele pairs, e.g. from *Aa* to *aa* (homozygous recessive) or from *Aa* to *aO* (hemizygous recessive), the recessive phenotype(s) will be seen in the mosaic individual. A detailed analysis of mutational mosaicism can provide information on the time of origin of the genetic change(s) involved.

In the large-scale studies on spontaneous and radiation-induced mutation rates in mice carried out at Oak Ridge National Laboratory, use has been made of the specific-locus method (Russel, 1951), in which wild-type irradiated and unirradiated mice were mated to animals homozygous for seven specific loci, five of which affected coat color. Thousands of F1 mice heterozygous for the seven loci have been analyzed. Of 112 mosaic mice reviewed by Russell (1964) at least 26 and probably as many as 40 were found to be somatic and germinal mosaics for one or another coat color locus for which the population was heterozygous. In the 20 mosaics for which the mutational evidence was most decisive, the percentage of germinal tissue carrying the mutation varied from 7% to 98%, the average being 49.5% based on 1815 progeny. When all 43 gonosomic mosaics reported by Russell in Table 1 are considered, the average percentage of mutated germinal tissue is 48.3% based on 3864 progeny. It is thus possible to interpret these results as showing that for 43 gonosomic mosaics about half the cells of the organism are mutant (Russell, 1964). This half-mutant state would imply participation in development of two blastomeres, one normal and the other mutant. As pointed out by Russell, the observations made on mosaic mice support the conclusion that the half-mutant state is the result of one of the following events: (i) mutation in one strand of a double-stranded gamete chromosome; (ii) mutation at the first cleavage or in the 2-cell stage. In the mouse, the first alternative cannot be taken as any more probable than the second (Russell, 1964).

In *Drosophila*, mutational mosaics have been studied using the same procedure as in the mouse: crosses of normal dominant males to recessive female and scoring mutations in which a dominant paternal gene had changed to a recessive either spontaneously or after treatment of the male with chemical mutagens. On the whole, the mosaics obtained seemed to present the mutation in about one-half of all cells. This result suggests that the mosaics were due to a mutation being already present in the sperm: possibility (ii) as discussed above for the mouse (Altenburg & Browning, 1961).

Another type of mosaicism which has been used extensively as a high-resolution analytical tool in developmental studies is the mosaicism resulting from somatic crossing-over. Somatic crossing-over, leading to gene recombination in dividing

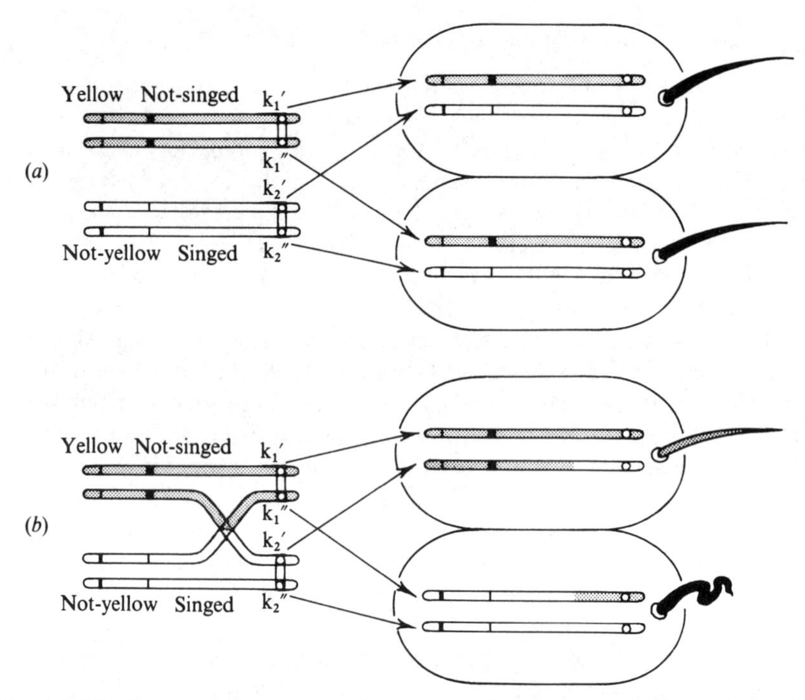

Fig. 4.4. Distribution of chromosomes heterozygous for two genes (yellow and singed) in *trans* position in *Drosophila*. On the left, the chromosomes are doubled and their sister kinetochores are labeled k_1' and k_1'' and k_2' and k_2'' respectively. On the right, the chromosomes are distributed to daughter cells. (*a*) no crossing-over. The daughter cells are alike genetically and they or their descendents may form not-yellow not-singed bristles; (*b*) crossing-over between the locus of singed and the kinetochore. The daughter cells differ genetically and may form a twin spot of yellow, not-singed and not-yellow, singed bristles (after Stern, 1968).

cells (mitotic recombination) was discovered by Stern (1936) in *Drosophila melanogaster*: it occurs between the two X-chromosomes and between pairs of homologous autosomes (in the Diptera, somatic pairing is an ever-present phenomenon) either at replication or during the replicated stage (G_2). Mitotic recombination takes place, at any one level, between only two of the four strands of a pair of homologous chromosomes. The two daughter cells resulting from somatic crossing-over differ genetically from the common mother cell (Fig. 4.4). Division of the two initial cells will lead to two adjacent genetically distinct cell clones (the so-called twin-spots) on a wild-type background. Since somatic crossing-over is greatly increased by ionizing radiations, larvae at different stages of development can be irradiated and the mosaicism resulting from somatic crossing-over is scored on the adult fly. Apart from cell death and possible developmental anomalies, which may occur at some radiation exposures, the clone size in the adult depends on the stage of irradiation, being the larger the earlier the irradiated stage. Analyses of cell clones arising from somatic crossing-over are very helpful in studies on cell lineages, morphogenesis, cell migration, cell division rates, the number of primordial cells in imaginal discs etc. (for extensive reviews on the use of mosaics in developmental studies in *Drosophila* see Stern, 1968; Nöthiger, 1972; Postlethwait & Schneiderman, 1973).

Chimeras from fused eggs or embryos

In 1927, Mangold & Seidel announced the development of chimeric newts following the fusion of early embryos of the same or two different species. More recently, chimeric mice have been produced by fusing two developing eggs, in the early cleavage stage, up to the 8-cell stage (Tarkowski, 1961, 1963–1964; Mintz, 1964, 1965). The eggs were deprived of the zona pellucida, either mechanically (Tarkowski) or enzymatically (Mintz), were fused and grown *in vitro* to the blastocyst stage and thereafter were implanted in the uterus of pseudopregnant foster mothers. The fully developed mice were born naturally or were delivered by Caesarean section. In rabbits, in which zona-free blastocysts fail to implant, chimeric animals with mixed coat colour have been obtained by the insertion of inner cell masses of blastocysts obtained from black rabbits into complete blastocysts with intact zona pellucida obtained from white rabbits (Gardner & Munro, 1974).

Chimeric mice have provided information on the fate of sex chimerism and on the distribution of cells during developments of the embryo and the adult. Fifty per cent of male–female chimeras (hermaphrodites) are to be expected from fused embryos. Both in Tarkowski's (1961, 1963) and Mintz's (1965) experiments, the hermaphrodites are rare and the males predominate. This led Tarkowski (1963) to suggest that the majority of potentially hermaphrodite chimeras may develop in a phenotypically male direction. This suggestion has received support from the observation that adult males with sex chromosome chimerism (XX/XY males) produce spermatozoa only from the genetically male component (as demonstrated by an analysis of their offspring) (Mystkowska & Tarkowski, 1968; Mintz, 1968). In their experiments on the fusion of embryos of genetically labeled strains (CBA/H–T6T6, darkly pigmented of the agouti type and CBA-p, light-colored coat and pink eye due to the small amount of pigment in the outer layer of the retina), Mystkowska & Tarkowski (1968) analyzed four adult and fully fertile chimeric males CBA-p/CBA-T6T6, with bicolor pigmentation of their coat; two proved to be sex chromosome chimeras (XX/XY). In the bone marrow, 51% and 19% of cells were of female origin, whereas in an adult incomplete hermaphrodite (one side of the reproductive system bisexual, the other typically male) recovered in the same experiment 96% of cells were of female origin. Thus, it appears that the majority of cells in the mouse embryo have to be of XX constitution before development is shifted in the female direction. Since the fertile XX/XY male mice produce sperm corresponding to the genotype of the XY component, the absence of XX germ cells must be due to an elimination phenomenon. It is not known, however, whether all XX germ cells in XX/XY individuals are eliminated soon after birth or some survive as spermatogonia but degenerate as soon as they enter meiosis (Mystkowska & Tarkowski, 1970).

The pink-eyed agouti chimeras studied by Tarkowski (1961, 1963, 1964) and Mystkowska & Tarkowski (1968) have provided information on the spatial distribution of the genetically different cell types in the retina and the coat of the animal. In the retinas, the chimerism was astonishingly fine-grained, with an abundance of small patches of pigmented and non-pigmented cells. This obviously implies an enormous amount of cell mixing during ontogeny. In the coat of the chimeras, the distribution of the dark and light areas was very irregular; the areas were irregular in outline, variable in size and their distribution did not exhibit any definite pattern. On the other hand, a definite pattern of distribution of melanocytes resulting in an alternate arrangement of stripes occurred in the chimeric mice obtained by Mintz (1967). This difference between Tarkowski's and Mintz's results is accounted for by the different starting materials used (see McLaren, 1969).

Graft chimeras in plants

In 1674, a Florentine gardener, Nati, described a strange form of orange, called 'bizzaria', which originated in 1644 in Firenze, Italy, as an adventitious bud from the swelling of an unsuccessful budgraft between bitter orange, *Citrus aurantium*, and citron, *C. medica*. The 'bizzarria' tree produced leaves, flowers and fruits of bitter orange and citron, and also compound fruits with the two kinds variously blended or segregated from each other. In more recent times, other synthetic chimeras consisting of tissues of two different species have arisen both spontaneously, generally as a result of an unsuccessful budgraft, and experimentally, following the pioneer works of Winkler (1907, 1910) (for a list of synthetic plant chimeras, see Brabec, 1965). A newly discovered graft chimera of two commonly cultivated *Camellia* species, *C. sasanqua* (2n = 90), and *C. japonica* (2n = 30), has been analyzed very accurately by Stewart, Mayer & Dermen, (1972).

Spontaneously arisen graft chimeras were defined to be 'graft hybrids' by many well-known botanists who assumed that, in the region of contact between the two parent tissues in the graft, cell fusion had occurred; this 'somatic hybridization' also explained, according to the same authors, the 'segregation' on the same individual of plant parts of each parent species. In 1907, however, Strasburger demonstrated that the chromosome number of the 'bizzarria' was not different from that of *Citrus aurantium* and *C. medica* (the correct chromosome number is 2n = 18, not 2n = 16 as counted by Strasburger): he also showed that *Laburnum adami*, a graft chimera known since 1830, had the same chromosome number (2n = 48) as the two parent species, *Laburnum anagyroides* and *Cytisus purpureus*. Strasburger concluded that neither the 'bizzarria' nor *Laburnum adami* were graft

TABLE 4.4. *Structure and behaviour of graft chimeras between species of Solanum lycopersicum* ($2n = 24$), *luteum* ($2n = 48$), *nigrum* ($2n = 72$). *The chromosome numbers of the other species were not counted by the authors, 1–4 from Winkler (1910), 5–10 from Jørgensen & Crane (1927). In the description of their graft chimeras, Jørgensen & Crane place second the species providing the exterior cell layer(s) of the shoot apex, either one (i) or two (ii)*

No.	Name	Structure of shoot apex			Fertility	Type of seed	Spontaneous reversion to
		L_I	L_{II}	L_{III}			
1	*S. koelreuterianum* H. Winkl.	Nigrum	Lycopersicum	Lycopersicum	None	—	Lycopersicum
2	*S. gaertnerianum* H. Winkl.	Nigrum	Nigrum	Lycopersicum	Low	Nigrum	Nigrum
3	*S. proteus* H. Winkl.	Lycopersicum	Lycopersicum	Nigrum	High	Lycopersicum	Lycopersicum
4	*S. tubingense* H. Winkl.	Lycopersicum	Nigrum	Nigrum	High	Nigrum	Nigrum
5	*S. lycopersicum-guineense* (i)*	Guineense	Lycopersicum	Lycopersicum	None	—	Lycopersicum
6	*S. lycopersicum-luteum* (i)*	Luteum	Lycopersicum	Lycopersicum	None	—	Lycopersicum
7	*S. nigrum-sisymbrifolium* (i)	Sisymbrifolium	Nigrum	Nigrum	High	(Not reported)	Nigrum
8	*S. lycopersicum-nigrum* (ii)	Nigrum	Nigrum	Lycopersicum	None	—	Twice *S. lycopersicum–nigrum* (i) once lycopersicum
9	*S. lycopersicum-luteum* (ii)	Luteum	Luteum	Lycopersicum	Low	Luteum	Luteum
10	*S. luteum-lycopersicum* (i)	Lycopersicum	Luteum	Luteum	Low	Luteum	Luteum

* Not studied cytologically.

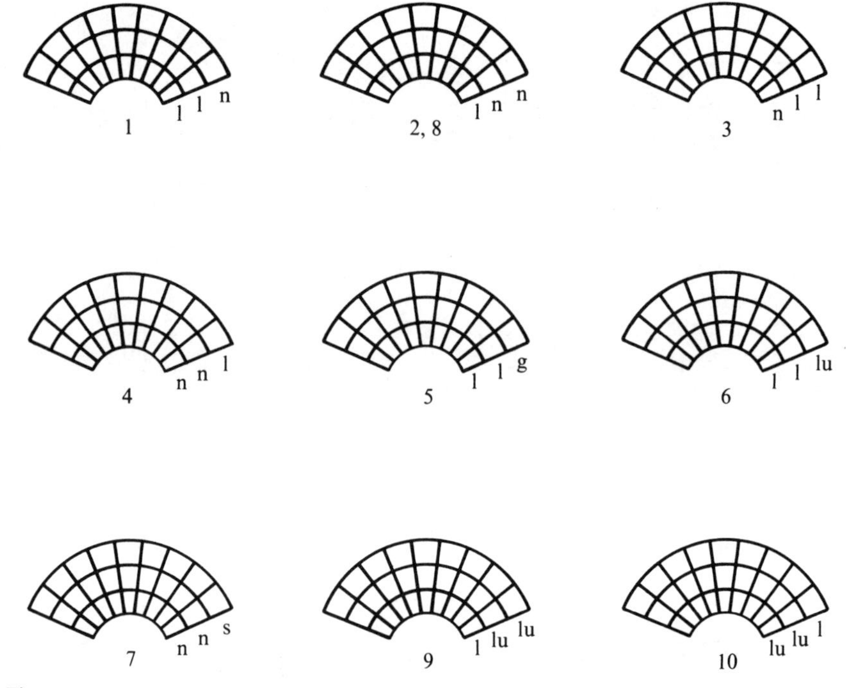

Fig. 4.5. Schematic representation of the distribution of the three outermost cell layers in the shoot apex of the *Solanum* graft chimeras, numbered as in Table 4.4. Species and varieties of *Solanum* are indicated as follows: g, *guineense;* l, *lycopersicum*; lu, *luteum*; n, *nigrum*; s, *sisymbrifolium*.

hybrids (if they were, they should have a chromosome number equal to the sum of the chromosome numbers of the two parent species); instead, they were to be considered as sexual hybrids. After the publication of Winkler's (1907) paper on chimeras, Strasburger (1909) concluded that the 'bizzarria' was a chimera. In the same year, Baur (1909) showed that the horticultural variety *albomarginata* of *Pelargonium* had a shoot apex consisting of two to three external layers of albino cells, lacking chloroplasts, covering normal green tissue. Baur assumed that this structure, which he called 'periclinal chimera',* was responsible for the origin of the graft chimeras, both the spontaneous ones ('bizzarria', *Laburnum adami*, etc.)

* Periclinal chimeras with only one dissimilar exterior cell layer in the shoot apex are called haplochlamydeous; those where the exterior component consists of two layers, diplochlamydeous (Meyer 1915). Meyer also calls the external and internal portion of the shoot apex in a periclinal chimera 'mantle' and 'core' respectively. This terminology might be useful to avoid confusion with the terms *tunica* and *corpus* which are widely used in describing the structure of shoot apices in plants (Clowes, 1961; Steeves & Sussex, 1972). In accordance with Satina, Blakeslee & Avery (1940) and most later authors, the cell layers in the dome of the shoot apex are numbered from the exterior to the interior: L_I (the outermost), L_{II} (the second) and L_{III} (the innermost). As will be seen below, in apple periclinal chimeras five apical layers (L_I–L_V) can be distinguished. Cell layers in shoot apices have been called primary histogenetic layers or apical layers by Dermen (1947).

and those produced experimentally by Winkler. That the 'bizzarria' is a periclinal chimera has been demonstrated by Tanaka (1927); demonstration of the periclinal nature of other graft chimeras is also available (references in Brabec, 1965). A decisive advance in the knowledge on the structure and behavior of graft chimeras was made possible by the classical investigations of Winkler (1907, 1908, 1910) who devised a method for the experimental production of graft chimeras in species of *Solanum*. Table 4.4. summarizes the most important observations made on *Solanum* graft chimeras, whose structure is schematically drawn in Fig. 4.5. The most relevant facts which emerge from the work of Winkler (1910) and Jørgensen & Crane (1927) are the following:

(i) despite the presence of tissues of two different species, the periclinal structure of the shoot apex makes possible the long-term maintenance of the chimera when propagated vegetatively. This shows the independence in development of the uppermost cell layers of the shoot apex which has been demonstrated for many angiosperms (see below);

(ii) spontaneous reversions, i.e. non-chimeric shoot apices (and plants) consisting of only one species occur in periclinal chimeras. Haplochlamydeous (i.e. monochlamydeous) chimeras (nos. 1, 4, 5, 6, 7, 10 in Table 4.4) give spontaneous reversions to the species which contributed tissues to the apex core (L_{II} and deeper tissue); diplochlamydeous (i.e. dichlamydeous) chimeras give spontaneous reversions preferentially to the species which contributed tissues to the apex mantle (L_I and L_{II}). Winkler has shown that by experimental production of adventitious buds from chimeras it is possible to obtain from a given chimera reversions to its parent species, due to the regeneration of adventitious buds from tissues of either species. The adventitious bud technique also leads, albeit infrequently, to the rearrangement of chimeras; for example, Winkler observed, among the regenerates of *S. proteus*, a few specimens of *S. koelreuterianum* and *S. gaertnerianum* (for the mechanism of layer rearrangement in chimeras, see below).

(iii) the sporogenous tissue of the flower and consequently the gametophytes (embryo sac and pollen) are in all cases derived from a well distinct cell layer in the shoot apex: L_{II}. This derivation is now well documented in many plant species (see below).

Periclinal chimeras and histogenesis in plants

Periclinal chimeras have an interesting application in studies on developmental processes: because they permit to trace the derivation from the apical meristems of the various tissues which form the aerial organs: stem, leaf and flower. To this end useful information has been provided by three types of periclinal chimeras: mutational (genetical), plastidial and cytological (cytochimeras). As seen in the analysis of graft chimeras in *Solanum*, reproduction of these chimeras through seeds shows that the gametes are derived from the second cell layer (L_{II}) in the shoot apex. Also, in the dichlamydeous plastidial chimera (two albino cell layers, L_I and L_{II}, surmounting a normal green core: AAG) of *Pelargonium zonale* var. *albomarginatum* studied by Baur (1909), it has been found that this variety, when reproduced by seed, gives rise exclusively to albino plants. In 1927, Chittenden concluded that the seed progeny from periclinal chimeras expresses the genetic condition of a well-defined cell layer (the sub-epidermal layer, now called L_{II}) in the angiosperm shoot apex.

In more recent times, periclinal chimeras have greatly helped in understanding the origin of different tissues from the shoot apex. According to the *tunica–corpus* theory propounded by Schmidt (1924) and accepted by many authors (details in Clowes, 1961; Steeves & Sussex, 1972), the shoot apex is made of an external portion consisting of one or more cell layers (*tunica*) covering the massive portion of the meristem (*corpus*). This structure is considered to result from the orientation of the planes of cell division in the meristem: exclusively or predominantly anticlinal (at right angles to the surface) in the *tunica*; anticlinal, periclinal (parallel to the surface) or oblique in the *corpus*. If the *tunica–corpus* concept is applied, the number of *tunica* layers in the angiosperm shoot apices ranges from one to five, with the majority of species having a two-layered *tunica*. When the rather rigid concept of *tunica* is accepted, difficulties arise in interpreting the anatomy of the shoot apex in many species of flowering plants; therefore, Popham (1963) has proposed four types of shoot apex organization in phanerogams, two of which apply to angiosperms. However, the most valid criterium of clearly defining the structure of a shoot apex is obviously its functional analysis; this is much facilitated by the use of periclinal chimeras.

Modifications in the planes of cell division, cell lethality, regenerative processes may occur spontaneously or can be induced experimentally in shoot apices; their developmental consequences are best analyzed in periclinal chimeras. The case of the 'pink' variety of *Poinsettia pulcherrima* can be taken as an illustrative example. This form with pink inflorescence bracts can be maintained indefinitely by vegetative propagation; it produces, however, plants with red bracts when adventitious buds are regenerated from roots of cuttings,* thus demonstrating its periclinal structure (Robinson & Darrow, 1929). From the pink *Poinsettia*, which is a monochlamydeous genetical chimera with 'white' (colorless) L_I, it has been possible to produce, by appropriate regeneration techniques, not only the two non-chimeric or 'homohistont' (Bergann, 1967) forms, red and white but also a new chimera with white-margined pink bracts, called Trebstii alba (Fig. 4.6). It must be concluded that the original form of *Poinsettia* is the one with red bracts, from which the pink form originated by somatic mutation (from red to white) which became stabilized as a monochlamydeous chimera, the starting material for reversions to red and white and for the origin of the dichlamydeous chimera, Trebstii alba (Bergann, 1967).

* In stem cuttings, the roots originate from the external layer of the stele (pericycle) which is derived from the cells of the third layer (L_{III}) or more deeply located cells in the shoot apex. The method of regenerating buds from roots was applied to the analysis of chimeras for the first time by Bateson (1916).

Fig. 4.6. Four types of *Poinsettia* and corresponding shoot apical structure (three outermost cell layers). The pink flowered monoectochimera Ecke's pink (apex: white–red–red) was used to produce the other three types of *Poinsettia*. One type is the diectochimera *Trebstii alba* (apex: white–white–red) with white margined pink flowers; another type is the homohistont red-flowered Imperator (apex non-chimeric; red–red–red), and the third type is the homohistont Ecke's white (apex non-chimeric: white–white–white). Following bud excision at the points shown by arrows the regenerated branches produce flowers which express the genetic constitution of the third apical cell layer (after Bergann, 1967).

By the regeneration of adventitious buds from the internal tissues of tubers (eye excision method), sometimes coupled with an analysis of seed progenies, Asseyeva (1927) was the first to show that in potato, a species long propagated vegetatively, spontaneous periclinal chimeras are not infrequent. This has been confirmed in more recent studies involving the eye excision method as well as the rearrangement of cell layers in shoot apical meristems resulting from treatment with ionizing radiations. Analyses have especially been done on potato ectochimeras bearing genes responsible for the coloration of the tuber skin (periderm) which is almost entirely derived from layer L_I of the apical meristem. In ectochimeras producing tubers with colored periderm, eye excision produces adventitious buds which give plants bearing tubers with colorless periderm, whereas irradiation may produce either this same type of tuber or tubers with a mosaic patterned (colored and colorless) periderm, depending respectively on the total or partial replacement of cells of L_I by cells of L_{II}. Cases of transformation of monochlamydeous into a dichlamydeous chimera by irradiation treatments have also been obtained in potato (review by Howard, 1967).

A well-documented case of a radiation-induced rearrangement of a chimera is that of the white-flowered carnation variety 'White Sim' which originated by somatic mutation from the red-flowered variety 'William Sim'. The monochlamydeous nature of 'White Sim' (L_I, 'white'; the remainder of the apex 'red') is demonstrated by two sets of experiments: (i) all plants raised from seeds of 'White Sim' have red flowers ('William Sim'); (ii) partial or total reversion to red is observed in flowers obtained from plants of 'White Sim' irradiated in the

Fig. 4.7. Reduplication in a three-layered chimeric shoot apex as a consequence of periclinal divisions in the first (left) or the second (right) cell layer (after Bergann, 1967).

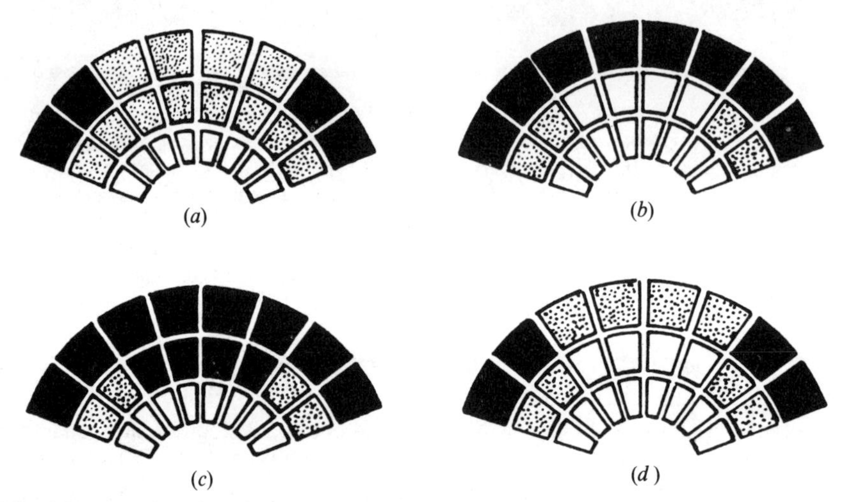

Fig. 4.8. Examples of perforation in a three-layered chimeric shoot apex: (*a*) perforation of L_I and 'filling' (replacement) with cells from L_{II}; (*b*) perforation of L_{II} and 'filling' with cells from L_{III}; (*c*) perforation of L_{II} and 'filling' with cells from L_I; (*d*) perforation of L_I and L_{II} and 'filling' of L_I with cells from L_{II} and of L_{II} with cells from L_{III} (after Bergann, 1967).

vegetative state. This reversion is due to the radiation-induced death of cells of L_I and their replacement by the genetically red, more deep-seated cells in the shoot apex (Sagawa & Mehlquist, 1957). That irradiation combined with adventitious bud techniques can lead to the isolation from a periclinal chimera of all expected forms is most beautifully demonstrated by the work of Bergann (1967) on the variety 'Madame Salleron' of *Pelargonium zonale*. This has been shown to be a triple chimera (trichimera), ABC, in which the external layer (A) consists of cells with green plastids, the second layer (B) is made of cells with albino plastids and the remainder of the apex (C) contains cells with green plastids and genes for the dwarf habit. From 'Madame Salleron' all nine possible types of shoot apices have been obtained: the three homohistonts (AAA, BBB, CCC) and the six heterohistonts (AAB, ABB, BCC, BBC, ACC, AAC).

The cellular mechanisms responsible for the rearrangement of cell layers in shoot apices have been discussed by several authors: e.g. Dermen (1960, 1965) and

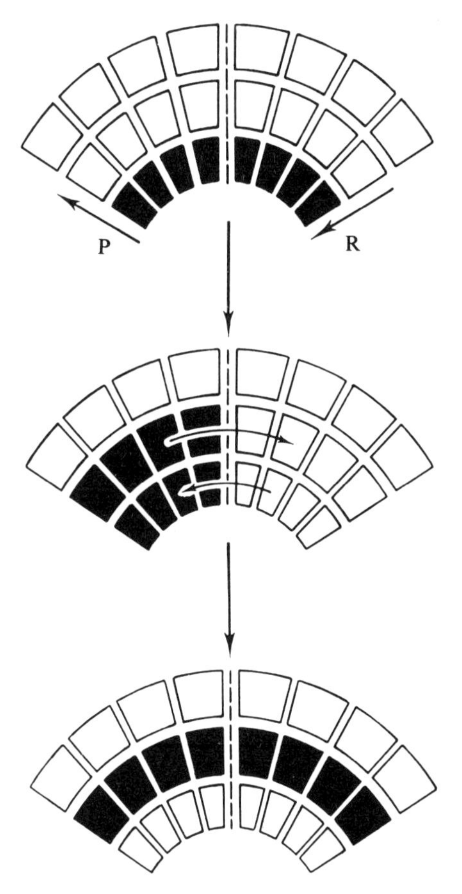

Fig. 4.9. Translocation from a diectochimera (white–white–green) to a white–green–white chimera. A perforation (P) in the left half of the apex leads to replacement of L_{II} (white) by cells of L_{III} and a reduplication (R) of L_{II} in the right half of the apex pushes the original L_{III} 'out' of the organogenic zone (i.e. into L_{IV}). Afterwards, anticlinal divisions produce an extension (transgression) to right of the green tissue and to left of the white tissue; the result is translocation of the green cell layer from the original L_{III} to L_{II} (after Bergann, 1967).

Bergann (1967). According to Bergann, the rearrangement of cell layers in shoot apices may result from three types of events, namely:

(i) reduplication (Fig. 4.7): cells of L_I, L_{II} or both may undergo a periclinal division, with the consequence that cells of L_{III} or the whole L_{III} are pushed outside of the 'active' histogenetic zone; i.e. they become L_{IV};

(ii) perforation (Fig. 4.8): cells of L_I, L_{II} or both are 'perforated' (an important cause may be cell lethality) and the resulting empty space is filled by cells of another layer which divide periclinally;

(iii) translocation (Fig. 4.9): a rare histogenetic anomaly which leads to an exchange between external and internal cell layers. These histogenetic anomalies, whose spontaneous incidence is low (in tobacco, in which the reproductive tissue of the flower originates from L_{II}, the frequency of participation of either L_I or L_{III} cells in the organization of the generative tissue ranges from 0.03 to 0.05%: Stewart & Burk, 1970), are much favored by irradiation. Irradiation not only leads to rearrangement of periclinal chimeras, but also helps in stabilizing, in the form of periclinal chimeras, radiation-induced genetic changes in shoot apices: this has been found in several vegetatively propagated species and studied most extensively in roses (Dommergues, Heslot, Gillot & Martin, 1967).

In periclinal cytochimeras the presence, in distinct histogenetic layers, of cells with different degrees of ploidy (generally diploid, and tetraploid) allows the tracing of cell lineages in the different organs through direct microscopic observation (chromosome counts, nuclear size). Following the pioneer developmental studies of Blakeslee and his associates (Satina, Blakeslee & Avery, 1940; Satina & Blakeslee, 1941, 1943; Satina, 1944, 1945) on colchicine-induced cytochimeras in *Datura*, investigations of a fundamental value on spontaneous and induced cytochimeras have been carried out by Dermen (1947, 1951, 1953, 1965), and Einset and his associates (Blaser & Einset, 1948, 1950; Einset, 1952; Einset & Pratt, 1954; Pratt, Ourecky & Einset, 1967). Our present knowledge on plant histogenesis as related to shoot apex organization can be briefly summarized as follows.

Histogenesis in the stem

In the apical dome of the shoot apex of cranberry, *Vaccinium macrocarpon*, there are possibly three independent histogenic layers (L_I, L_{II}, L_{III}) as suggested by obtaining, through colchicine treatment, the following chimeras: 4–2–2,* 2–4–4, 2–4–2, 2–2–4. In the stem of these chimeras Dermen (1947) found that the epidermis was derived from L_I (indicating that cell division in this layer is anticlinal), the hypodermis and sometimes part of the cortex were derived from L_{II} and the remainder of the stem originated from L_{III}. Since in cranberry chimeras no distinction can be made between cells of L_{III} and subjacent cells of the meristem because of their identical chromosome number, nothing can be learned on histogenesis in the central cylinder (stele) of this species. The situation is much better in the apple in which, as first shown by Dermen (1951), the dome of the shoot apex shows five distinct apical layers (in some apices more than five: Fig. 4.10). In the stem of a spontaneous 2–2–2–2–4 chimera Dermen found that the epidermis and five to seven layers of cells in the outer cortex were diploid (2x) and the rest of the stem tissue was tetraploid (4x). The evidence in favor of an hypothesis of generally five apical layers (Dermen, 1951) in the apple promeristem which contribute to the primary tissues of the stem in a fairly regular pattern has been

* For brevity, periclinal cytochimeras are usually designated by numerals referring to the degree of ploidy of the apical layers, from the exterior to the interior: 2 = 2x (diploid), 4 = 4x (tetraploid).

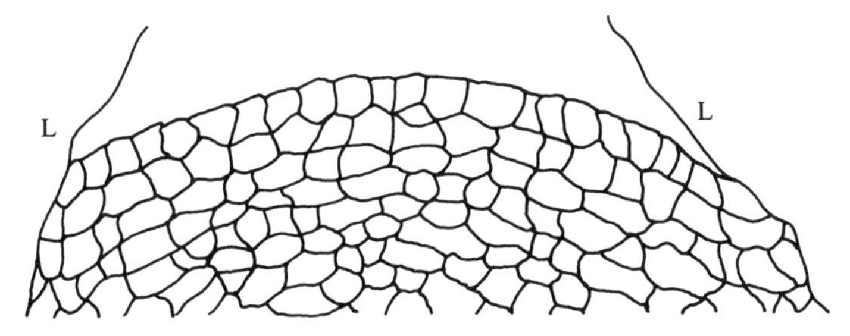

Fig. 4.10. Longitudinal section of the apical dome of an apple shoot apex showing the arrangement of cells in seven or eight layers. L: leaf (drawn from a microphotograph kindly provided by Dr Haig Dermen).

presented by Pratt *et al.* (1967): L_I developed into epidermis; L_{II} and L_{III} formed the outer and inner cortex; L_{III}, L_{IV} and L_V contributed to the vascular tissue in varying degrees, differing with the cytochimera and the individual bud; L_V (and probably more internal cells of the promeristem) formed the pith tissue (Table 4.5). Although the apical layers in the apple are independent enough to produce certain primary tissues in the same bud, they are not strictly self-perpetuating, as is demonstrated by the occurrence of layer replacements (Table 4.6).

Histogenesis in the leaf

From their analysis on *Datura* cytochimeras Satina & Blakeslee (1941) concluded that the epidermis of the leaf was derived from L_I, while the mesophyll and the vascular tissue (procambial strands) were derived from L_{II} and L_{III} respectively. Contrary to the situation in *Datura*, in cranberry both L_{II} and L_{III} can contribute to the formation of the mesophyll, and from both these layers the conductive tissue of the leaf may arise: moreover, in the parts of the cranberry leaves where the mesophyll was 2x the conductive tissue (veins) was also 2x; in the parts where the mesophyll was 4x, the conductive tissue was 4x (Dermen, 1947).

The most comprehensive studies on leaf histogenesis have been made in the apple using a variety of cytochimerical types (Blaser & Einset, 1948; Dermen, 1951; Pratt *et al.*, 1967). In the apple, as in *Datura*, the interior of the leaf including the vascular tissue is formed from L_{II} and L_{III} (2–4–2–2–2 and 2–2–4–4–4 chimeras in Table 4.5); only in one case and at the base of leaves, the participation of L_{IV} in vascular tissue formation has been observed (2–2–2–4–4 chimera in Table 4.5). In any case, the relative contribution of L_{II} and L_{III} to the tissues forming the interior of the leaf is quite variable so that, at the extremes, the internal tissues may be derived entirely or almost entirely from either L_{II} or L_{III}.

TABLE 4.5. *Distribution of diploid (2x) and tetraploid (4x) metaphase figures in meristematic primary tissues of apple cytochimeras* (from Pratt *et al.*, 1967)

		Stem			Leaf		Flower			
							Floral tube		Flower part†	
Apical pattern*	Epi-dermis	Cortex	Vascular tissue	Pith	Interior	Vascular tissue	Interior	Vascular tissue	Interior	Vascular tissue
4–2–2–2–2	4	2	2	2	2	2	2	2	2	2
2–4–2–2–2	2	4	2	2	4	4	4	2	4	2
		2	4	—	2	2	2	4	—	4
2–4–2–4–4	2	4	4	4	4	4	4	4	4	4
		2	2	—	2	2	2	2	2	2
2–4–4–2–2	2	4	4	4	4	4	4	—	4	4
		—	2	2	2	—	—	—	—	—
2–4–4–4–4	2	4	4	4	4	4	4	4	4	4
2–2–4–4–4	2	2	4	4	2	2	2	4	2	2
		4	—	—	4 (base)	4 (base, middle)	4	—	—	4
2–2–2–4–4	2	2	4	4	2	2	2	4	2	2
		—	—	—	—	4 (base)	4	—	—	4
2–2–2–2–4	2	2	2	4	2	2	2	2	2	2
		—	4	—	—	—	—	4	—	—

Abbreviations: 2 = 2x; 4 = 4x.
* Based on relative sizes of cells, nuclei and mitoses in protomeristem.
† Sepals, petals, stamen and style.

TABLE 4.6. *Spontaneous and radiation-induced variation in apple cytochimeras as determined by cytological examination of growing buds* (from Pratt *et al.*, 1967)

Predominant type (L_I to L_V)	Variation (L_I to L_V)
2–4–2–2–2	4–2–2–2–2
	2–4–4–2–2
	2–4–4–4–4
	2–2–4–4–4
	2*
2–4–4–4–4	2–2–4–4–4
	2–2–2–4–4
2–2–4–4–4	2–4–2–2–2
	2–4–4–4–4
	2–2–2–4–4
	2–2–2–2–4
	2*
	4*
2–2–2–4–4	2–2–4–4–4
	2–2–2–2–4
	2*

* Non-chimeric (homohistont).

Histogenesis in the flower

The most comprehensive studies on ontogenesis of the floral tissues have been made by Blakeslee and his associates (Satina & Blakeslee, 1941, 1943; Satina, 1944, 1945) who used periclinal cytochimeras of *Datura*. These studies demonstrated that

(i) the microsporogenous tissue in the anther and the archesporial cell (megaspore mother cell) in the ovule were derived from L_{II};
(ii) histogenesis of petals and sepals is identical to that of leaves (L_I gives rise to the epidermis, L_{II} and L_{III} give rise to the internal tissues of the petal and sepal);
(iii) the stamen, apart from the microsporogenous tissue, results mainly from the activity of L_{III};
(iv) the carpel is derived mostly from L_{III} and in part (upper portion of the carpel wall and septa) from L_{II};
(v) all three apical layers participate in the production of the ovule, but in each of its components (funicle, nucellus and integuments) the cells derived from one of the three layers, L_{III}, L_{II} and L_I, predominate in number over the others.

Further work on cytochimeras has shown that the sporogenous tissue of the flower in other dicotyledons is derived from L_{II}; e.g. in cranberry (Dermen, 1947), grapes (Einset & Pratt, 1954), apple (Blaser & Einset, 1950; Dermen, 1953; Pratt *et al.*, 1967) (Table 4.5) and peach (Dermen & Stewart, 1973).

Visible markers and organogenesis in plants

A common method for the induction of mutations in plants is treatment with radiations or chemicals of whole plants or plant parts (seeds, cuttings, tubers, bulbs, rhizomes) in which multicellular shoot apices, sometimes consisting of hundreds or thousands of cells, are present. When the treated shoot apices, which are now complex mosaics, grow to form the new plant (M_1: M, initial of mutagen), each mutated cell becomes the progenitor of a cell lineage making a sector in the apex. If an inflorescence or a flower-bearing branch is formed from this sectorial apex, among the seeds produced by selfing (M_2), one or several will give rise to plants showing the mutation (in general, induced mutations are recessive). In these progenies (spike progeny, panicle progeny, branch progeny, etc.), the relative proportion of mutated and normal plants can give an estimate of the number of initial cells which participated in the formation of the inflorescence or branch (more properly, the sporogenous tissue in the flowers of such organs). Thus, Anderson, Longley, Li & Retherford (1949), who made an analysis of M_2 progenies raised from atom bomb irradiated seeds of maize, concluded that the sporogenous tissue of the male inflorescence in their material should have originated from an average number of seven to eight initial cells. Also Kaukis & Reitz (1955), using radiation-induced chlorophyll-deficient mutations, obtained information on the number of initial cells responsible for the organization of the sporogenous tissue of a panicle in *Sorghum*.

Chlorophyll-deficient mutations, which offer the great advantage of being analyzable at the plantlet stage, have been extensively used in the outstanding investigations of Gaul (1959, 1961) on the chimeric nature of plants raised from irradiated seeds. In the barley seed embryo, the apical meristem comprises the primordia of the first five culms (and spikes); the size of the primordia reflects their ontogenetic history in that the first developed culm originates from the largest primordium and the following culms originate from progressively smaller primordia. Moreover, during further growth of the plant, other culms bearing spikes (tillers) may be formed: they are not preformed as primordia in the embryo, and are developed from a few or even a single cell of the embryo. When seeds are irradiated, mutation from dominant to recessive (from *AA* to *Aa*, heterozygous) or loss of a dominant allele (from *AA* to *AO*, hemizygous) may involve genes responsible for chlorophyll formation; in the M_2 progeny (barley is an autogamous species) chlorophyll deficient mutants will occur. If one spike were produced by a single mutated cell, the recessive mutants in this particular spike progeny would be expected to occur with the frequency typical of a mendelian segregation, i.e. 25% (25% *AA*+50% *Aa*+25% *aa*). In case of the origin of a spike from two cells, one of which is mutant, the mutant frequency in the spike progeny would be half that value, i.e. 12.5%. Since these 'segregation' ratios are to be ascertained statistically, difficulties may arise in connection with the rather small population size of a barley spike progeny which, at some radiation exposures, is reduced by

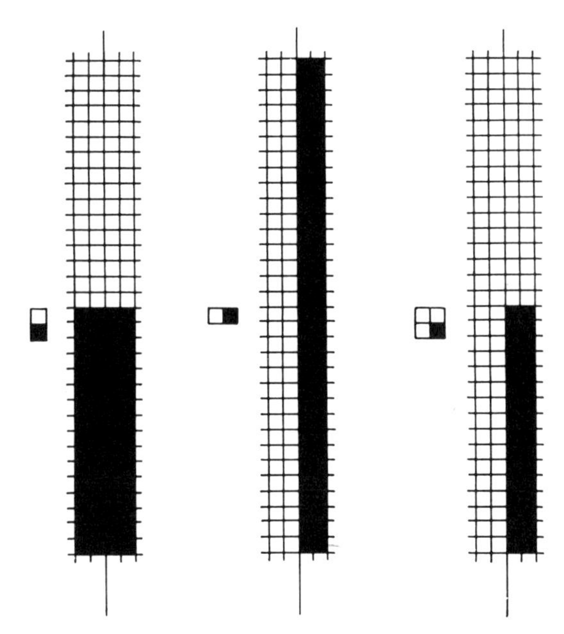

Fig. 4.11. Diagrammatic representation of the origin of a *durum* wheat (*Triticum durum*) spike from two or four initial cells one of which mutated (in black). In the first case, the spike sector bearing mutants covers either the transversal or the longitudinal half of the spike, depending upon whether the two initials are superposed or juxtaposed. In the case of four initials, the spike sector bearing mutants covers one quarter of the spike (lower right quarter in the example drawn in this figure). Based on the 'topographical' analysis of spikes produced by plants deriving from irradiated seeds (D'Amato, 1965*a*).

induced sterility. Despite these limitations, Gaul has presented convincing evidence that:

(i) the frequency of mutations is maximal in the spikes of the first five culms (primordia already present in the embryo) and progressively decreases with increasing ontogenetic order of the culms:

(ii) the M_2 mutant 'segregation ratio', indicative of the size of the mutated spike sector, is clearly smaller in spikes from preformed primordia (first to fifth) than in later formed spikes. When spikes are analyzed in two separate groups, the first two spikes and the remaining spikes in each plant, it is calculated that, on average, two cells produce one spike in the first group whereas in the second group only one cell produces one spike, or even two or more spikes.

Gaul's observations have been confirmed and extended in *Triticum durum* whose spike offers the advantage over barley of a clearly greater (by some 50%) spike progeny size. In this material, the statistical analysis of 'segregation ratios' has been supplemented by a new method of visual analysis ('topographical method') which has proved of decisive value: it consists of germinating whole (unthreshed)

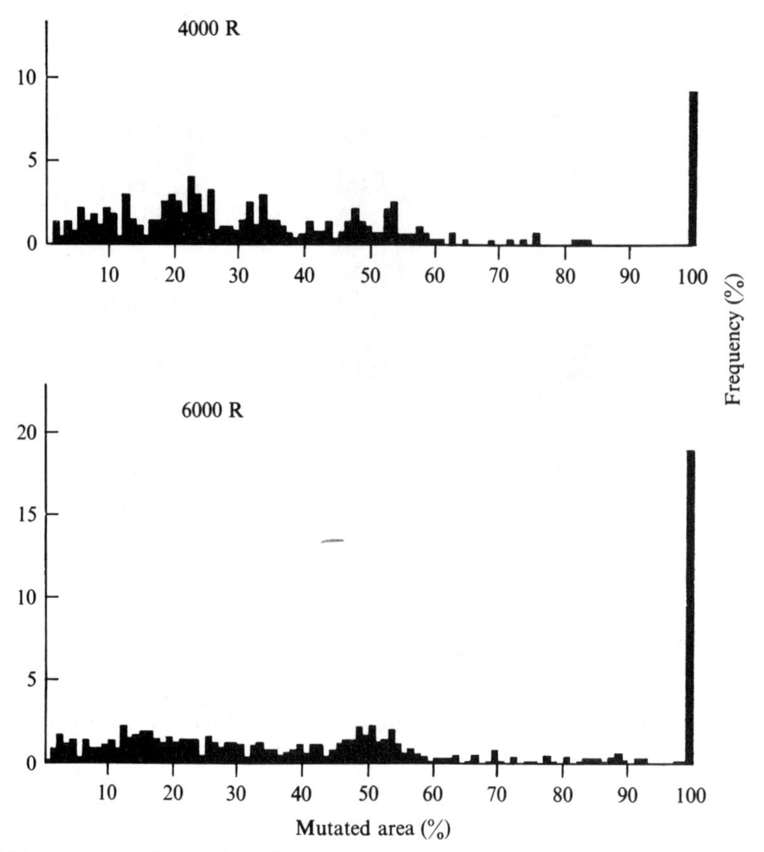

Fig. 4.12. Frequency distribution of mutated tepal area (as a percentage of total tepal area) in flowers of gladiolus produced from corms irradiated with 4000 or 6000 R. The mutated area gives an estimate of the number of initial cells that formed the tepal (e.g. 100% = one initial; 50% = two initials) (after Buiatti *et al.*, 1969).

M_1 spikes and taking note of the localization of the mutants within the spike (D'Amato, 1965*a*). Since the *Triticum durum* spike is distichous (two rows of flowers, and seeds, along the rachis of the spike), the spike can be easily divided into two longitudinal and two transversal halves. It has been demonstrated that, when a spike originates from a single mutated cell, mutants are distributed at random within the spike: when the spike originates from two cells, one of which mutant, the mutant plantlets are distributed within one half, either longitudinal or transversal, of the spike; when the spike originates from four cells, one of which mutant, the mutant plantlets are distributed within one quarter of the spike (Fig. 4.11).

Another case of ontogenetic analysis of a distichous spike is worth considering: the inflorescence of gladiolus. For this study, use was made of varieties heterozygous for flower color, which allowed analysis of distribution of color changes directly in M_1; that is, in spikes produced from irradiated corms (Buiatti, Tesi & Molino, 1969). Since in the gladiolus spike flowers develop with a regular sequence from the base (first developed flowers) to the top (later formed flowers) this study has not only verified expected facts (e.g. lower mutation frequency and greater mutant sector size in later formed flowers), but also permitted the number of initial cells per flower and per tepal in a given flower to be estimated. Thus, for example, an analysis of mutated sector areas in percent of total tepal area demonstrated a preponderance of tepals formed by a single initial cell (100% mutated area = wholly mutated tepal) followed by two classes of tepals formed by two (50% mutated area) and four (25% mutated area) initial cells respectively and by cases in which tepals started development from more than four initials (Fig. 4.12).

An elegant analysis on the ontogeny of the stem and leaf in *Epilobium*, based on the use of spontaneous and induced chlorophyll deficient markers, has been published by Michaelis (1967).

Germ line and soma

From the works on graft chimeras and periclinal chimeras discussed above and from other information available in literature (references in Popham, 1963) the conclusion can be drawn that in dicotyledons the sporocytes (micro- and mega-sporocytes) and ultimately the reproductive cells in the flower are derived from the second cell layer (L_{II}) in the shoot apex. In monocotyledons (only a few species have been studied so far) the megasporocytes are derived from L_{II} and the microsporocytes are derived from L_{III} or from L_{II} and L_{III}. Thus, in angiosperms, as in most if not all plant species, no separation exists between germ line and soma, any somatic cell being a potential progenitor of a new individual (Brink, 1962). Whether or not a somatic cell enters the reproductive lines strictly depends on its localization in a defined apical layer at the time of phase change, i.e. the passage of the apex from the vegetative to the reproductive phase (Brink, 1962). The cellular and molecular aspects of the transition of the shoot apex from the vegetative to the reproductive condition are not well understood: there are indications, however, that nucleic acid and protein metabolism in the meristem play a role in the floral transition process. Thus, e.g. in *Sinapis alba* a single 22-hour, long-day exposure induces two-month-old plants to flower, the first flower buds being initiated at about 58 hours from the beginning of the long day. From the diagram of Fig. 4.13 (Jacqmard, Miksehe & Bernier, 1972) it is seen that at the presumed time of arrival of the floral stimulus at the apex there is an increase in total RNA synthesis and chromatin-dependent RNA synthesis followed by preferential mitotic stimulation of G_2 cells in the apex, so that at the end of the first mitotic peak a synchronization of cell population at about 30 hours is achieved. At about 40 hours nuclear DNA synthesis begins, accompanied by an increase in

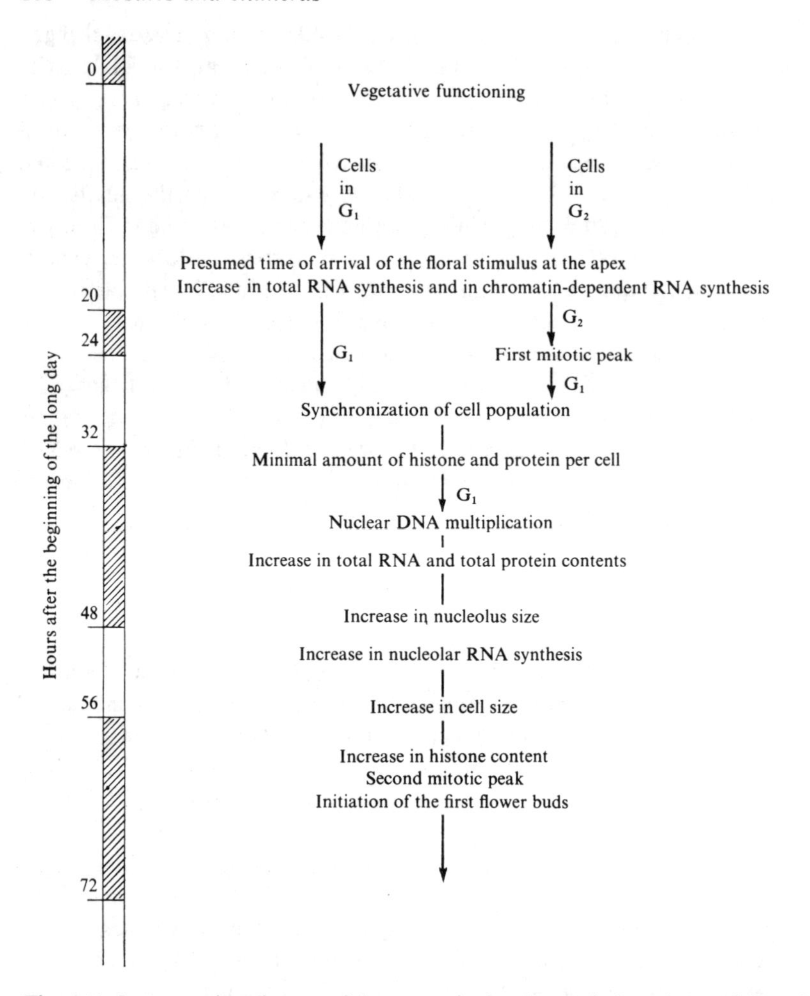

Fig. 4.13. Interpretative diagram of the events in the shoot apical meristem of *Sinapis alba* in transition from the vegetative phase to flowering (after Jacqmard *et al.*, 1972).

total protein, total RNA and nucleolar RNA. The second mitotic peak at 58–60 hours marks the initiation of flower buds in *Sinapis*.

Although no separation between germ line and soma occurs in plants, it must be realized that the two types of apical meristem (the root and the shoot apex) have different prospective values: the root apex provides for the development and growth of the root system; the shoot apex, besides providing for the development and growth of the aerial organ system, is destined to enter the reproductive line in the flower at the time of phase change. A difference in prospective values between the two meristem cell lines would imply that decision for separation of one cell line from the other is taken at the time, during embryogenesis, when shoot and root meristems are differentiated. In this connection, the case of *Xanthisma texanum*, a diploid species (2n = 8), is instructive. In this species, types have been found which show one, two, three or four supernumerary chromosomes, generally called B-chromosomes (Battaglia, 1964), in the cells of their shoots and in the sporocytes ('germ cells'), while the root cells have the normal chromosome complement, 2n = 8 (Berger & Witkus, 1954; Witkus, Ferschl & Berger, 1959). In these types, B-chromosomes are present in all the cells of the young embryo up to the late spherical stage; at the transition from this stage to the heart-shaped embryo, when root and shoot tissues are differentiated, the B-chromosomes are eliminated through non-congression at metaphase or anaphasic lagging in the root cells (Berger, McMahon & Witkus, 1955; Berger, Feeley & Witkus, 1956; Witkus, Ferschl & Berger, 1959). The two nuclear conditions which originated during embryogenesis keep quite stable during plant development: adventitious roots produced from cuttings (these roots originate from the stem pericycle) have been found to have cells with the same chromosome complement as the cells of the original shoot. Clearly then, in *Xanthisma* there is some controlled difference with respect to genome between true root tissue and true shoot tissue (and 'germ line') (Semple, 1972). In other plant species, a.o., *Sorghum purpureo-sericeum* (Darlington & Thomas, 1941), *Poa timoleontis* (Nygren, 1957), *Haplopappus gracilis* (Östergren & Fröst, 1962), the number of B-chromosomes originally transmitted by the zygote is maintained in the 'germ line' (shoot, pollen mother cells), while the B-chromosomes are eliminated from the root. The exact time of somatic elimination of the B-chromosomes in these species is not known, however.

In the majority of animals, in contrast with plants, separation of the germ line from soma occurs during early embryogenesis. There are various lines of evidence that the factor ultimately responsible for cells becoming germ cells resides in the cytoplasm. In many eggs, this germ-cell determinant (germinal cytoplasm) occupies the basophilic cytoplasm at the vegetal pole; in scyphozoans, chaetognaths, rotifers, insects, crustaceans and amphibians, the germinal cytoplasm includes deeply stained granules. During the first cleavage divisions the granular cytoplasm becomes aggregated at one pole of the mitotic spindle. In the copepod *Cyclops*, segregation of the granular cytoplasm occurs at the third cleavage division. The single blastomere which receives the germinal cytoplasm is the source of the germ line (Austin, 1965; Blackler, 1970).

As a rule, the karyotype of the somatic mitoses closely agrees with that found in the germ line. But cases of gonadosomic mosaicism are known. The classical example is the horse ascaris, *Parascaris equorum*, of which two 'races' exist, one diploid (2n = 2) and the other tetraploid (2n = 4). *Parascaris*, first studied by Boveri (1887), has polycentric chromosomes, each of which consists of six segments, each segment having one centrally located centromere (Serra & Picciochi, 1961). In *Parascaris*, the four cells formed by the second cleavage division, identified as A, B, C and P, show a different nuclear behaviour. In cells A, B and

C, the chromosomes dissociate into their component monocentric chromosomes and into acentric fragments detached from the chromosomal ends (chromosome diminution). At the third division, one of the daughter cells of P undergoes chromatin diminution and the same happens at the fourth division, so that only one cell, P_4, retains the intact chromosomes, two in *univalens* and four in *bivalens*. It is from P_4 that all the germ cells eventually develop (Austin, 1965). In *Ascaris lumbricoides*, the nuclear DNA of somatic cells was compared by hybridization with germ cell (spermatid) DNA (Tobler, Smith & Ursprung, 1972; Tobler, Zulauf & Kuhn, 1974). It was found that the eliminated DNA consists of repetitious and unique sequences in a ratio of about 1:1 (in the germ cells, this ratio is about 1:9) and also contains ribosomal cistrons (the number of these cistrons per haploid genome was calculated to be about 350 in the germ line and about 255 in the somatic lines).

Another type of gonadosomic mosaicism results from the differential elimination of chromosomes from nuclei of somatic cells during the early cleavage mitoses: it occurs in three dipteran groups, the Orthocladinae, the Cecydomidae and the Sciaridae. In the fourteen species of Orthocladinae studied by Bauer & Beermann (1952), the somatic cells contained four (two species) or six chromosomes (S-type) while in the germ line cells, in addition to the S-type chromosomes, there were chromosomes whose number varied from 2 to 52. In the Cecydomidae, which have been studied by many authors (references in Camenzind, 1966), the number of chromosomes which are eliminated from the somatic cells (E-type chromosomes) varies from 8 to 67. In the Sciaridae, the elimination involves the paternal X-chromosome and some large chromosomes which are confined to the germ-line and are referred to as 'limited chromosomes' or L-chromosomes (Metz & Schmuck, 1931). In male and female embryos of *Sciara coprophila* at 60–71 hours after oviposition, chromosomes are eliminated from each non-dividing germ cell nucleus. A paternal X-chromosome and a variable number (0–3) of L-chromosomes apparently pass through the intact nuclear membrane into the cytoplasm where they disappear after a few days. Following elimination, 2 limited and 8 regular chromosomes remain (Rieffel & Crouse, 1966).

5

The chromosome complement of differentiated cells

Diploidy

In their analysis of the cell cycle in the intestinal epithelium of the mouse, Quastler & Sherman (1959) assumed that, shortly after mitosis, the cell 'takes a decision' whether to divide or differentiate. This meant that the cell cycle, in addition to G_1, S, G_2 and M, might include another phase, or period, between the end of mitosis and the beginning of differentiation or a new mitotic cycle. Quastler & Sherman (1959) calculated that in their material this period lasted up to 4 hours after mitosis. Oehlert, Lauf & Seemayer (1962) designated this period as a 'reversible post-mitotic cell condition' and Bullough (1963) called the period following mitosis and preceding G_1 'dicophase'. As pointed out by Epifanova & Terskikh (1969), the phase of 'taking the decision' of Quastler & Sherman (1959), the 'reversible post-mitotic cell condition' of Oehlert *et al.* (1962) and the 'dicophase' of Bullough (1963) are all synonyms of G_0, the true resting stage, or stage of 'no cell cycle', whose concept has been developed by Lajtha (1963, 1964) and Quastler (1963).

In considering the early events in mammalian liver regeneration following a partial hepatectomy, Lajtha (1963–4) has called attention to the fact that DNA synthesis starts about 15 hours after the operation and is the prerequisite for mitosis, which occurs some 10 hours later. In Laitha's reasoning, if there were a very long G_1 period of months or years, and if it were shortened to 15 hours, then, instead of the delayed DNA synthesis peak, an immediate increase should take place. Since this is not the case, liver cells have a true resting stage (G_0), from which they can be triggered into a cell cycle at any time. Thus, cells in the G_0 period form a cell population with unlimited proliferative potential: after mitosis, the cell may immediately enter into the mitotic cycle or into the G_0 state and at any time it may be triggered into the cycle again (Fig. 5.1). Quastler (1963), on the other hand, has pointed out that pure proliferative populations are rare in the mammalian organism; as a rule, they are embedded in a system which also contains one or several kinds of mature cells and, sometimes, one or more populations of potentially proliferative cells (G_0). Then, according to Quastler, a cell proliferation system is made up of at least six compartments (Fig. 5.2).

G_0 is not the only 'resting' or 'nul' period in the cell cycle. The investigations of Gelfant (1962, 1963, 1966) on the patterns of cell proliferation in stimulated mouse epidermis first demonstrated that the epidermis comprised two distinct cell

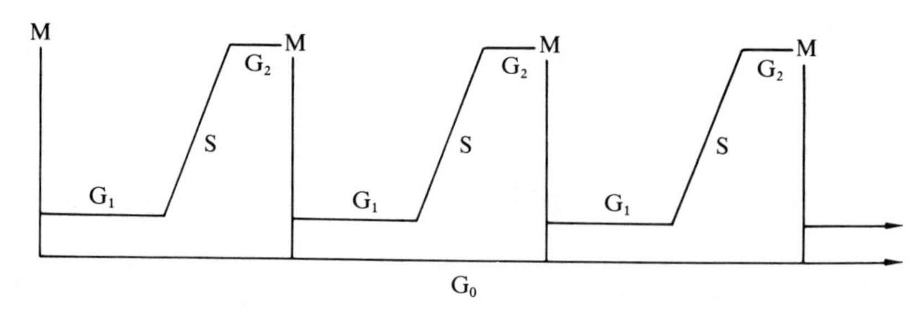

Fig. 5.1. Concept of the G_0 period. Following mitosis (M) a cell may immediately enter into the cell cycle (G_1–S–G_2–M) or into a state of no cycle (G_0). From G_0 the cell may be at any time triggered into the cell cycle (after Gilbert & Lajtha, copyright © 1965, The Williams and Wilkins Co., Baltimore.)

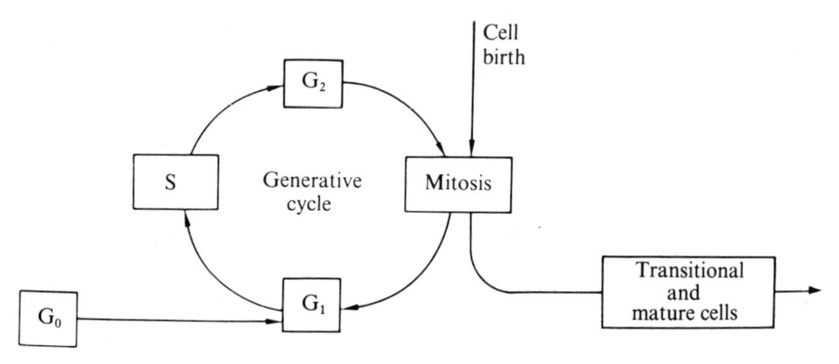

Fig. 5.2. Compartments of a cell proliferation system in the interpretation of Quastler (after Quastler, 1963).

types with regenerative capacities: a population of cells which have not synthesized DNA prior to stimulation (G_1) and a population of cells which have completed DNA synthesis (G_2) and can enter mitosis under an appropriate stimulus. A situation similar to that in mouse epidermis has been found by Gelfant and other workers in several animal and plant tissues (review in Gelfant, 1966). Thus, the cell cycle can be regarded as a system of alternating periods of rest and active cell proliferation (Fig. 5.3). The cell may move out of the cell cycle either after mitosis and prior to DNA synthesis (2C DNA content) or after the completion of DNA synthesis (4C DNA content). From either state, which have been called resting period 1 (R_1) and 2 (R_2) because 'it is inexpedient to denote resting periods with symbols of the "gap" nomenclature' (Epifanova & Terskikh, 1969), cells are able to re-enter the mitotic cycle at any time.

Since R_1 and R_2 are 'periods of decision' for mitosis or differentiation, it must be expected that cell differentiation in the diploid state – i.e. in the absence of chromosome endoreduplication, a common concomitant of the process of cell differentiation (see below) – may occur either in the 2C (R_1) or in the 4C (R_2)

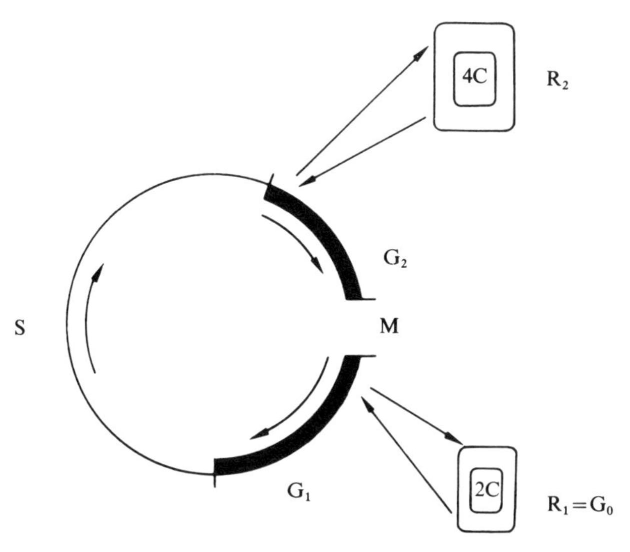

Fig. 5.3. Diagram of the diploid cell cycle illustrating the alternation of resting periods (R_1 and R_2) and periods of active proliferation. According to this interpretation, the cell may rest either in the pre-DNA synthetic phase, 2C(R_1 = G_0), or in the post-DNA synthetic phase, 4C (R_2) (after Epifanova & Terskikh, 1969).

condition.* In both plants and animals, however, diploid cell differentiation in 2C is much more widespread than differentiation in 4C. In plants, the nuclear conditions in differentiated tissues have been extensively studied and hundreds of species have been investigated. Taking the literature on differentiated root tissues only, Tschermak-Woess (1956) reported that in a total of 179 species investigated, the majority (140) showed both diploidy and endopolyploidy (chromosome endoreduplication) while the rest (39) were characterized by the exclusive occurrence of diploidy. An interesting aspect which has emerged from the extensive analyses on the nuclear cytology of root, stem and leaf tissues is that among angiosperms the ability to differentiate tissues in the diploid state occurs preferentially in some genera and families: e.g. *Crinum*, *Lilium*, *Paeonia*, Compositae (D'Amato, 1950; Holzer, 1952; Fenzl & Tschermak-Woess, 1954). Most striking is the accumulation of species with diploid cell differentiation in the family Compositae. Taken as a whole, the available information on differentiated plant cells supports the view that the nuclear cytology of differentiation is strictly bound to the genetic make-up of taxonomic groups, thus appearing in general as an aspect of phylogenesis (D'Amato, 1964*a*, 1965*b*; Brunori & D'Amato, 1967).

Since most studies on differentiated plant tissues have been based on an analysis of mitoses induced by wounding or hormone treatment, it has not been possible to ascertain whether the nuclei in the diploid differentiated tissues of the

* Dewey and Howard (1963) use the symbol g_2 to denote the category of cells which differentiate in the 4C nuclear condition.

species investigated are all held in R_1 (2C) or part in R_1 (2C) and part in R_2 (4C). The first possibility is documented by a few investigations in which the karyological analyses have been supplemented with DNA cytophotometric and autoradiographic data. A well-studied case is that of Jerusalem artichoke, *Helianthus tuberosus* (Compositae). In this species, all differentiated tissues consist of 2C cells only (Partanen, 1959; Adamson, 1962). When explants of tuber tissue are cultured *in vitro*, the first two or three divisions are approximately synchronous (Yeoman & Evans, 1967); at 25 °C in the dark, DNA synthesis, which begins at 25 h, lasts 14 h and is followed almost immediately by mitosis which lasts for 2 h (Mitchell, 1967). Obviously, the exclusive occurrence of 2C nuclei in the differentiated tissues depends on a strict control of the sequence of DNA synthesis and mitosis which allows only one resting period in the cell cycle, the G_0 (R_1). Cogent evidence for this conclusion comes from the observation that Jerusalem artichoke tissue remains indefinitely diploid both under the influence of a wide range of concentrations of the hormone naphthalene acetic acid (NAA) and during tumorous growth induced by the crown gall bacterium, *Agrobacterium tumefaciens* (Partanen, 1959).

In differentiated plant tissues consisting of cells with diploid and cells with endoreduplicated (endopolyploid) nuclei, it can be demonstrated that the diploid cell population comprises both nuclei in R_1 and, less frequently, in R_2 (4C). Tobacco pith tissue contains nuclei with 2C, 4C and 8C DNA contents. When this tissue is explanted *in vitro* in the presence of kinetin and indole acetic acid (IAA), both diploid and endopolyploid nuclei are triggered into mitosis; feeding the culture with tritiated thymidine shows that the majority of diploid mitoses are labeled and the rest are unlabeled, thus demonstrating the occurrence in pith tissue of differentiated diploid cells resting in either R_1 (2C) or R_2 (4C). Other 4C cells, however, undergo DNA synthesis and enter an 8C (endotetraploid) mitosis (Pätau, Das & Skoog, 1957; Pätau & Das, 1961). Results very similar to those on tobacco pith have been obtained with segments of pea root which also contain diploid and endoreduplicated nuclei: 2C, 4C and 8C (McLeish & Sunderland, 1961). When pea root segments are cultured on a medium containing auxin and kinetin with tritiated thymidine added, most diploid and polyploid mitoses are found to be labeled; at 52 h after the beginning of culture only 36% diploid mitoses are unlabeled (cells in R_2 (4C) at the time of excision of the root segments) (Matthysse & Torrey, 1967). The tobacco and pea cases exemplify a situation which frequently occurs during cell and tissue differentiation both in plants and in animals: R_2 (4C) is the stage from which the 'resting' nucleus embarks upon the endoreduplication course, (R_2) 4C $\rightarrow$ 8C $\rightarrow$ 16C, etc., producing chromosome polyteny of different degree.

In summary, differentiation of a tissue in a pure diploid state or in a mixed diploid–endopolyploid state most probably reflects the characteristics of the cell cycle of the proliferating system which gave rise to the tissue with one resting period – R_1 (2C) – or with two resting periods – R_1 (2C) and R_2 (4C) – respectively.

Binucleation, multinucleation and nuclear fusion

Binucleation and/or multinucleation, sometimes combined with nuclear fusion, endomitosis and chromosome endoreduplication occur in some animal and plant tissues (reviews in D'Amato, 1952*b*; Pera, 1970). The production of binucleate and multinucleate cells by nuclear divisions not followed by cytokinesis (coenocyte formation) is to be regarded as a process of somatic polyploidization (D'Amato, 1952*b*). Thus, for example, a cell containing two diploid nuclei as a result of acytokinetic mitosis must be regarded genetically and functionally as a tetraploid cell, by analogy with the diploidization process in higher fungi, for which definite evidence has been obtained that a cell with two haploid nuclei (dikaryon) is a diploid cell (Buller, 1941; Raper & Flexer, 1970).

The occurrence in the mammalian liver of binucleate and multinucleate cells and of cells with nuclei of different sizes has been a matter of interest since the early twenties. From a very accurate study on the rat liver, Beams & King (1942) concluded that binucleate cells arise from uninucleate cells by mitotic division not followed by cytoplasmic cleavage, and that larger nuclei arise when the two nuclei of a binucleate cell enter mitosis synchronously and fuse their spindles into a single spindle. If cytokinesis occurs, two tetraploid cells are produced; if not, a binucleate cell with two tetraploid nuclei remains. In agreement with Beams & King, Wilson & Leduc (1948) stated that in the mouse liver binuclearity arises by acytokinetic mitosis; they also assumed that fusion of mitotic spindles (or spindle poles) in binucleate cells is the most probable mechanism of formation of polyploid nuclei. Wilson & Leduc (1948) suggested, however, that in the adult animal cases of liver cell polyploidization by endomitosis may occur, as indicated by some of the observed nuclear stages. Swartz (1956) also regards endomitosis as one of the mechanisms of polyploidization in the liver.

Although multinucleate cells occur, sometimes with great frequency, in the liver of adult individuals of different mammals (Wilson & Leduc, 1948, and the literature here reported), binucleation is clearly the distinguishing feature in the development and differentiation of this organ. It seems probable that binucleation is also important in the development and differentiation of other mammalian tissues in which the incidence of binucleate cells is high (endometrium, salivary gland, adrenal cortex and medulla); but knowledge on the nuclear cytology of these tissues is scanty. There is considerable variation in the incidence of binucleation in liver parenchymal cells* among mammals (Table 5.1): the mouse has a particularly high level (55–58%) whereas the guinea pig has a comparatively low level (6%). During liver development in mice and rats, changes in binucleation and polyploidy occur; throughout adult life in mammals, including man, the degree of polyploidy continues to increase (a.o. St. Aubin & Bucher, 1952; Swartz, 1956;

* Parenchymal cells constitute approximately 90–95% of the total hepatic cellular mass, but only 60–65% of the population. The remaining cell population (35–40%) is made of diploid nonparenchymal cells (Bucher, 1963, and references there cited).

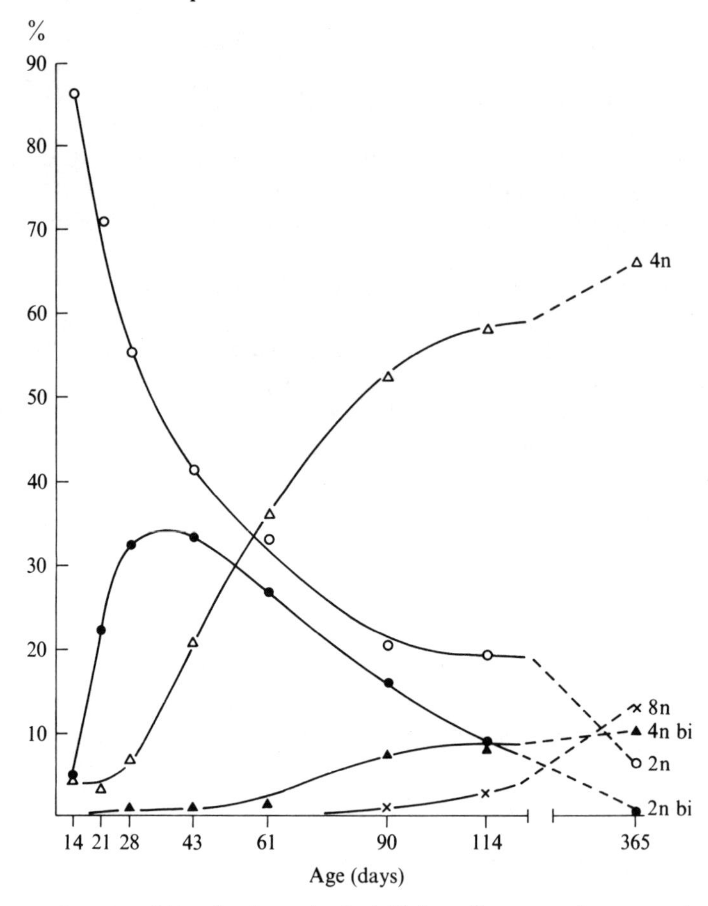

Fig. 5.4. Development of binucleation and polyploidy in rat liver parenchyma at various ages. 2n: diploid; 2n bi: cells with two diploid nuclei; 4n: tetraploid; etc. (after Alfert & Geschwind, 1958).

Alfert & Geschwind, 1958; Doljanski, 1960; Epstein, 1967). A study of the development of the complete ploidy pattern in female rats up to one year of age has shown that the liver cell population shortly after birth is predominantly diploid and cells of increasing degree of polyploidy appear in sequence and rise in frequency with age. Moreover, the frequency distribution of binucleation supports the view that binucleate cells contribute to the polyploidization process (Fig. 5.4). In the mouse, a clear relationship has been observed between body weight and the incidence and degree of polyploidy in the liver (Fig. 5.5). Neither in the rat, nor in other mammals, are age or body weight important factors *per se* of cell polyploidization in the liver; rather, this process seems to be under hormonal control.

The essential role of binucleation in the production of polyploid nuclei in the

TABLE 5.1. *Comparison of the incidence of binucleation in various mammalian livers by cell separation studies* (from Wheatley, 1972)

Species	Strain	Number of animals	Age (d)	Binucleation Average (%)	Binucleation Range
Rat	Sprague-Dawley	46	50–200	26.1	21.0–39.2
	Hooded Lister	4	52	32.1	29.9–33.2
	Ash/CSE	14	58	35.8	32.0–38.6
Mouse	Hairless	5	49	57.8	52.6–63.0
	Theiler's Original	39	60	55.1	44.2–66.2
Hamster	Golden Syrian	5	100	35.0	25.6–38.4
Guinea pig	Hartley	6	35	6.4	5.2–6.8

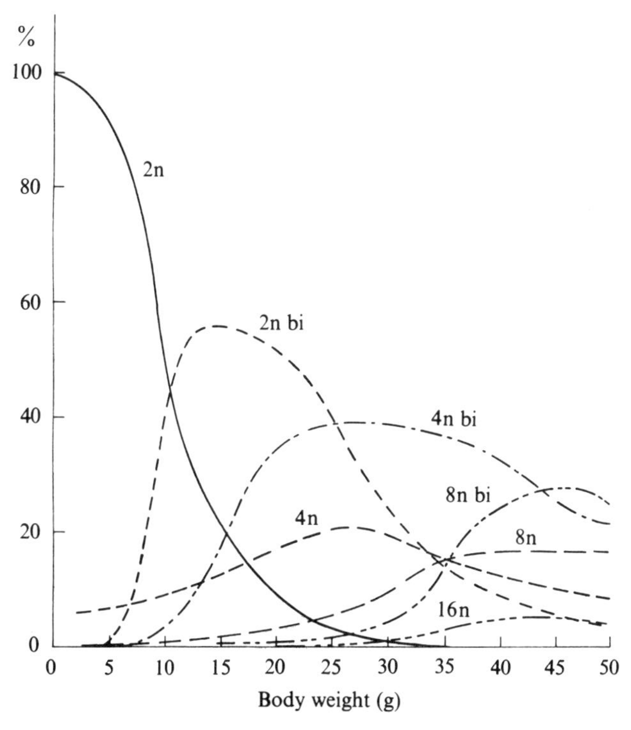

Fig. 5.5. Relationships of the frequencies of various classes of liver parenchyma cells to body weight in the Swiss mouse. 2n: diploid; 2n bi: cell with two diploid nuclei; 4n: tetraploid; etc. (after Epstein, 1967).

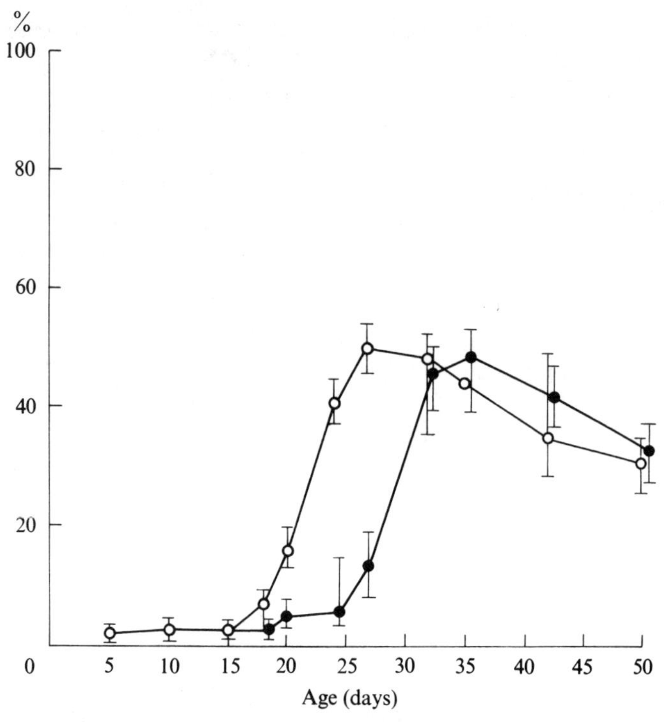

Fig. 5.6. Development of binucleate cells in the liver of adolescent rats weaned at 15 days of age (early-weaned, empty circles) or at 25 days of age (late-weaned, solid circles) (after Wheatley, 1972).

mammalian liver* is well demonstrated by the recent investigations of Nadal & Zajdela (1966*a*, *b*) and Wheatley (1972) on rat and mouse livers. In the rat, the period of intense production of binucleate cells corresponds approximately with weaning. A very clear correlation indeed exists between the time of weaning and the development of binucleate cells in the liver (Fig. 5.6). Binucleation always precedes the appearance of nuclei of the next ploidy level (e.g. a 2n binucleate is followed by a 4n uninucleate): this is particularly striking at the fifth and sixth weeks of age (Fig. 5.7). The essential role of mitosis for the production of polyploids from binucleate cells is most clearly documented in the cytological analyses on liver regeneration (Chapter 9).

* In contrast with the mammalian liver, the frog liver contains uninucleate cells, which in some species have been found to remain diploid and in others to comprise different ploidy classes (Bachmann & Cowden, 1965*a*, *b*; Bachmann, Goin & Goin, 1966). Since diploid, tetraploid and octoploid nuclei, as determined by DNA cytophotometry, in the liver of *Pseudacris* species all contain a single nucleolus (Bachmann & Cowden, 1967), the most probable mechanism of polyploidization seems to be chromosome endoreduplication.

An extensive literature exists on DNA cytophotometry on liver cells; a list of references can be found in the recent papers of Dutt (1970, 1971).

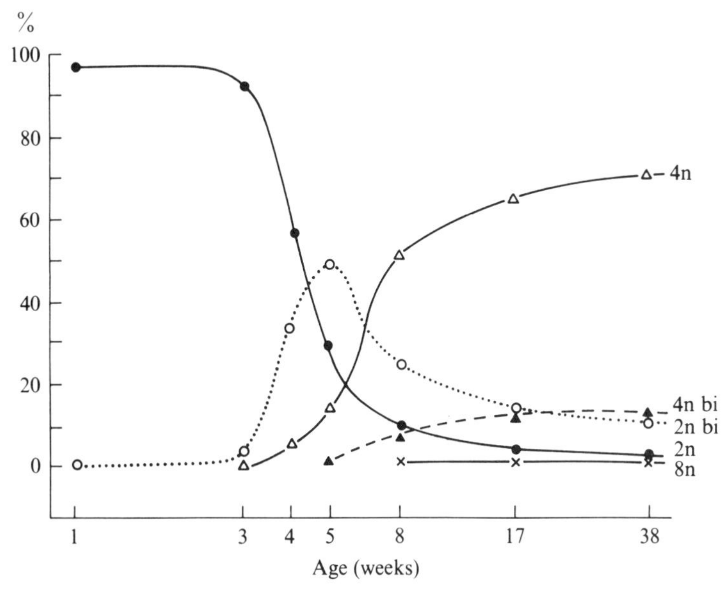

Fig. 5.7. Frequencies of various types of cells in rat liver at various ages (in weeks). 2n: diploid; 2n, bi; cells with two diploid nuclei; 4n: tetraploid, etc. (from Nadal & Zajdela, 1966*a*).

Nuclear polyploidization via binucleation, as occurring in the mammalian liver, can be obtained experimentally in root meristems treated with caffeine. This drug produces binucleate cells by inhibiting cytokinesis (Gosselin, 1940; Kihlman & Levan, 1949). When the binucleate cell divides, the two nuclei enter mitosis synchronously (bimitosis). Parallel bimitoses, through fusion of spindles, give rise to two tetraploid nuclei (and cells); coaxial bimitoses, through fusion of two telophase groups (the upper and the lower pole of the spindles), gives rise to three cells in a tier: diploid, tetraploid, diploid (López-Sáez, Giménez-Martin & González-Fernández, 1965).

Polyploidization of nuclei via bimitosis also occurs in plants. For example, in the three antipodals of the embryo sac of *Caltha palustris* four mitoses occur: the first mitosis produces a binucleate cell with two haploid nuclei, and the other three mitoses are constantly characterized by the fusion of the two spindles at metaphase, so that the antipodals contain two octoploid nuclei each (Grafl, 1941). Binucleation and multinucleation, combined with other polyploidization mechanisms (nuclear or spindle fusion, restitutional mitosis, endomitosis and chromosome endoreduplication), are of common occurrence in such ephemeral plant tissues as the embryo suspensor, the endosperm and the microsporangial tapetum, especially the anther tapetum of angiosperms. A very extensive literature has been accumulated on the nuclear cytology of these tissues; for reviews and recent papers on the subject, reference is made to Brink & Cooper (1947), D'Amato (1952*b*),

Chiarugi (1954), Wunderlich (1954), Tschermak-Woess (1956), Enzenberg (1961), Nagl (1962*a*, *b*) and Carniel (1963). The variety of the cytological mechanisms which operate in the polyploidization of the anther tapetum, a generally single-layered tissue which envelops the pollen mother cells, is worth considering. The following types of development of the anther tapetum may be distinguished:

(i) the tapetal cells remain uninucleate from the time of their formation to their degeneration. Nuclear growth by chromosome endoreduplication in the uninucleate tapetum is rather common: the number of endoreduplication cycles in an uninucleate tapetum does not seem to exceed three;

(ii) the tapetal cells undergo one or more (up to four) nuclear divisions not followed by wall formation. Due to several mitotic disturbances (fusion of resting or prophase nuclei, fusion of mitotic spindles or telophasic chromosome groups, chromosome stickiness), nuclei of different ploidy occur in the tapetum. An interesting feature of the multinucleate type of tapetum is the increasing degree of irregularities in successive mitotic cycles;

(iii) both endomitosis and chromosome endoreduplication are common in the multinucleate-type tapetum, sometimes producing an extraordinary nuclear heterogeneity in the tissue. This situation is clearly exemplified by the anther tapetum of potato, *Solanum tuberosum* (2n = 4x = 48), which develops by three mitotic divisions which occupy the time interval from the differentiation of the microsporogenous tissue to metaphase I in the microsporocytes (Fig. 5.8).

Endopolyploidy and polyteny

In both plants and animals, the commonest nuclear change accompanying cell differentiation is the repeated duplication of the chromosomes within an intact nuclear envelope and without spindle formation. The process of endonuclear chromosome duplication can take two main forms: (i) endomitosis (Geitler, 1939*a*) which leads to endopolyploidy and (ii) supernumerary chromonemal reproduction (Lorz, 1947) – more commonly called endoreduplication (Levan & Hauschka, 1953) – which leads to polyteny (Fig. 5.9). The subject of endopolyploidy and polyteny has been reviewed on several occasions (Lorz, 1947; Geitler, 1948, 1953; D'Amato, 1952*b*, 1965*b*; Tschermak-Woess, 1956, 1963, 1971; Beermann, 1962; Partanen, 1965; Stange, 1965; Catarino, 1968).

Endomitosis

In various bugs of the order Heteroptera, a.o. the pondskaters of the genus *Gerris*, the resting nuclei of different tissues contain chromosomes individualized in a form not much different from that of mitotic chromosomes: this condition makes the determination of ploidy level of nuclei possible (Geitler, 1937). In *Gerris lateralis*, the ploidy level of the resting nuclei can be established by counting the number of the X-chromosomes which keep heteropycnotic; in *Lygaeus saxatilis* males,

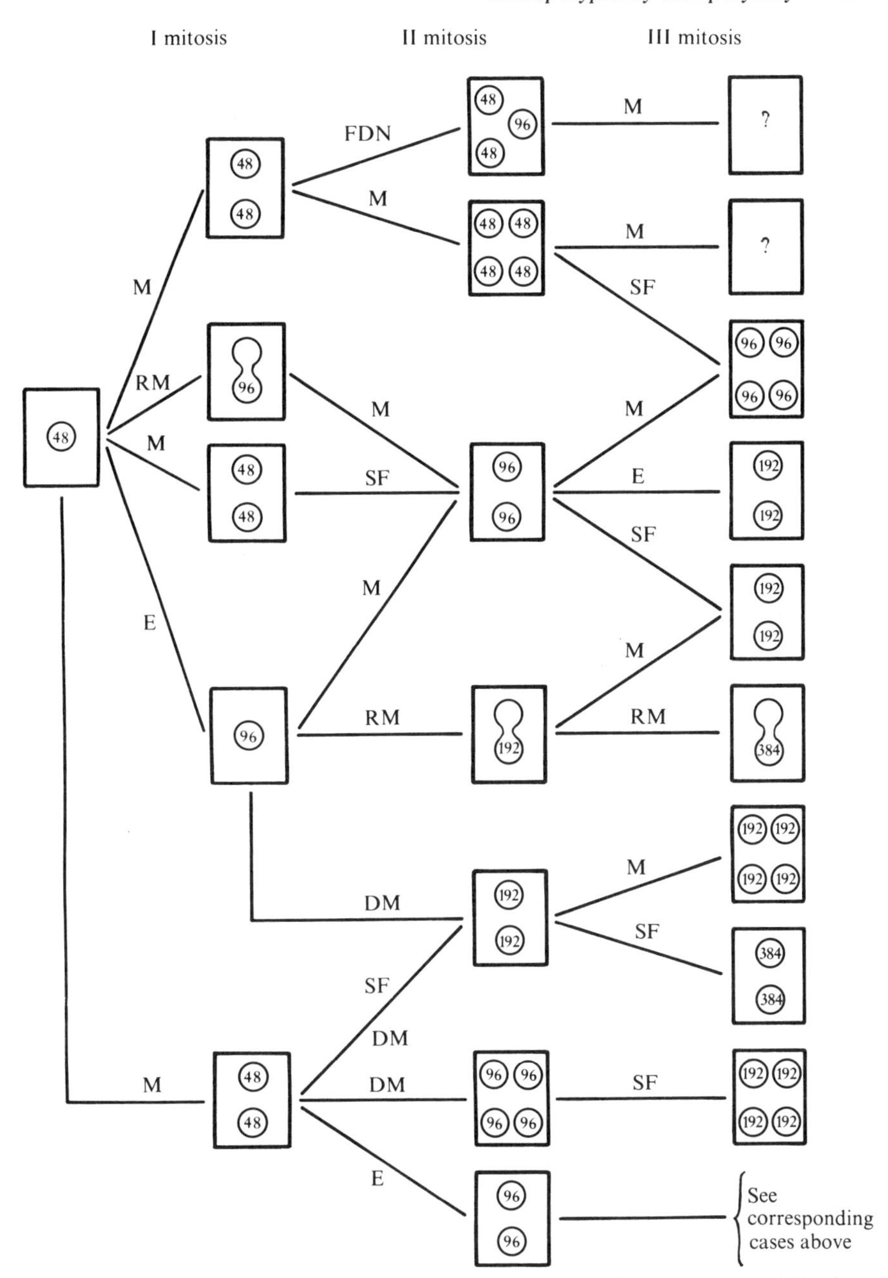

Fig. 5.8. Schematic representation of the mechanisms of polynucleation and polyploidization in the cells of the anther tapetum of potato, *Solanum tuberosum* ($2n = 4x = 48$). DM: diplochromosome mitosis; E: endomitosis; FDN: fusion of daughter nuclei; M: mitosis; RM: restitutional mitosis; SF: spindle fusion; SF/DM: spindle fusion in diplochromosome mitoses (condensed from Avanzi, 1950).

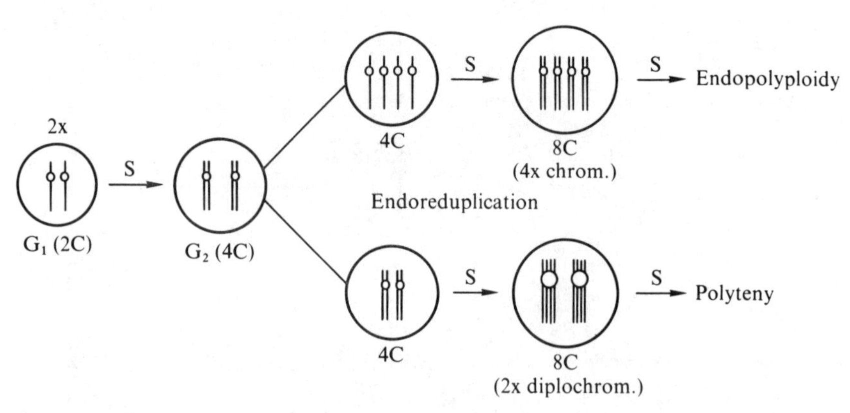

Fig. 5.9. Comparison of endomitosis and chromosome endoreduplication starting from an initial diploid (2x) nucleus (only one pair of chromosomes is shown).

heteropycnotic Y-chromosomes serve the same scope (Geitler, 1939*a*, *b*). The cytological analysis shows that, in *Gerris*, different ploidy degrees occur in different tissues of the adult (imago); in the giant nuclei of the salivary glands 1024-ploidy (1024x) has been ascertained and, based on nuclear size, a maximum ploidy level of 2048x has been estimated. Geitler (1938, 1939*a*, *b*) has studied the origin of the polyploid nuclei by analyzing larvae of different ages. For example, many nuclei of the Malpighian tubules are diploid in the first larval stage and become 64x in the imago; the nuclei of the salivary glands are already at least 16x in the first larval stage. Polyploidization occurs by endomitosis: the chromosomes enter a prophase stage (endoprophase), then contract further (endometaphase), their chromatids separate parallel to each other (endoanaphase) and finally decondense to take the resting nuclear structure. The essential difference between endomitosis and mitosis lies in the absence of the mitotic spindle, so that endomitosis may be likened to a colchicine-induced mitosis (C-mitosis: Levan, 1938), apart from the much lower degree of chromosome contraction in endomitosis. The culminating point of an endomitosis corresponds to the stage of late prophase of mitosis (further details in Geitler, 1953).

Endomitosis, as discovered and extensively investigated by Geitler in Heteroptera, also occurs, sometimes with variations, in nematodes, gastropods, crustaceans, collembolans and homopterans (references in Geitler, 1953). Endomitosis in Malpighian tubules of the homopteran species *Planococcus citri*, the mealy bug, has been recently described by Nur (1968). Since endomitosis in the mealy bug does not resemble very closely the descriptions of endomitosis in the heteropterans *Gerris lateralis* (Geitler) and *Corixa punctata* (Lipp, 1953), Nur (1968) has reported a stage by stage comparison between endomitosis in the mealy bug and the two heteropterans. Nur (1966) has also described an interesting aspect of endomitosis in male mealy bugs (2n = 10), which have the paternal chromosome set in a permanent heterochromatic (H) state from early embryogeny (lecanoid chromosome system: Hughes-Schrader, 1948). In the nuclei of the testis sheath and of most of the oenocytes, the maternal euchromatic (E) chromosomes undergo several endomitotic cycles while the paternal chromosomes do not generally reduplicate: this results in nuclei with 5H chromosomes and 10, 20, 40, 80, or 160 E chromosomes.

In his reviews, Geitler (1948, 1953) has included under the collective term of 'endomitosis' both endomitosis *sensu stricto* and the process of endonuclear chromosome duplication which takes place in the absence of any obvious condensation and decondensation stages: in other words, endomitosis is said to include chromosome endoreduplication, the essential mechanism of polytenization. The term 'endomitosis' in this wide sense is also used by John & Lewis (1968) and by Geitler's associates (Tschermak-Woess & Hasitschka, 1953; Tschermak-Woess, 1959; Nagl 1965*b*, 1970*a*, *b*). By concentrating on 'angiosperm endomitosis', which occurs without any obvious nuclear changes, Geitler and his associates have provided evidence for the occurrence of a 'dispersion stage' which follows DNA replication and is similar to the dispersion stage which was interpreted by Heitz (1929) as the initial portion of mitotic prophase. Thus, the angiosperm type of endomitosis is regarded as a short-cut of mitosis in the very early prophase and, according to Nagl (1970*a*, *b*), should be distinctly differentiated from the concept of endoreduplication. However, as pointed out by D'Amato (1954), the crucial difference between endomitosis and endoreduplication is chromosome polytenization, the commonest process of endonuclear chromosome duplication to be found in animals and plants, including angiosperms. The need for keeping endomitosis separate from polytenization (endoreduplication) has been stressed by many others: Huskins (1947), Levan & Hauschka (1953), Chiarugi (1954), Hsu & Moorehead (1956), Pätau & Das (1961), Levan & Hsu (1961), Van Parijs & Vandendriessche (1966).

Chromosome endoreduplication (polyteny)

Chromosome polyteny of a different order can be demonstrated in a variety of tissues and cells in both plants and animals.

In plants, polyteny of low order (one to a few endoreduplication cycles) is of especial developmental significance because of the extraordinary regenerative capacity of the plant cell, which will be discussed in another section of this book. In 1939, Grafl described the polyploid mitoses which occurred in the differentiated tuber and leaf tissues of *Sauromatum guttatum* under the effect of wounding; she demonstrated that the tetraploid (4x), octoploid (8x) and 16-ploid (16x) mitoses were characterized respectively by 2n 'pairs', 2n 'groups of four' and 2n 'groups of eight' chromosomes. These mitoses – which we now call diplochromosome, quadruplochromosome and octuplochromosome mitoses – were correctly taken

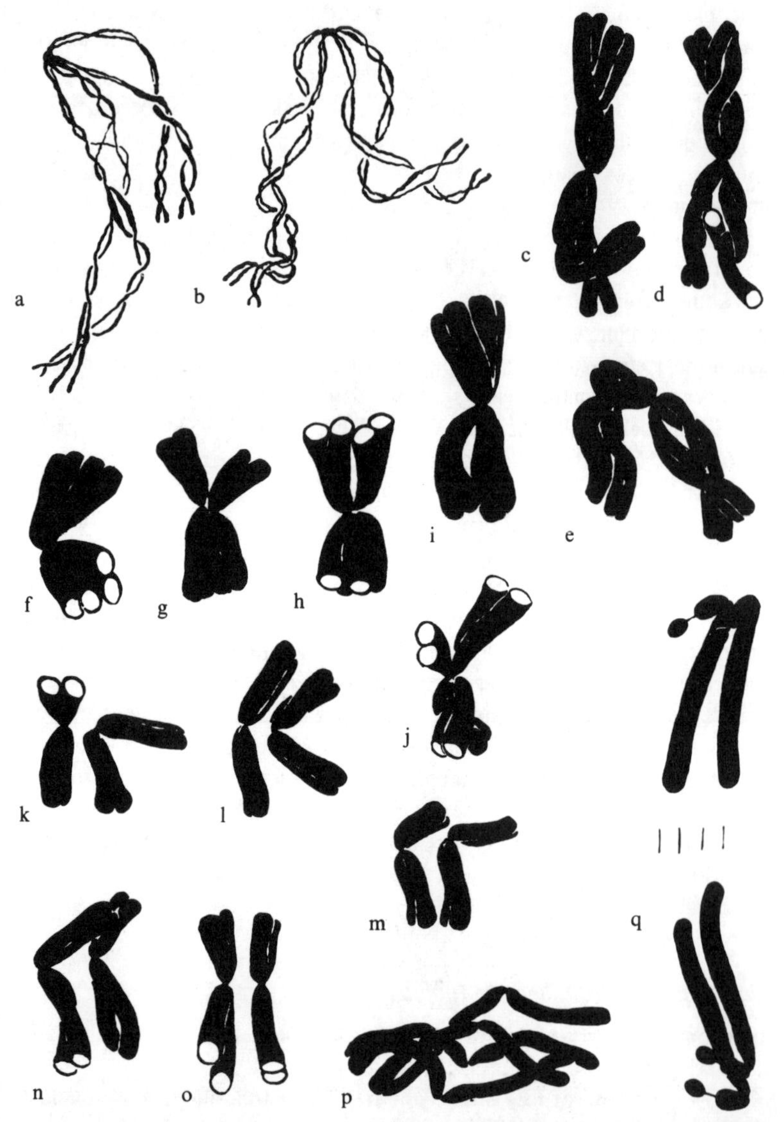

Fig. 5.10. Course of diplochromosome mitosis in cells of the differentiated root zone of *Allium cepa* stimulated into division by treatment with phytohormones (after Levan, 1939).

by Grafl as demonstration of the occurrence in the differentiated tissues of *Sauromatum* of nuclei which had undergone 'endomitosis' (endoreduplication). In the same year, 1939, Levan gave a masterly description of the diplochromosome mitoses found in the elongation (differentiated) zone of onion roots treated with phytohormones (Fig. 5.10). Levan's conclusion that the double chromosome

TABLE 5.2. *Two examples of karyological anatomy in plants (phytohormone-induced mitoses)*

Organ	Tissue	Level of ploidy	References
Bean stem	1. Procambium	2x	Dermen, 1941
	2. Cambium	2x	
	3. Cortex including endodermis		
	4. Phloem	4x, 8x and 16x	
	5. Medullary rays	4x, 8x and 16x	
	6. Pith	4x, 8x and 16x	
Onion root	1. Pericycle	2x with isolated 4x* cells	D'Amato & Avanzi, 1948
	2. Procambium	2x	
	3. Epidermis	2x and 4x	
	4. Cortex including endodermis	Mainly 4x cells, some 2x or 8x†	
	5. Stele	4x with 8x xylem cells	

* 4x: 2n diplochromosomes (4 chromatids each).
† 8x: 2n quadruplochromosomes (8 chromatids each).

reproduction leading to diplochromosomes was induced by phytohormones was disproved by three groups working independently on the nuclear cytology of differentiated root issues (Berger, 1946; Berger & Witkus, 1948; D'Amato, 1948*a*, *b*; D'Amato & Avanzi, 1948; Huskins, 1947; Huskins & Steinitz, 1948*a*, *b*); they presented several lines of evidence that endoreduplicated nuclei already exist in the differentiated tissues and phytohormone treatment is effective in stimulating them into mitosis.

The ease with which endoreduplicated plant nuclei (generally up to 16x and 32x) are induced to divide by either wounding or phytohormone treatment has greatly expanded our knowledge on 'karyological anatomy' (Geitler, 1952). The two first described cases of karyological anatomy in plants are reported in Table 5.2. Extensive data on the karyological anatomy of plants are available (Lauber, 1947; D'Amato, 1952*b*; Holzer, 1952; Geitler, 1953; Tschermak-Woess & Dolezal, 1953; Fenzl & Tschermak-Woess, 1954; Tschermak-Woess, 1956; Enzenberg, 1961; Carniel, 1963; Erbrich, 1965; Hesse, 1968). From Table 5.2 it is apparent that meristematic tissues (pericycle, procambium, cambium) remain diploid amid tissues whose cells quite commonly undergo endoreduplication. Undoubtedly diploidy in the meristem cell line is ensured by a strict control of the sequence of DNA synthesis and mitosis; this situation is now documented for the pericycle of the onion root (Corsi & Avanzi, 1970).

A few more morphogenetic aspects which have emerged from the karyological anatomy of plants are worth considering. Firstly, a relation has been found in

TABLE 5.3. *Incidence (%) of chromosome endoreduplication in the cortical parenchyma of thin roots and pull-roots of Bellevalia romana (mitosis stimulation by 2,4-D)* (from D'Amato, 1952*a*)

Type of root	Number of mitoses analyzed	2x	4x and 8x*
Thin	130	39.2	60.8
Thick (pull-roots)	1005	2.8	97.2
Very thick (pull-roots)	2647	0.9	99.1

* All mitoses with endoreduplicated chromosomes (diplochromosomes or, very rarely, quadruplochromosomes).

parenchyma between succulence and an increased level of chromosome endoreduplication. In *Bellevalia romana*, as in other bulbous plants, each bulb bears both thin roots and thick fleshy roots, the latter being known as 'pull-roots'. Mitosis stimulation by root treatment with sodium 2,4-dichlorophenoxyacetate (2,4-D) demonstrates a far higher incidence of chromosome endoreduplication in the cortical parenchyma of pull-roots as compared with that of thin roots (Table 5.3). Succulent plants with developed water-storing parenchymata are classical examples of high levels of endoreduplication (Czeika, 1956).* In *Kalanchoe blossfeldiana*, very succulent leaves can be induced by growing plants under short-day conditions; the increased succulence is due to further nuclear growth of the mesophyll cells by endoreduplication as shown by an analysis of wounding induced mitoses: 16x and 32x in short-day leaves and 4x and 8x in long-day leaves (Witsch & Flügel, 1952). The effect of short-day conditions on succulence and chromosome endoreduplication has been found in other Kalanchoideae (Resende & Catarino, 1963; Resende, Linskens & Catarino, 1964; Catarino, 1968). In *Bryophyllum crenatum*, the mesophyll cells of the short-day (succulent) leaves reach 32-ploidy while the maximum level of endoreduplication in the mesophyll of long-day leaves is 8x. Since short-day plants of *B. crenatum* contain 20 times less growth-promoter (IAA) than the long-day plants and in addition contain a growth-inhibitor, Catarino (1965, 1968) has investigated the effect on endoreduplication of a well-known growth-inhibiting factor, salinity. In *Phaseolus*, salinity reduces cell division and the rates of RNA and protein syntheses, but does not appreciably affect DNA synthesis (Nieman, 1965). A non-succulent species, *Lobularia maritima*, occurs as a succulent in areas near the sea in Portugal; in the succulent leaves, which are also obtained when plants are grown under the effect of sodium chloride, DNA values up to 64C and 32x mitoses (stimulation

* In Compositae, in which tissue differentiation in a diploid state is common, no correlation is found between succulence and endoreduplication. Among nine succulent species of Compositae investigated by Czeika (1956) only one had endoreduplicated nuclei (up to 32x).

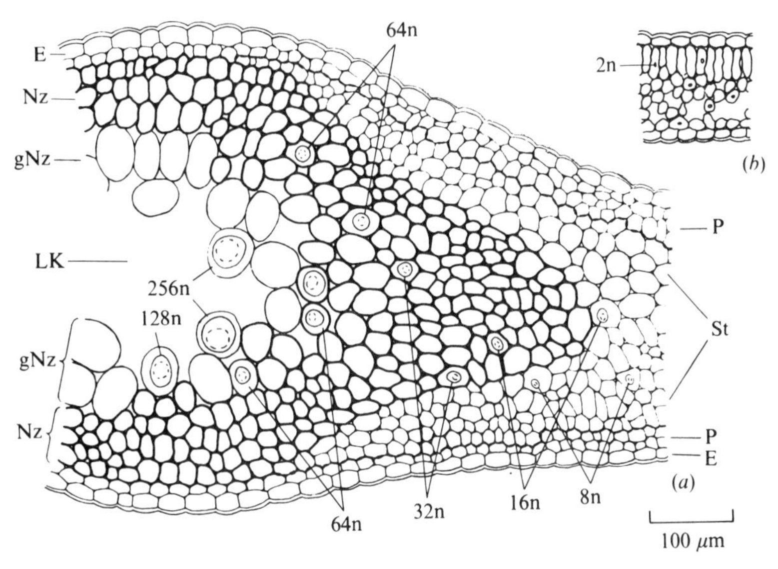

Fig. 5.11. Distribution of different levels of ploidy in cells of a fully developed gall formed by *Neuroterus numismalis* on leaves of *Quercus robur* (a transverse section of normal leaf is shown in (*b*). E: epidermis; P: gall parenchyma; St; starch containing cells; Nz: nutritive cells; gNz: large protein-containing nutritive cells; LK: chamber housing the larva (after Hesse, 1968).

by wounding) have been demonstrated in the mesophyll cells (in the non-succulent leaves, the corresponding values are 16C and 8x). According to Catarino, all external and internal factors reducing overall plant growth favour chromosome endoreduplication.

Another aspect brought about by the karyological anatomy of plants is the gradient of endoreduplication levels in some organs and tissues. The most striking example is offered by leaves which transform into galls under the action of gall-forming insects belonging to Cynipidae, Acarina, Diptera. The degree of endoreduplication commonly increases from the periphery towards the larval chamber surrounded by the largest nutritive cells (Fig. 5.11); these are estimated to be 1024x in the galls of *Andricus marginalis* (Hesse, 1968). Another example of a tissue where gradients in polyploidization (by endoreduplication and/or defective mitosis) are frequently encountered is the endosperm. In some species, there is an increase in the level of polyploidization from the periphery to the central portion; a well-studied case is maize endosperm, in which the largest nuclei are estimated to be 384x (Duncan & Ross, 1950; Tschermak-Woess & Enzenberg-Kunz, 1965). In other species, the endoreduplication activity is highest in cells or cell groups of the endosperm taking haustorial function; in the species studied by Erbrich (1965), the maximum number of endoreduplication cycles is estimated to vary from 7 (384x) to 13 (24 576x). A third example is the embryo suspensor of some plant species, which generally have under-developed endosperm; the most striking case is the bean suspensor (*Phaseolus coccineus* and *P. vulgaris*), an elongated structure whose nuclei undergo a progressively increasing number of endoreduplication cycles from the proximal to the basal (micropylar) region consisting of giant cells reaching a DNA content of 4096C (Nagl, 1962*a*, *b*, 1965*a*, 1967*a*, 1970*c*) and, in exceptional cases, of 8192C (Brady, 1973). These giant cells contain polytene chromosomes in a diploid number ($2n = 22$),* which are condensed and banded (like the salivary gland chromosomes of Diptera) in the inactive state and granular or fibrillar in the active state (Nagl, 1965*a*, 1967*a*, 1969*a*, *b*, *c*, 1970*c*). The giant *Phaseolus* cells are the only material so far available for cytochemical, cytophotometric, autoradiographic and biochemical investigations on plant polytene chromosomes (Nagl, 1969*a*, *b*, 1970*c*, *d*; Avanzi, Cionini & D'Amato, 1970; Avanzi, Durante, Cionini & D'Amato, 1972; Cionini & Avanzi, 1972; Brady & Clutter, 1972, 1974; Brady, 1973; Lima-de-Faria *et al.*, 1975).

Recently, Nagl (1973) has likened the plant embryo suspensor to the mammalian trophoblast. In the rat and the mouse trophoblast, the nuclear DNA content (cytophotometry) ranges from 128C to 4096C: the chromatin is homogeneously distributed in the nucleus or, only occasionally, may be organized in the form of polytene chromosomes (Zybina, 1970; Nagl, 1972*a*; Barlow & Sherman, 1974). DNA cytophotometry of the nuclei of mouse trophoblast has given lower values than in rat trophoblast (between 512C and 1024C) (Barlow & Sherman, 1972). Moreover, CsCl density gradient centrifugation and renaturation studies on the DNA extracted from these giant nuclei have shown that the proportion of satellite DNA to total DNA is similar to that found in the DNA extracted from embryos: i.e. 10% in both materials. This result demonstrates that during nuclear growth in the trophoblast both the satellite and the main band DNA replicate to the same extent (Sherman, McLaren & Walker, 1972). This situation differs markedly from that of polytene chromosome cells where the nuclei have much less than the full complement of satellite DNA (under-replication, see below).

Before turning to a discussion on dipteran polytene chromosomes it seems pertinent to summarize briefly some recent work on the experimental induction of diplochromosome and quadruplochromosome mitoses in animals and plants. In cultured cells of the marsupial *Potorous tridactylus* (Walen, 1965) and in cultured human blood cells (Schwarzacher & Schnedl, 1965, 1966; Herreros & Giannelli, 1967), the origin of diplo- and quadruplochromosomes has been studied by autoradiography following ^{3}H-thymidine feeding during the first DNA synthesis phase, that intercalated between G_1 (2C) and G_2 (4C). It has been found that in

* Polytene chromosomes, which occur in antipodals, synergids and andosperms of some angiosperms are found only occasionally (Tschermak-Woess, 1963; Erbrich, 1965), but in the chalazal haustorium of *Rhinanthus alectorolophus* they are always present. In all cases, the polytene chromosomes are unbanded and unpaired (n in antipodals and synergids, 3n in endosperms). The highest level of polytenization in *Rhinanthus* is estimated to be seven endoreduplications (Tschermak-Woess, 1957).

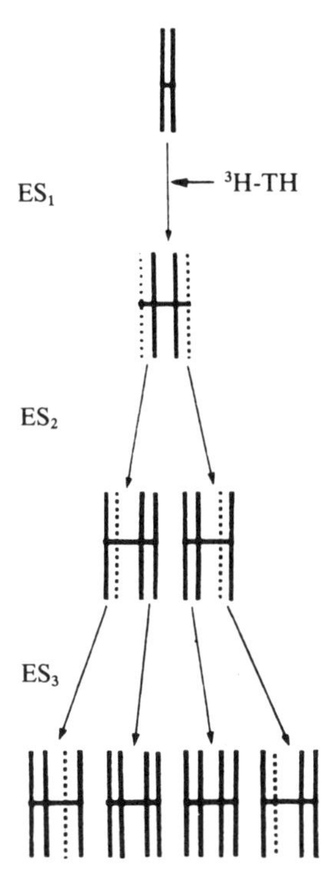

Fig. 5.12. Schematic representation of the spatial relationship of old and new chromatid sub-units during successive replication cycles leading to diplochromosomes (after ES_2) and quadruplochromosomes (after ES_3). Each chromatid is represented by two sub-units which are drawn as solid or broken lines to indicate whether they are labeled or unlabeled. The centromere is represented by a transverse line. ES_1, ES_2, ES_3: respectively, first, second and third DNA synthesis periods. The two chromatid sub-units which are labeled when ^{3}H-thymidine is present during ES_1 are distributed as shown in the diagram in the diplochromosome (after ES_2) and the quadruplochromosome (after ES_3) (after Herreros & Giannelli, 1967).

practically all diplochromosomes (97% in Herreros & Giannelli's experiments) only the outer chromatids are labeled; in the quadruplochromosomes, it is the inner chromatids of the outer chromosomes that are always labeled (Fig. 5.12). These results are consistent with the view that during endoreduplication the newly synthesized DNA strand is on the outside. In the botanical literature, there is only one case reported of induction of diplo- and quadruplochromosomes. These occur in high frequency in apical root meristems of pea, *Pisum sativum*, recovering in water after various treatments with 8-azaguanine (Nuti-Ronchi, Avanzi & D'Amato, 1965). DNA cytophotometric and autoradiographic analyses have shown that 8-azaguanine blocks cells in G_2 (4C) and that the majority (90%) of nuclei destined to give diplochromosomes make the whole or part of their first S (2C$\rightarrow$4C) before 8-azaguanine treatment, the remainder (10%) enter DNA synthesis in the first hours of treatment. Evidence has been presented that, in the pea root, 8-azaguanine triggers the endoreduplication cycle in 4C cells belonging to the differentiating cell line of the meristem, e.g. the R_2 period in Epifanova & Terskikh's (1969) terminology.

Polytene chromosomes of Diptera

As seen in the preceding section, the polytene chromosomes in the giant cells of *Phaseolus* are present in the diploid number (the homologues are not paired). Unpaired polytene chromosomes, with a banded structure similar to that of dipteran polytene chromosomes, have been described in the salivary gland cells of some species of Collembola (Prabhoo, 1961; Cassagnau, 1968). Banded polytene chromosomes also occur during the development of the macronuclear anlage in some species of Ciliates (Ammermann, 1964, 1965, 1971; Golikowa, 1965; Alonso & Perez-Silva, 1965); in *Stylonychia mytilus*, it has been possible to ascertain that the polytene chromosomes are unpaired (Ammermann, 1971).

The unpaired condition of the polytene chromosomes in the organisms listed above strikingly contrasts with the close homologous association (somatic synapsis, somatic pairing) which characterizes the polytene chromosomes which occur in two-winged flies (Diptera). Homologous pairing of dipteran polytene chromosomes, discovered by Heitz & Bauer (1933), Painter (1933, 1934) and King & Beams (1934), reflects the condition of close homologous association (somatic pairing) which occurs in the diploid nuclei and mitoses of Diptera (Metz, 1916; Dobzhansky, 1936; Cooper, 1948). Since these initial discoveries, knowledge of polytene chromosomes has been reviewed from time to time. Recent reviews either cover the whole subject (Beermann, 1962) or offer an extensive survey of a particular aspect (Swift, 1962; Clever, 1968; Pavan & Da Cunha, 1969; Ashburner, 1970, 1972; Beermann, 1972; Berendes 1972, 1973; Rudkin, 1972). Polytene chromosomes have been reported to occur in a wide variety of families of Diptera (Ashburner, 1970). They are most common in larval tissues (esophagus, salivary glands, gut, gastric ceca, Malpighian tubules, fat body, tracheary wall cells, muscle, certain cell types in nerve ganglia), but they also occur in the pupa (certain cells in developing bristle sockets and in the epidermis of the footpads) and in the adult (Malpighian tubules, nurse cells of the ovary). Quite recently, polytene chromosomes have been found in the abdominal pericardial cells of the adults in some species of *Drosophila* (Kambysellis & Wheeler, 1972). Dipteran polytene chromosomes have a typical banded structure which reflects in an enormously magnified but faithful picture the chromomeric structure of the eukaryotic chromosomes (Beermann, 1972, review).

Undoubtedly the most intensively investigated polytene chromosomes are those of the salivary glands. In *Drosophila melanogaster*, the two salivary glands arise as invaginations from the blastoderm near the middle of the embryonic period (8 hours after egg laying) and each consists of about 125 initially diploid cells which

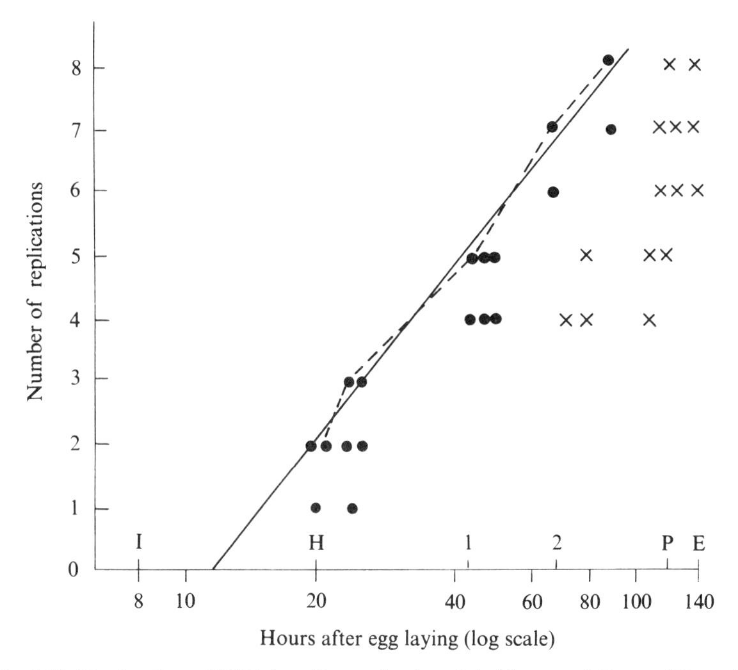

Fig. 5.13. Replication of DNA in salivary gland nuclei of larvae of *Drosophila melanogaster* at different times after egg deposition at 25 °C. I: invagination of the salivary gland plaque; H: hatching from the egg; 1 and 2: first and second larval molts; P: puparium formation: E: eversion of the imaginal disks. Based on data from two laboratories (after Rudkin, 1972).

enter endoreduplication before the larva hatches (Fig. 5.13). The maximal level of polytenization is attained at puparium formation and eversion of the imaginal discs (eight endoreduplication cycles); this value is one step below the maximum value estimated by Swift (1962) for the larval salivary gland chromosomes of *D. melanogaster* (1024C, corresponding to nine endoreduplications). This level of polytenization is lower than that occurring in other species of Diptera. In *D. virilis*, salivary gland cells may go through ten endoreduplication cycles, whereas salivary gland cells of *Sciara coprophila* may go through twelve, and salivary gland cells of *Chironomus tentans* through thirteen endoreduplication cycles (Rasch, 1966, 1970*b*; Daneholt & Edström, 1967).

The ultimate level of polyteny attained by a certain cell type may be influenced by several factors. In salivary glands of sciarid flies infested by protozoa (one gregarine and at least three species of microsporidia) and by polyhedrosis virus, the infected cells undergo further DNA doublings; in *Rhynchosciara angelae*, microsporidian infection may lead to chromosomes with as much as 32 (and exceptionally 128) times the DNA of chromosomes of non-infected cells of the same gland (Roberts, Kimball & Pavan, 1967; Pavan & Da Cunha, 1968). Extra duplications in polytene chromosomes over the normal maximal value are also obtained in salivary glands transplanted to adult abdomens (Hadorn, Ruch & Staub, 1964) and in glands of a mutant of *Drosophila melanogaster* with an extended larval period (Rodman, 1967). The ultimate level of polyteny is also tissue-specific. For example, in *Drosophila virilis*, salivary gland cells may go through ten replication cycles, whereas, in the same period of development in the same larva, cells in the gastric cecum may attain maximal polyteny after nine replication cycles and cells of the Malpighian tubules may undergo only six replication cycles (Rasch, 1970*b*).

During polytenization some chromosome regions either do not replicate or lag behind during replication (under-replication). The polytene chromosome cells of some *Drosophila* species possess a chromocenter which consists of the associated kinetochore regions of the chromosomes, the heterochromatic arms of the X chromosomes and the entire Y. The chromocenter comprises two types of heterochromatin: an extremely condensed chromatin (α-heterochromatin) is generally located within or associated with a loosely arranged particulate chromatin (β-heterochromatin) (Heitz, 1933, 1934). It was suggested by Heitz that in the salivary glands of *Drosophila virilis* the heterochromatic regions do not attain a level of polyteny in salivary glands equivalent to the level of polyteny of the rest of the chromosome complement. Heitz's suggestion has been confirmed by DNA cytophotometric measurements in polytene chromosomes of *Drosophila melanogaster* and *D. hydei* (Rudkin, 1965, 1969; Berendes & Keyl, 1967; Mulder, Van Duyn & Gloor, 1968). It has been found that most of the DNA present in the chromocenter replicates only once or a few (2–3) times, whereas the DNA of thc euchromatic segments of the chromosome complement undergo eight to ten replications. The differential replication behavior of heterochromatin and euchromatin is also demonstrated by biochemical analyses and cytological DNA–RNA hybridization experiments. *Drosophila virilis* and *D. melanogaster* possess repetitive DNA sequences which can be isolated as satellites from CsCl gradients (three satellites in *D. virilis*, one in *D. melanogaster*); the satellites constitute respectively 41% and 8% of the total DNA isolated from diploid tissues. The satellites make up only a minute fraction of the DNA isolated from polytene nuclei. Complementary RNA (cRNA), transcribed *in vitro* from the largest satellite of *D. virilis*, hybridized primarily to α-heterochromatin in the chromocenter of polytene nuclei and to the centromeric heterochromatin of mitotic chromosomes. The degree of hybridization in the two classes of nuclei was similar, thus demonstrating that α-heterochromatin either does not replicate or else very rarely replicates during polytenization. Evidence was also found that β-heterochromatin, or at least a part of it, comprises repetitive DNA sequences which replicate during polytenization (Fig. 5.14). Similar observations were made in *D. melanogaster* (Gall, Cohen & Polan, 1971). In *Drosophila hydei* also, repetitive DNA sequences are under-represented in polytene chromosomes: 5% of total DNA, instead of 20% present in embryo or pupal DNA, indicating that 25% of the repetitive sequences are

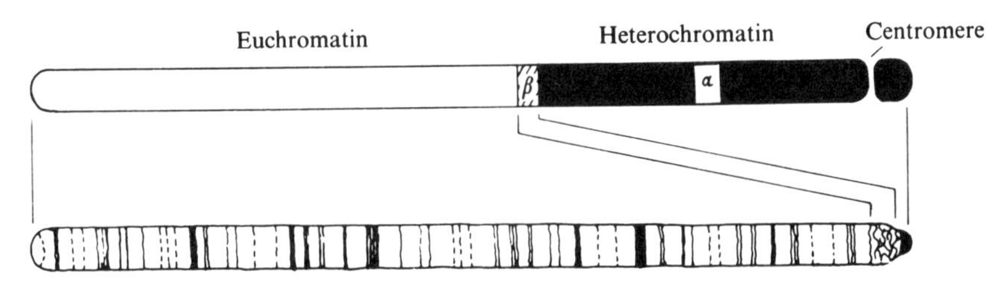

Fig. 5.14. Comparison of mitotic (upper diagram) and polytene chromosomes of *Drosophila*, based on data from *D. virilis* and *D. melanogaster*. Three regions can be distinguished: (1) α-heterochromatin: highly repetitive DNA sequences (satellites), probably not replicated during polytenization; (2) β-heterochromatin: repetitive sequences of which at least some are replicated during polytenization; (3) euchromatin: largely unique sequences with a small percentage of widely distributed repetitive sequences: both unique and repetitive sequences are replicated during polytenization (after Gall *et al.*, 1971).

replicated during polytenization. In addition, a heavy DNA satellite (1.713 g/cm^3), forming 2%–3% of embryo and pupal DNA, was not observed in any of the salivary gland DNAs (Dickson, Boyd & Laird, 1971; see also Hennig, Hennig & Stein, 1970). Using ^{3}H-thymidine electron microscope autoradiography, DNA replication in the chromocentre of polytene chromosome cells of *Drosophila melanogaster* has been studied. It has been shown that the α-heterochromatin does not replicate at all from after hatching till the third instar while the β-heterochromatin replicates as much as the euchromatin (Lakhotia, 1974). Another dipteran species in which under-replication of the most repetitive sequences in the salivary gland chromosomes has been described is *Rhyncosciara angelae* (Balsamo *et al.*, 1973).

DNA–RNA molecular hybridization experiments have also demonstrated the under-replication in polytene chromosomes of the DNA sequences (rDNA, ribosomal cistrons) that code for ribosomal RNA (rRNA). Saturation hybridization experiments in which rRNA of *Drosophila hydei* was hybridized with total DNA from polytene cells revealed that the rDNA is about three steps behind euchromatin in polytenization (Hennig & Meer, 1971). Comparable results have been obtained in *D. melanogaster* (Sibatani, 1971; Spear & Gall, 1973). In this species, in which the rDNA is located at or near the nucleolus organizer of the X and Y chromosomes (Ritossa & Spiegelman, 1965), the rDNA content of diploid and polytene cells of XO and XX larvae has been studied. It has been found that, whereas in diploid cells the rDNA content is proportional to the number of nucleolus organizers, in polytene cells the rDNA amount is independent of this number (XO and XX flies have the same rDNA amount in their polytene chromosome cells). This rDNA compensation is due to independent polytenization of the rDNA leading to a change in the number of lateral copies of the original nucleolus organizer without any variation in the tandem number of rDNA genes (Spear & Gall, 1973; Spear, 1974). Under-replication of ribosomal cistrons has been found in the polytene chromosomes of four species of *Rhynchosciara* (Gambarini & Meneghini, 1972; Gambarini & Lara, 1974) and in *Phaseolus coccineus*, the only plant species investigated so far (Lima-de-Faria *et al.*, 1975). An interesting aspect of under-replication during the polytenization process has emerged from a recent investigation by Blumenfeld & Forrest (1972). DNA preparations from embryos and adults of *Drosophila virilis* are resolved, in agreement with the determinations of Gall *et al.* (1971), into a main band (I: 1.700 g/cm^3) and three less dense satellite bands (II: 1.691 g/cm^3; III: 1.688 g/cm^3; IV: 1.669 g/cm^3). Bands II, III and IV represent 23, 9 and 8% respectively of embryo DNA preparations; they represent 16, 4 and 8% respectively of adult DNA preparations. Thus, the representation of bands II and III is about 50% less in adults, whereas the representation of band IV is not reduced in adults. Since the reduced representation of satellite DNAs reflects polyteny in adult cells, different satellite sequences must be differentially under-replicated during polytenization (Blumenfeld & Forrest, 1972).

Another aspect of the autonomy in replication behaviour in replicating units is the accumulation of DNA at the level of some bands producing so-called DNA puffs in polytene chromosomes of some sciarid flies. These structures were discovered in *Rhynchosciara angelae* by Breuer & Pavan (1955) who suggested that they produce excessive amounts of DNA. Their suggestion has been confirmed by the spectrophotometric studies of Rudkin & Corlette (1957) on *R. angelae* and of Swift (1962), Crouse & Keyl (1968) and Rasch (1970*a*) on *Sciara coprophila*. DNA puff formation involves a local geometric increase in DNA content, indicating complete rounds of DNA duplications. The sudden development of many DNA puffs in cells of a late fourth instar larva of *Sciara* leads to a significant deviation of the nuclear DNA content from that expected on the basis of complete rounds of replication of the entire genome. The amount of 'extra' DNA synthesized during DNA puff formation is 10–12% of total nuclear DNA. Due to its significance in some phases of development, the process of extra DNA synthesis (amplification) will be discussed in detail in the next chapter.

Polynemy

In some species of animals and plants chromosome size has been found to vary between cells within the individual even when differences in DNA values are not involved. For example, in the early embryo of *Schistocerca gregaria* a fixed number of very large neuroblasts occur, all of which undergo a series of unequal cell divisions; the larger daughter cells, which result from these unequal divisions and retain the morphological and the physiological characteristics of the neuroblasts, contain much larger chromosomes than the smaller daughter cells, although both cell types have the same DNA content (John & Lewis, 1968). In *Arvelius albopunctatus*, as in other species of pentatomid bugs with seven lobes in the testis, the spermatocytes in all but lobes 4, 5 and 6 are of normal size, whereas those in lobes 4 and 6 are much larger and those in lobe 5 are much smaller. The nuclear

TABLE 5.4. *Chromosome volume and DNA content measured photometrically in two replicates of twelve Feulgen stained cells of known 4C DNA content in different meristems of* Vicia faba (from Bennett, 1970)

Meristem	Age of plant (d)	Replicates in arbitrary units	Mean in arbitrary units	Chromosome volume (μm^3)
Main root	7	21.66 20.83	21.24	637
Lateral root of first order, grown up	21	21.45 21.54	21.50	398
Shoot	38	20.95 20.85	20.90	273
Lateral root of first order, just emerged	38	21.08 21.08	21.08	474
Lateral root of second order, just emerged	38	20.83 21.41	21.12	303

volumes in the three types of spermatocytes are about 200, 400 and 1600 μm^3, but their DNA content is the same and so is the course of meiosis; however, in the largest class of cells both the nucleolus and the cytoplasm contain a greater amount of protein and RNA (Schrader & Leuchtenberger, 1950). Equivalent situations exist in plants as well. Berrie (1959) has found that in the cycad *Encephalartos berteri* the chromosomes at the second pollen grain mitosis are smaller than at the first mitosis. Baetcke, Sparrow, Naumann & Schemmer (1967) have compared the chromosome volumes of shoot and root meristem cells in thirty plant species and have shown great differences in several of them. A large natural variation in chromosome size also exists between cells in meristems of different age in a given species; in the three closely investigated species (*Secale cereale*, *Allium cepa* and *Vicia faba*) the uniformity in nuclear DNA content of the meristematic cells investigated was ascertained cytophotometrically (Bennett & Rees, 1969; Bennett, 1970). In *Vicia faba* the difference in chromosome volume between meristems can be greater than 2 times (Table 5.4) and the chromosome volume shows a positive linear correlation with each of the nuclear components measured: RNA, total nuclear protein and nuclear histone (Bennett, 1970).

Changes in chromosome volume of the sort described above might differ from two recently described cases in which increased chromosome volume at mitosis has been claimed to be associated with an increase in nuclear DNA content. In the female gametophyte of *Pinus silvestris* all mitoses are haploid (n = 12) and no indication is found of diplochromosome formation (endoreduplication). Haploid mitoses with especially large chromosomes, which occur in the micropylar region of the gametophyte, have shown a DNA content twice that of the corresponding phases of the haploid mitoses which occur in other parts of the gametophytes (Nagl, 1965*b*, 1967*b*). This condition is comparable to the one found during the development of the ganglionic cells (neuroblasts) of *Drosophila melanogaster.* Two-wavelength Feulgen microspectrophotometry has revealed that the DNA value of metaphases in neuroblasts of third instar larvae was twice that of metaphases in neuroblasts of first instar larvae; all mitoses measured were diploid ($2n = 8$) and did not show any indication of diplochromosome formation (the four chromatids of each diplochromosome would be expected to fall apart at anaphase to form two tetraploid daughter nuclei) (Gay, Das, Forward & Kaufman, 1970). Although some process of extra DNA synthesis cannot be excluded, as discussed by Nagl (1967*b*), both Nagl (1967*b*) and Gay *et al.* (1970) are inclined to conclude that the condition found in *Pinus* and *Drosophila* is best interpreted as evidence of an increased polynemy in chromosomes. Quite recently, however, a reinvestigation of the *Drosophila* stock used by Gay *et al.* (1970) by the very precise method of scanning microspectrophotometry (Zeiss UMSP) has demonstrated the same DNA value (4C) for metaphases with different chromosome size in first instar and third instar larvae (Van de Flierdt, 1975). This observation raises doubts on the possibility that intra-individual differences in mitotic chromosome volume may reflect differences in strandedness.

Reduced chromosome number and aneuploidy

In a previous section of this chapter we have seen that cells with endoreduplicated nuclei can enter mitosis either spontaneously or under experimental influences; proneness to divide is most common in cells which have undergone one or two endoreduplication cycles (mitoses with diplochromosomes and quadruplochromosomes respectively). Since these endoreduplicated nuclei divide with a normal bipolar spindle, which segregates sister chromatids at opposite poles, mitoses lead to the formation of daughter nuclei with a polyploid chromosome number: tetraploid (4x) and octoploid (8x) from a diplochromosome and a quadruplochromosome mitosis respectively. Chromosome endoreduplication followed by mitosis in a regular sequence is a normally occurring phenomenon in the development of apical meristems, especially in root apices, of a number of plant species: a phenomenon called 'polysomaty' by Langlet (1927). The classical, and most extensively studied, example of polysomaty is that of spinach, *Spinacia oleracea* (De Litardière, 1923; Lorz, 1937, 1947; Gentcheff & Gustafsson, 1939; Berger, 1941*a*). In the root apex of spinach ($2n = 12$), the ploidy levels progressively increase from the promeristem to the base of the meristem in the following sequence: 12 chromosomes – 12 diplochromosomes – 24 chromosomes – 24 diplochromosomes – 48 chromosomes – 48 diplochromosomes – perhaps 96 chromosomes (for a review on polysomatic meristems, see D'Amato, 1952*b*).

Spontaneous mitoses either do not occur or are exceptional in systems, both plant and animal, in which nuclei have undergone several endoreduplication cycles.

A remarkable exception to this rule are the cells of the iliac epithelium in mosquito (2n = 6) larvae. The cells may undergo up to five endoreduplication cycles (64x) and, when the pupal metamorphosis begins, they acquire the ability to divide by mitosis (Berger, 1938, 1941*b*; Grell, 1946). At first prophase in these cells, it can be seen that in each nucleus there are three bundles of chromonemata which are due to repeated endoreduplication in a situation of somatically paired homologous chromosomes. As these cells approach metaphase, the three pairs of homologous bundles fall apart to form six bundles of sister chromosomes; at anaphase there is an apparent polar movement of chromosomes, but not of chromatids (somatic reduction). This first division is frequently followed by one or more mitoses with a short interphase during which, it is assumed, no chromosome replication occurs; moreover, each mitosis is characterized by somatic pairing. Thus, the number of cells in the iliac epithelium is greatly increased, their size is progressively diminished and their chromosome number is reduced to tetraploidy (4x) or octoploidy (8x). The substitution of somatic pairing for chromosome replication in interphase as a mechanism of somatic chromosome reduction in mosquitoes (Grell, 1946) is an interesting phenomenon worth being investigated thoroughly with appropriate cytochemical and autoradiographic techniques.

Another mechanism of somatic chromosome reduction has been observed in the root apical meristems and ovules of a number of colchicine-induced tetraploid (2n = 4x = 32) plants of *Ribes nigrum* (Vaarama, 1949). Whereas most mitoses in the young ovules showed 16 chromosomes (variation from 15 to 18), the chromosome number in the root meristem cells varied between 4 and 32 (modal frequency: 17). The reduction of the tetraploid chromosome number resulted from the organization of two separate and independent spindles which resulted in the production of four cells from the original cell; the autonomy of the separately dividing chromosome groups was seen as early as interphase in the form of a constricted nucleus. The presence of two separate spindles in one cell is also the condition responsible for the wide variation in chromosome numbers (from 23 to 83) in the root tips of *Hymenocallis palathinum* (Snoad, 1955). That prophasic chromosome segregation followed by the organization of 'twin spindles' in a cell occurs normally, though with very low frequency, in the root apices of several plant species has been shown by Huskins and co-workers (Huskins, 1948; Wilson & Cheng, 1949; Huskins & Cheng, 1950; Pätau, 1950). They have also shown that the frequency of reductional groupings and twin spindle formation can be increased experimentally, thus allowing a detailed analysis of the process to be made. In cytologically favorable material, it has been demonstrated that segregation of the diploid chromosome complement into two homologous groups, which organize separate spindles (formation of four haploid nuclei from an initial diploid nucleus) occurs with much greater frequency than would be expected if the distribution of the chromosomes were a random process. Somatic reduction in diploid tissues is responsible for the formation of haplo-diploid chimeras in which the size of the haploid sector(s) is related to the time of occurrence of somatic segregation. For

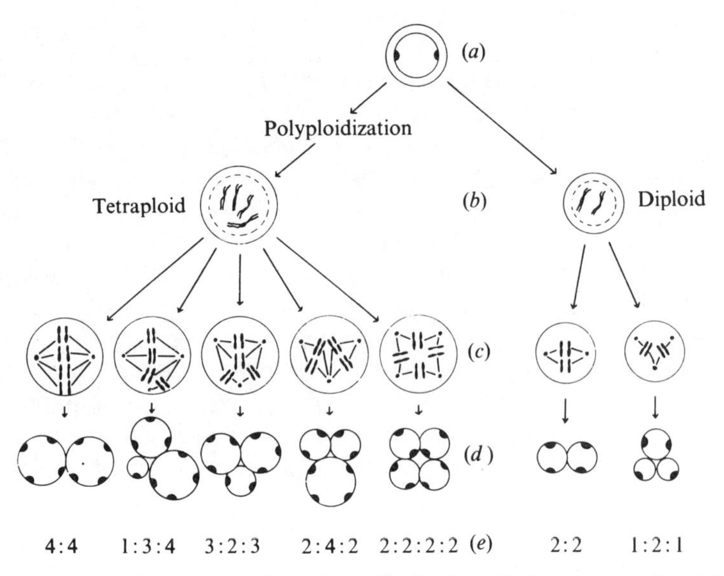

Fig. 5.15. Schematic representation of the distribution of chromosomes in diploid and tetraploid cells of *Microtus agrestis* through bipolar and multipolar mitoses. (*a*) original diploid cell with two chromocenters; (*b*) prophase; (*c*) anaphase; (*d*) daughter nuclei (one chromocenter denotes a haploid chomosome set); (*e*) ploidy of daughter nuclei (after Pera, 1970).

example, the two haplo-diploid plants of *Haplopappus gracilis* (2n = 4) studied by Lima-de-Faria & Jaworska (1964) differed widely in the percentage of haploid mitoses in the leaf primordia: 1.5% and 77.4% respectively.* In an interspecific *Sciara* hybrid Metz (1942) found a mosaic salivary gland in which some of the nuclei had chromosomes of only one of the parental species; the segregation had apparently occurred a few cell generations before the final mitosis in the gland. Other examples of somatic chromosome reduction, both in plants and in animals, are reviewed by Pera (1970).

A third mechanism of somatic chromosome reduction is multipolar mitosis. A scrutiny of the zoological and botanical literature (references in Pera, 1970) shows that multipolar mitoses both *in vivo* and *in vitro* are rather common in polyploid nuclei, whereas multipolar spindles are exceptionally formed in diploid mitoses. Indications have been found by several workers that multipolar mitoses sometimes lead to segregation of whole genomes. Detailed analyses on multipolar tetraploid and diploid mitoses in tissue cultures of *Microtus agrestis* (Pera, 1969; Pera & Rainer, 1973; Schwarzacher & Pera, 1969) have shown that daughter nuclei with different ploidy levels may result from multipolar mitoses (Fig. 5.15).

In addition to segregation of whole genomes (which, as stated above, also

* Since *H. gracilis* reproduces sexually, the haplo-diploid chimeras in this species cannot originate by semigamy.

TABLE 5.5. *Frequency distribution (per cent) of the various types of genome segregation (one genome: n = 21) in the dividing diploid, triploid and tetraploid cells of the liver in newborn and 6-day-old rats* (data from Gläss, 1957)

	2n = 42	3x = 63		4x = 84	
Age of rats	21:21	42:21	21:21:21	42:42	42:21:21
Newborn (12 hours)	7.9	13.3	13.3	38.0	9.6
6 days	9.2	(44.4)*	(11.2)	(40.0)	(20.0)

* Values in parenthesis are based on 5–9 mitoses observed.

results from twin spindle formation) multipolar mitoses also produce, with higher frequency, aneuploid chromosome complements. This condition is best exemplified in the liver of newborn and very young rats (2n = 42) in which cells are still dividing actively (Gläss, 1957, 1958; Marquardt & Gläss, 1957). Of particular interest is the process of genome segregation in this material. The proliferating cell population consists of haploid (n = 21), diploid (2n = 42), triploid (3x = 63) and tetraploid (4x = 84) cells; genome segregation may occur in diploid and in tetraploid cells (Table 5.5).

Reductional grouping followed by a regular meiosis has been found to occur in pollen mother cells of induced tetraploids and octoploids in tomato (Gottschalk, 1958). It seems probable that a sequence reductional grouping-meiosis may be responsible for the occasional segregation of diploids from experimental tetraploids in plants.

Age-associated changes in nuclei

Some cells die as they differentiate, e.g. xylem cells in plants and keratinized epidermal cells in animals; some other cells, as the sieve tubes (phloem) in most, but not all, plant species (Evert, Davis, Tucker & Alfieri, 1970) and red blood cells in mammals, lose their nuclei. Sieve elements are the only cases known of enucleate cells (they contain cytoplasm only) that can survive and remain functional (transport of organic solutes and synthesis of particular substances as slime and callose) for long periods of time; up to 50 years in the palm *Sabal palmetto* (Parthasarathy & Tomlinson, 1967). Sieve elements originate from the longitudinal unequal division of a meristematic (procambial or cambial) cell which produces a larger cell destined to differentiate into a sieve element and a smaller companion cell. In other cases (e.g. secondary phloem), the prospective companion cell divides once or twice; each sieve element is then associated with two or four companion cells. When the sieve element/companion cell(s) complex attains full

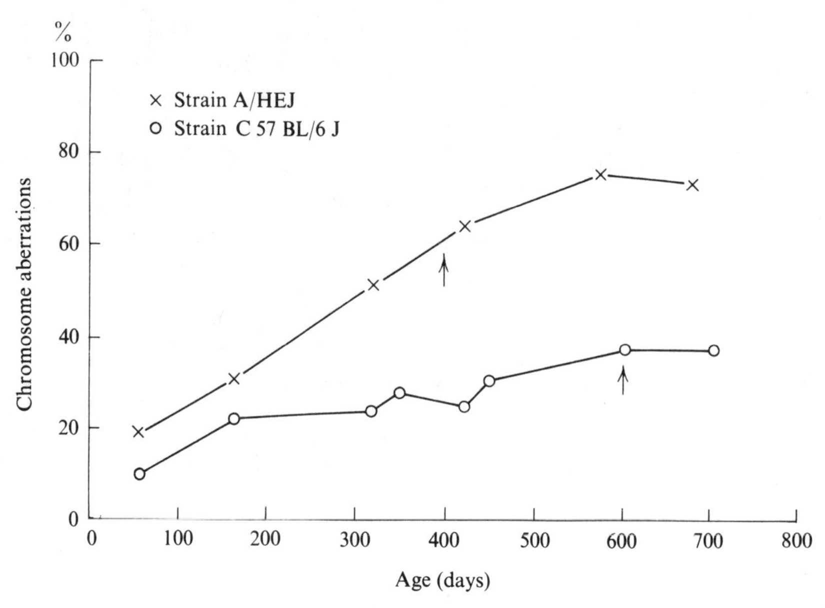

Fig. 5.16. Percentage of chromosome aberrations as a function of age in liver cells of two different inbred strains of mice. The median life of each strain is indicated by the arrows (after Curtis, 1966).

differentiation, the nucleus in the sieve element is resorbed whereas its cytoplasm persists, due to the many plasmodesmata that connect the sieve element with the companion cell(s). The functional relationship between the enucleate sieve element and the nucleate companion cell(s) is very close: in the senile secondary phloem of some taxa, cessation of function of sieve elements is concomitant with death of the closely associated companion cells (Esau, 1971, 1973). A detailed cytological study of phloem differentiation in *Vicia faba* has shown that the companion cell which is associated with the sieve element undergoes chromosome endoreduplication; this fails to occur (i.e. the nuclei remain diploid) when the sieve element is associated with either two or four companion cells. It seems as though the lack of a nucleus in the sieve element is compensated by a 'polyploid' condition of the companion cell(s); this is achieved by chromosome endoreduplication in the case of a single companion cell and by multicellularity in the case of two or four companion cells (Resch, 1958).

If we now turn to the variety of differentiated animal and plant tissues whose cells do not divide (and hence do not undergo renewal), we can state that a large body of evidence now exists on the accumulation with age of genetic changes in tissues. For recent reviews on this subject, reference is made to D'Amato (1964*b*), Curtis (1966, 1971) and Burnet (1974). Among animals, the best-known case is that of the mammalian liver which was investigated by Curtis and his associates (see

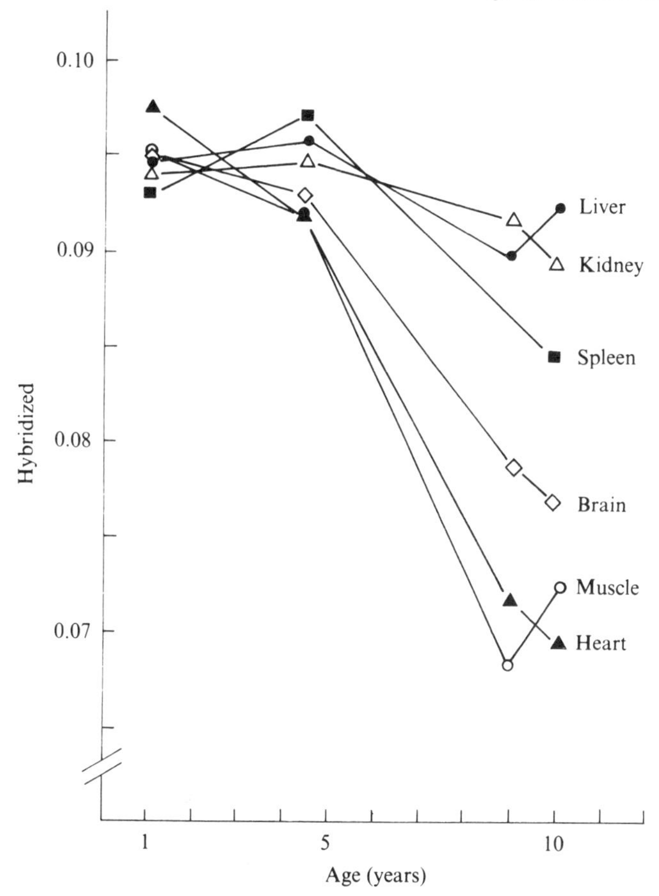

Fig. 5.17. Average percentage hybridization between DNA and ^{3}H-rRNA plotted against age for various tissues in the dog (after Johnson *et al.*, 1972).

Curtis, 1966, 1971). Liver cells of mice and dogs of different ages were scored for chromosome structural changes at anaphase following mitosis induction by partial hepatectomy (mitoses are quite rare in intact liver). It was shown that in mice the frequency of chromosome aberrations (deficiencies and translocations, the two types of aberrations which are easy to score at anaphase) steadily increased with age and the proportion of aberrations rose more rapidly in a short-lived than in a longer-lived strain (Fig. 5.16). It was also shown that the frequency of chromosome aberrations in liver cells rose more rapidly in mice than in dogs, whose life span is six times as long. These experiments strongly support the concept that mutations in somatic cells are connected with aging; but there are exceptions to this general rule (Curtis, 1971).

It is a reasonable assumption that in animals, as in plants (see below), the progressive increase in frequency of chromosomal aberrations in non-dividing somatic cells with increasing age is paralleled by a progressive increase in gene mutations. Analyses of changes in nucleoprotein (review by Hahn, 1970) and DNA (references in Johnson, Chrisp & Strehler, 1972) during cell aging do not help in detecting the mutation of specific genes in somatic cells. To this end, the technique of RNA–DNA hybridization *in vitro* seems promising. Using this technique, Johnson *et al.* (1972) have found a selective loss of rRNA genes during the aging of non-dividing tissues in the dog. The decrease in rDNA is most significant in heart and skeletal muscle (Fig. 5.17). The occurrence of a loss of rRNA genes in aging somatic cells implies that other subtle and previously undetected damage may occur to other genes present in the aging cell (Johnson *et al.*, 1972).

In plants, seeds provide outstanding material for studies of the age-associated genetic changes (reviews by D'Amato & Hoffmann-Ostenhof, 1956; Sax, 1962). Seeds may age under natural conditions or, more commonly, during prolonged storage. The genetic changes which occur during seed aging in the quiescent apical meristems of the embryonic root and shoot can be studied both immediately upon seed germination (chromosome structural changes in the first mitotic cycle post-germination) and in the progeny of plants raised from aged seeds (segregating gene mutations). In confirmation of the classical investigations of Navashin (1933) on aged *Crepis* seeds, it has been shown in a variety of plant species that: (i) with increasing age of seeds there occurs a progressive increase in both chromosome aberrations and gene mutations and (ii) the mutation rate does not increase arithmetically year after year but rises very much more rapidly in later years, that is, during senescence. As pointed out by Navashin himself, the lack of a linear relation between the duration of storage and frequency of spontaneous mutations excludes the effect of natural radiation or other continuously acting factors and points to the importance of cell metabolism in the process. Nowadays, there is ample evidence, in both plants and animals, that spontaneous mutation is an endogenous process; it has been said that it is the result of 'automutagenesis' (D'Amato & Hoffmann-Ostenhof, 1956) or 'intrinsic mutagenesis' (Burnet, 1974).

6
Differential DNA replication

The under-replication of some portions of the genome during chromosome polytenization, which has been discussed in the preceding chapter, is but one aspect of the differential replication of DNA. The investigations carried out in the last few years have shown that differential DNA replication is more common than previously suspected and that it may play a major role in differentiation and development. The best examples of differential DNA replication studied so far are those of the preferential replication of specific DNA sequences defined as gene amplification or, more generally, DNA amplification. For reviews on gene amplification, reference is made to Tobler (1975) and Buiatti (1977). Buiatti's review presents direct and indirect evidence for gene amplification in plants. Our discussion on differential DNA replication will first consider the amplification of ribosomal DNA (rDNA) in oocytes.

Amplification of rRNA genes in animal ovaries

Amplification of the DNA sequences coding for ribosomal RNA (rDNA) is found to occur in the ovaries of several species of amphibians and insects and in a few other animal species (Tobler, 1975), but it is best understood in amphibian oocytes.

Amphibia

In species of *Rana* and *Bufo*, Brachet (1940), Painter & Taylor (1942) and Guyenot & Danon (1953) demonstrated that the very numerous (hundreds) nucleoli which are formed in the nuclei (germinal vesicles) of oocytes at meiotic prophase are not attached to the chromosomes; they contain DNA which may take the form of one or occasionally two or more laterally located granules. In *Bufo valliceps*, these nucleoli (peripheral, multiple or extrachromosomal nucleoli as they are termed) lie against the inside of the nuclear envelope with their DNA granules always oriented toward the membrane and accumulate in finger-like protrusions of the nuclear envelope. Guyenot & Danon (1953) also described, in mature oocytes, a 'DNA enrichment' of chromosome regions in the form of Feulgen-positive spherules sometimes containing a 'vacuole'. More recent work on oocytes has led to a better definition of the distribution of DNA in multiple nucleoli by

TABLE 6.1. *Number of extrachromosomal nucleoli in oocytes of twenty-two species of amphibia* (from Buongiorno-Nardelli *et al.*, 1972)

Anura		Urodela	
Species	Number of nucleoli per oocyte	Species	Number of nucleoli per oocyte
Bombina variegata	1002	*Ambystoma mexicanum*	502
Bufo americanus	977	*Ambystoma opacum*	523
Bufo bufo	1017	*Ambystoma talpoideum*	1040
Bufo viridis	1015	*Desmognathus fuscus*	246
Bufo woodhousei fouleri	545	*Necturus maculosus*	1054
Hyla cynerea	253	*Plethodon jordani melaventris*	253
Hyla versicolor	970	*Pseudotriton ruber schenkii*	248
Rana agilis	526	*Triturus cristatus*	1025
Rana catesbeiana	1035	*Triturus taeniatus*	1042
Rana clamitans	570		
Rana esculenta	516		
Rana pipiens	562		
Xenopus laevis	1080		

means of the specific binding to DNA of ^{3}H-actinomycin D (Ebstein, 1969) and to the isolation of the DNA cores of multiple nucleoli of amphibia as circular molecules of DNA with their associated nucleoproteins (Miller, 1964, 1966; Kezer, 1965). This work has culminated in the visualization at the electron microscope of the genes coding for rRNA in *Triturus viridescens* and *Xenopus laevis* (Miller & Beatty, 1969). Evidence for a massive extra replication of DNA in amphibian oocytes was also provided by the observation that the germinal vesicle of *Triturus viridescens* contains four times more DNA than the expected 4C value (the DNA value of a meiotic prophase nucleus) (Izawa, Allfrey & Mirsky, 1963) and the germinal vesicle of *Rana pipiens*, the leopard frog, contains about 150 times more DNA than the 4C content (Haggis, 1966).

Using molecular DNA–RNA hybridization, Brown & Dawid (1968), Evans & Birnstiel (1968) and Gall (1968) were the first to show that the formation of multiple nucleoli in amphibian oocytes is the result of the amplification of the rRNA genes which are located at the nucleolus organizing region (for details on the localization of rDNA at the nucleolus organizing regions of chromosomes, see Chapter 7). The degree of amplification of the ribosomal cistrons in oocytes is different in different species of amphibia (Brown & Dawid, 1968); also the number of multiple nucleoli in oocytes differs from species to species (Table 6.1). In *Xenopus laevis*, the most extensively studied amphibian species, the extra synthesis of rDNA, which can be separated as a satellite with buoyant density of 1.723

g/cm³ (buoyant density of main band: 1.699 g/cm³) starts in zygotene nuclei less than 7 days after the end of the premeiotic S phase and is completed 18 to 24 days later, toward the end of pachytene. The rDNA which is synthesized during this 24-day period corresponds to about 3400 nucleolus organizers starting from the initial four (one for each chromatid of the nucleolus organizing chromosome pair) (Bird & Birnstiel, 1971; Coggins & Gall, 1972). The extra rDNA accumulates as granules around the nucleolus and at the end of amplification it appears as a compact mass (cap) attached to the pachytene chromosome mass; it labels very heavily when hybridized with radioactive 18S and 28S RNA (Gall & Pardue, 1969; John, Birnstiel & Jones, 1969). The cap rDNA eventually segregates into the hundreds of extrachromosomal nucleoli which synthesize rRNA in preparation for the massive accumulation in the oocyte cytoplasm of ribosomes to be used during embryogenesis until the onset of gastrulation when embryos are composed of about 30000 cells (Brown & Littna, 1964). It is at this time that the synthesis of rRNA, which had stopped in the mature egg, starts again (Brown, 1966).

An interesting aspect of rDNA amplification has been observed in the oocytes of F_1 hybrid females from a cross between *Xenopus laevis* and *X. mulleri*: only the nucleolus organizer of *X. laevis* is amplified (the rDNAs of the two species have 'spacer' regions of about the same length but very different in sequence), whereas the rDNA of *X. mulleri* is not amplified to any appreciable extent. The *X. laevis* chromosomal nucleolus organizer is invariably 'dominant'; moreover, the oocytes of the hybrid female produce as much extra rDNA as is made in normal *X. laevis* oocytes, which have four *X. laevis* nucleolus organizers (Brown, Wensink & Jordan, 1972; Brown & Blackler, 1972). This result is not surprising because it was already known that *X. laevis* females heterozygous for a deletion of the nucleolus organizer (one organizer instead of two) amplify their rDNA to the same level as normal females (two organizers) (Perkowska, Macgregor & Birnstiel, 1968). It is generally assumed that if one of the two organizers is lost by deletion, the oocyte is in some way able to compensate for the loss, possibly by undergoing an additional round of replication of rDNA over the number of replications needed to attain the *ca.* 1000-fold amplification observed in the oocytes of normal females. This theory of compensation has been challenged by Macgregor (1973) who proposes that only one of the two nucleolus organizers (that is, the nucleolus organizer of only one of the two chromosomes of the nucleolar bivalent) is ever involved in amplification. That this may be so is indicated by studies on the spermatocytes of *Plethodon cinereus cinereus*, the red-backed salamander (Macgregor & Kezer, 1973). As determined by rRNA–DNA hybridization *in situ*, two chromosome pairs (numbers 7 and 14) in this species carry clusters of ribosomal cistrons, but only the chromosome pair 7 organizes a nucleolus. In spermatocytes hybridized *in situ* with rRNA it is found that the nucleolus organizer of one of the two homologues in bivalent 7 is much more heavily labeled than the other. Obviously, the most plausible explanation for this condition is the differential involvement of the organizers in the amplification process.

Amplification of rDNA in oocytes is not the first increase in the number of ribosomal cistrons to occur in the germ line of *Xenopus laevis* (Gall, 1969; Gall & Pardue, 1969; Kalt & Gall, 1974). An amplification of rDNA equalling 20–40 nucleolus organizer equivalents occurs in premetamorphic oogonia and spermatogonia. In male germ cells the amplified rDNA is lost at the onset of meiotic prophase; instead, the loss of amplified rDNA at the onset of meiotic prophase is partial in oocytes, in which rDNA amplification will start again in zygotene (Bird & Birnstiel, 1971; Coggins & Gall, 1972).

Models have been proposed to account for the amplification of rDNA. Tocchini-Valentini and his associates (Crippa & Tocchini-Valentini, 1971; Tocchini-Valentini, Mahdavi, Brown & Crippa, 1974) have proposed that the first template rDNA molecules are synthesized from specific RNA molecules, which must contain spacer as well as the 18S and 28S sequences, through the intervention of an RNA-dependent DNA polymerase ('reverse transcriptase'). They have found evidence for the occurrence, at early stages of amplification in *Xenopus*, of a DNA–RNA hybrid molecule, sedimenting at 27S, from which a 45S RNA molecule can be separated by heat denaturation. They have also found in young *Xenopus* ovaries an enzyme with reverse transcriptase characteristics. On the basis of their results, Tocchini-Valentini *et al.* (1974) propose that amplification starts with the transcription of a 45S RNA precursor from the full rDNA unit; this RNA is reverse-transcribed to form extrachromosomal rDNA molecules which are then replicated repeatedly during amplification possibly by a rolling circle intermediate (see below). Tocchini-Valentini *et al.* (1974) have emphasised the importance of a precise timing of the development of the ovaries for an analysis of the involvement of reverse transcriptase in the initial phases of amplification; they therefore think that the inability of Bird, Rogers & Birnstiel (1973*b*) to find the class of 27S RNA–DNA hybrid molecules in *Xenopus* oocytes could be due to the use of frogs which were at a later stage of the amplification process.

Other theories for rDNA amplification in *Xenopus* are based on the more direct concept of DNA replication from a DNA template. Hourcade, Dressler & Wolfson (1973, 1974) and Bird, Rochaix & Bakker (1973*a*) studied at the electron microscope the amplifying rDNA of *Xenopus laevis*. They found that whereas most (90–95%) of the rDNA molecules were linear, the majority of the remaining molecules were in the form of circles and rolling circles (Fig. 6.1). The lengths of circles and rolling circles was not random: the smallest ones were clearly grouped at lengths corresponding to 1, 2, 3, 4 and 5 rDNA repeat units (the length of the largest rolling circle corresponded to about 16 rDNA repeats). Bird *et al.* (1973*a*) also studied the rDNA after labeling with ^{3}H-thymidine and provided evidence that rDNA rolling circles are involved in replication. Thus, the extra rDNA is synthesized on numerous replicating molecules which are not attached to the ribosomal cistrons on the chromosomes; but the mode of formation of the first extrachromosomal rDNA templates remains to be clarified. Although a reverse transcriptase might be involved (Tocchini-Valentini *et al.*, 1974), it can also be postulated

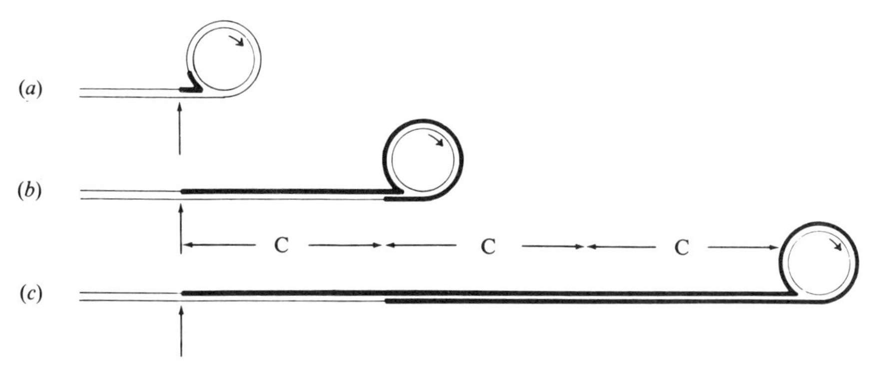

Fig. 6.1. Diagrammatic representation of progressive stages in the labeling of a rolling circle. Vertical arrows mark the beginning of a pulse with a tritiated DNA precursor; 'C' is the circumference of the circle. (*a*) the fork has replicated only a small part of the circle and the two labeled strands lie close together, each labeled in one strand of the duplex; (*b*) the fork has now copied slightly more than one circumference of the circle. Most DNA is labeled in one strand but, since the start of the second cycle, a small region of doubly-labeled duplex has appeared; (*c*) the fork has now traveled around the circle three times. Since the first cycle was completed, all the DNA appearing in the tail has been doubly-labeled. The circle remains labeled in one strand as does a section at the distal end of the tail equal in length to 'C' (after Bird *et al.*, 1973*a*).

(Hourcade *et al.*, 1974) that the first step of amplification is the creation of an extrachromosomal rDNA circle. The first DNA circle could arise either by making a replica of the chromosomal rDNA or by the excision of one or more rRNA genes from the chromosome (Brown & Blackler, 1972; Hourcade *et al.*, 1974). Buongiorno-Nardelli, Amaldi & Lava-Sanchez (1972) have suggested that rDNA units are excised from the chromosomes at meiotic prophase as monocistronic circles which replicate to form circles of doubled circumference at each replication. This model is, however, contradicted by the observations of Hourcade *et al.* (1973, 1974) and Bird *et al.* (1973*a*) which demonstrate that the size of circles and rolling circles corresponds to an arithmetic rather than a geometric progression.

Chromosomal rDNA and amplified rDNA of *Xenopus laevis* have been found to be very similar if not identical. The slight differences in their physical characteristics have been attributed to the fact that chromosomal rDNA contains about 4.4% of its bases as 5-methyl deoxycytidylic acid whereas amplified rDNA is unmethylated (Wensink & Brown, 1971).

An extraordinary feature of the growing oocytes of the tailed frog *Ascaphus truei* is that all of them have eight germinal vesicles because the eight nuclei produced in the last three oogonial mitoses remain in the same cell. The eight oocyte nuclei undergo meiosis in synchrony, but late in oogenesis seven of them disintegrate; when reaching their maximal size, the germinal vesicles contain about the same number of extrachromosomal nucleoli (about 100). During pachytene there occurs amplification of nucleolar DNA up to 5 pg per germinal vesicle (the diploid nuclear DNA content of *Ascaphus* is about 8 pg). These observations suggest that an *Ascaphus* oocyte attains the same level of amplification of nucleolar DNA as an oocyte of *Xenopus* or other amphibian species, but it does so by first becoming 8-nucleate and then accomplishing only a relatively modest amplification of its nucleolar DNA (Macgregor & Kezer, 1970).

Insects

In insects two types of ovarioles occur, panoistic and meroistic ones. The former type is the more primitive; there are no nurse cells and oogenesis is characterized by the lampbrush state of the chromosomes at meiotic prophase and the formation of extrachromosomal nucleoli. In the meroistic type of ovary the oocyte nucleus, as a rule, no longer supplies the ooplasm with RNA. The germinal vesicle incorporates only small amounts of RNA precursors; its chromosomes are in a condensed state and the nucleoli are inactive. In the meroistic ovary the function of the germinal vesicle has been taken over by the nurse cells (non-functional oocytes). Nurse cells are polyploid or polytene; they actively synthesize RNA that is transferred to the growing oocytes through intercellular cytoplasmic bridges. There is cytological and cytochemical evidence that the nurse cells undergo amplification of nucleolar DNA (Bier, Kunz & Ribbert, 1967; Ribbert & Bier, 1969). Among insect species with meroistic ovaries, water beetles (Dytiscidae) and craneflies (Tipulidae) seem to deviate from this general pattern; they have both the DNA amplification mechanism in nurse cells and the amplification of rDNA in oocytes.

In 1901, Giardina described a conspicuous 'chromatin ring' (*anello cromatico*) in oogonia and oocytes of *Dytiscus*. This and similar chromatin bodies occurring in the oocytes of other insects including *Tipula* were later shown to contain DNA (Bauer, 1933). In recent years cytochemical, autoradiographic and electron microscope studies (Urbani & Russo-Caia, 1964; Kato, 1968; Urbani, 1969) have shown that the chromatin body, which includes the nucleoli, contains a high amount of DNA (from 20 to 40 times the DNA value of diploid cells), synthesizes DNA and RNA independently of the chromosomes and disintegrates during oocyte growth. The condition found in *Dytiscus* is very similar to that occurring in the oocytes of *Tipula oleracea* which also show a DNA body (Lima-de-Faria & Moses, 1966); in both cases, all evidence indicates that the DNA body represents an amplification of nucleolar DNA. Indeed, amplification of rDNA has been demonstrated in the oocytes of three species of Dytiscidae; *Colymbetes fuscus*, *Dytiscus marginalis*, and *Agabus bipustulatus* (Gall, Macgregor & Kidston, 1969). The DNA of these species comprises a main band DNA (mDNA) and a high density satellite DNA, which can be shown to hybridize with *Xenopus* rRNA. In *Colymbetes*, somatic DNA contains satellite DNA amounting to 3.4% of total DNA as opposed to 23% of total DNA in ovariole tips; as shown by DNA hybridization with *Xenopus* rRNA, the amplification of the high density satellite in the ovariole tips is accompanied by a parallel amplification of rDNA. In *Dytiscus*, the high density satellite DNA, which makes an average 24% of total DNA in somatic tissues, is amplified about two-fold in ovariole tips. The level of amplification found in ovariole tips does not give a good estimate of the degree of amplification of rDNA in oocytes (the oocytes constitute a very small fraction of the cells contained in an ovariole tip). So, the amount of high density satellite per oocyte must be very great.

The DNA–rRNA hybridization data suggest that Giardina's body contains much DNA which is not complementary to rRNA (Gall *et al.*, 1969). Electron microscope analyses have shown that a high percentage of the rDNA in the oocyte nuclei of dytiscid beetles occurs as circles falling into size classes that are integral multiples (1, 2, 3, 4 etc) of a unit circle. The unit circle probably contains the coding sequences for one precursor rRNA plus accompanying spacer sequences (Gall & Rochaix, 1974; Trendelenburg, 1974).

In the last few years, much information on gene amplification in insect oocytes has been gained from extensive investigations on the house cricket, *Acheta domesticus*. Initial cytochemical and autoradiographic studies (Allen & Cave, 1968; Lima-de-Faria *et al.*, 1968; Cave & Allen, 1969) showed that the DNA body, which is already present in oogonia, undergoes further DNA synthesis during interphase and early meiotic prophase, when its DNA content amounts to about 25% of the total nuclear DNA. The DNA body encloses the nucleoli and at pachytene and diplotene it becomes 'puffed' at its periphery, where RNA is actively synthesized, and forms a shell of RNA around the DNA body. The DNA body disintegrates at late diplotene, releasing its RNA, DNA and histone into the nucleoplasm.

In 1969, Lima-de-Faria, Birnstiel & Jaworska showed that the DNA of *Acheta*, when analyzed in the analytical ultracentrifuge, comprises besides the main band DNA (buoyant density: 1.699 g/cm^3) a G–C rich satellite (1.716 g/cm^3) which specifically hybridizes with rRNA of *Acheta*. They then demonstrated that the DNA body is the site of a differential amplification of ribosomal cistrons, the amount of rDNA in ovaries at the time of maximal size of the DNA body being 18 times higher than in testes (no detectable amplifaction). These observations were confirmed and extended by Cave (1972) who showed that the amplified rRNA genes are localized in the DNA bodies of early prophase oocytes and that, as the cells proceed through diplotene, the DNA complementary to rRNA is gradually incorporated into the developing nucleolar mass. More recently, Cave (1973) and Pero *et al.* (1973) have shown that the complementary RNA (cRNA) transcribed *in vitro* from the satellite DNA can also be localized in the DNA body by hybridization *in situ*. Since the amount of DNA non-homologous to rRNA present in the satellite is approximately 92.3% (Pero *et al.*, 1973), it must be concluded that, as first suggested by Gall *et al.* (1969) for dytiscid beetles, the amplified DNA present in the DNA bodies of *Acheta* contains, besides rDNA, a large proportion of 'spacer' DNA and other sequences which may have other functions (Cave, 1973; Pero *et al.*, 1973). Trendelenburg, Scheer & Franke (1973) have recently

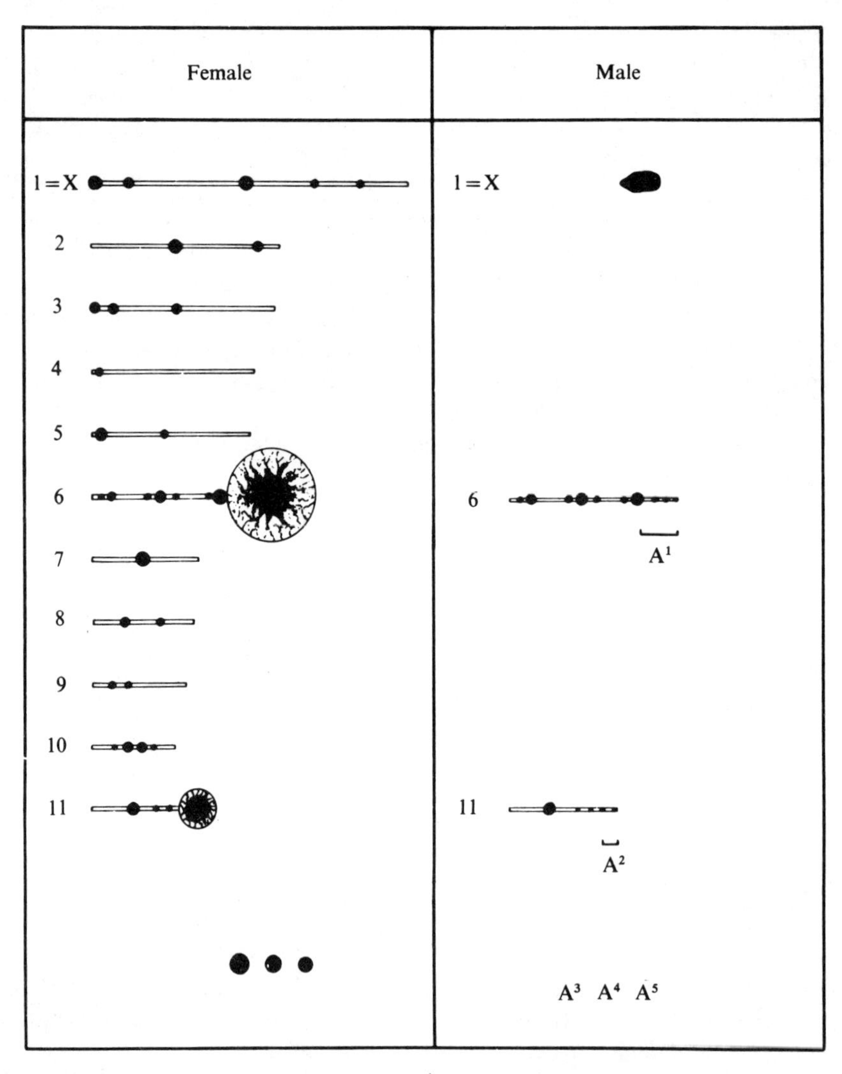

Fig. 6.2. Schematic representation of the pachytene chromosomes of *Acheta domesticus* in the oocytes and the spermatocytes. Only the chromomeres used as markers are shown. Three chromosomes have different chromomere patterns in the male: the X, chromosome 6 and chromosome 11. Chromosomes 6 and 11 have major chromomeres in the female. Since these major chromomeres are absent in the male, the chromomeres involved in their formation (the amplicons) can be accurately localized (A^1 and A^2). The other three smaller major chromomeres (A^3, A^4 and A^5) cannot be localized in specific chromosomes (after Lima-de-Faria, 1974).

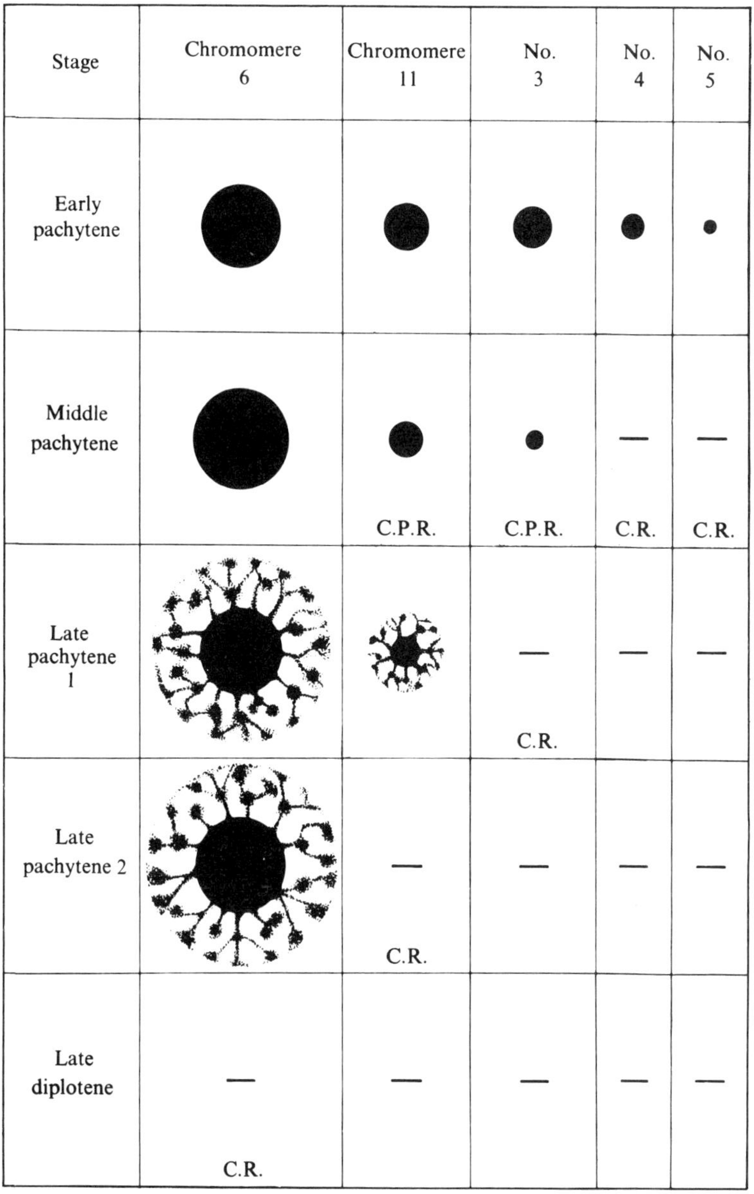

Stage	Chromomere 6	Chromomere 11	No. 3	No. 4	No. 5
Early pachytene	●	●	●	●	●
Middle pachytene	●	● C.P.R.	● C.P.R.	— C.R.	— C.R.
Late pachytene 1			— C.R.	—	—
Late pachytene 2		— C.R.	—	—	—
Late diplotene	— C.R.	—	—	—	—

Fig. 6.3. Schematic representation of the major chromomeres of *Acheta domesticus* from early pachytene to late diplotene in oocytes. The two largest major chromomeres can be localized in chromosomes 6 and 11, the other three are described as Nos. 3, 4, 5. C.P.R.: copies partially released; C.R.: copies released. In the case of the major chromomeres 6 and 11 the release of copies is preceded or accompanied by a 'puff' formation (after Lima-de-Faria *et al.*, 1973*c*).

spread the germinal vesicles of *Acheta* oocytes at diplotene using the Miller technique and have made molecular weight determinations of the mature *Acheta* rRNA and their precursors. They found that the amount of DNA homologous to mature rRNA corresponds to 21% of the repeating unit. This would leave a 'spacer' of about 80%.

Since *Acheta* has chromosomes which can easily be recognised at pachytene by their size and chromomere markers, Lima-de-Faria and his associates have been able to localize DNA amplification in distinct chromomeres belonging to different chromosomes (Lima-de-Faria *et al.*, 1973*a*, *b*, *c*; Ullman, Lima-de-Faria, Jaworska & Bryngelsson, 1973; Lima-de-Faria, 1974). In *Acheta* oocytes there are five amplification sites (amplicons) (Fig. 6.2). Of these two are major chromomeres, or DNA bodies, localized in chromosomes number 6 and 11 and the other three are smaller chromomeres which are only present at very early pachytene (the entanglement of chromosomes at this stage does not permit their localization in chromosomes). The amount of DNA in the major chromomeres is 1.8×10^9 (chromosome 6), 1.4×10^9 (chromosome 11) and 0.66×10^9 (three smaller chromomeres) nucleotide pairs. The amount of DNA in the major chromomeres of chromosome 6 (amplicon formed by three chromomeres) and chromosome 11 (amplicon formed by two chromomeres) is in both cases about 100 times greater than the DNA amount of the initial amplicons (haploid). Saturation rRNA–DNA hybridization experiments show that in ovaries the degree of amplification of the genes for 18S and 28S RNA (which are in a 1:1 ratio) is between 4 and 5 times the amount of ribosomal cistrons in somatic tissues. This is, however, an underestimate of the actual degree of amplification because the ovarian DNA consists of oocyte and follicle cell DNA. This implies that as the number of follicle cells increases during ovarian development, the oocyte DNA is diluted by the follicle cell DNA. Fig. 6.3 summarizes the behavior of the five major chromomeres from early pachytene to late diplotene. The major chromomeres of chromosomes 6 and 11 form 'puffs' at different times during the meiotic process and they release their copies into the nucleoplasm at different times: at late pachytene the major chromomere of chromosome 11 and at late diplotene the major chromomere of chromosome 6. At that time, this structure becomes differentiated into a main nucleolus and a secondary nucleolus which, in the electron microscope, are seen to consist of various classes of particles resembling preribosomes. Both the main and the secondary nucleolar components cross the nuclear envelope and pass into the cytoplasm (Jaworska & Lima-de-Faria, 1973; Jaworska, Avanzi & Lima-de-Faria, 1973).

Other animals

Amplification of ribosomal cistrons in oocytes has been demonstrated for the teleost *Roccus saxatilis* (Vincent, Halvorson, Chen & Shin, 1969), the clam *Spisula solidissima* and the worm *Urechis caupo* (Brown & Dawid, 1968).

As pointed out by Gall (1969) and Tobler (1975), DNA amplification is of common occurrence though not universal during oogenesis in animals. In many cases, however, the evidence rests only on cytochemical and/or autoradiographic data.

Amplification of rRNA genes in somatic tissues

So far, the only case of massive amplification of ribosomal cistrons in somatic tissues is provided by the differentiating metaxylem cells in the root of *Allium cepa* (Innocenti & Avanzi, 1971; Avanzi, Maggini & Innocenti, 1973; Innocenti, 1973). In the onion, as in a number of plant species, the cells destined to differentiate as the centrally located lignified elements (metaxylem) of the root vascular system begin differentiation just behind the apical meristem. Prospective metaxylem cells in the onion root comprise the axial file of cells and the immediately adjacent five cell files. During early differentiation these cells undergo a conspicuous growth in both diameter and length, so that at 15 mm from the apical meristem they are 3 times longer than parenchymatic cells in the root cortex (916–977 μm and 311–320 μm respectively). Cytochemical, cytophotometric and electron microscopical analyses have shown that the early differentiation involves a defined sequence of stages during which chromosome endoreduplication, differential DNA replication and extrusion of nuclear material from the nucleus into the cytoplasm take place (Table 6.2). Before DNA release from the nucleus, the extra synthesis of DNA, which can be easily detected by labeling with ^{3}H-thymidine, is mostly restricted to nucleolar and perinucleolar chromatin and leads to the formation of a conspicuous DNA shell around the nucleolus. This DNA is specifically labeled by hybridization *in situ* with homologous ^{3}H-rRNA. DNA extracted from root segments containing differentiating metaxylem cells shows a 4.5 to 6-fold higher saturation plateau with rRNA than DNA extracted from root apical meristems. Since metaxylem cells comprise only a very small fraction of all root cells, the degree of amplification per individual metaxylem cell must be very high; Avanzi *et al.* (1973) estimate an amplification level of between 1125 and 3000.

In contrast with the situation in differentiating metaxylem cells in the onion root, the degree of rDNA amplification which has been observed in a few animal somatic tissues is very low. For example, Collins (1972) reported a 60% amplification of rRNA genes during Wolffian lens regeneration in the iris of *Triturus viridescens* (see Chapter 9). Koch and Cruceanu (1971) noted an increase of 80% in the amount of rDNA in response to the hormone triiodothyronine in cultured human liver cells. The validity of this observation has been contested by Ryffel, Hagenbüchle & Weber (1973) on account of the very high concentration of triiodothyronine used (a concentration 100-fold that of the less effective thyroxine *in vivo*). They have shown that the number of rRNA genes is not changed in liver and tail muscle of *Xenopus laevis* larvae during thyroxine-induced metamorphosis.

Tobler (1975) has pointed out that no uncontested proof exists for the occurrence

TABLE 6.2. *Sequence of stages during the differentiation of metaxylem cells in the root of* Allium cepa (from Avanzi *et al.*, 1973)

Nuclear DNA content	Stage	Distance from root tip (mm)
4C–8C	Chromosome endoreduplication	—
8C–(8C–16C)	Differential DNA synthesis	—
(8C–16C)–8C	Extrusion of the nucleolus and release of extra DNA	2
8C–16C	Chromosome endoreduplication	
16C–(16C–32C)	Differential DNA synthesis	5
(16C–32C)–16C	Release of extra DNA	
16C	Chromatin condensation	10

of gene amplification in animal somatic tissues and that, even when accepting the instances of low-level amplification, it is doubtful whether this has biological significance. It must be stressed, however, that in the search for rDNA amplification, and gene amplification in general, the sole analysis of saturation curves in RNA–DNA hybridizations *in vitro* is not adequate. Such an analysis must be supplemented with two other analyses – RNA–DNA hybridization *in situ* and nuclear DNA cytophotometry – which alone can give information on the frequency and types of cells which might undergo amplification within a tissue or a tissue system. In this context, the example of the differentiating metaxylem in the onion root (Avanzi *et al.* 1973) is very instructive. The metaxylem cells, which comprise 0.1 to 0.2% of the total cell population, are the only cells that undergo a massive rDNA amplification in the subapical portion of the root.

rRNA gene magnification and compensation

Ritossa & Spiegelman (1965) were the first to show that in *Drosophila melanogaster* the 28S and 18S rRNA cistrons are clustered in equal numbers at the nucleolus organizer (NO) located in the proximal heterochromatin of the X chromosome and in the short arm of the Y chromosome. Ritossa, Atwood & Spiegelman (1966*b*) then demonstrated that the *bobbed* (*bb*) mutation (which in homozygous flies is expressed by slow growth, short bristles and thin cuticles) is due to a partial deficiency of rDNA. When further studying the *bobbed* phenomenon, Ritossa (1968) discovered an increase (magnification) in rRNA cistrons in males homozygous for the *bobbed* mutation. He found that strongly *bobbed* males X^{bb}/Y^{-bb} ($-bb$ is a severe deletion for rDNA) and X^{bb}/O, although highly sterile, could be mated to $\widehat{XX}/Y^{-bb}$ females; the male progeny, which had the same sex chromosome constitution of the male parent, were no longer extreme *bobbed*. The change produced in this generation was carried over into the next so that in three generations the progeny were phenotypically wild-type. Hybridization experiments between rRNA and the DNA extracted from adult flies showed that this reversion from *bobbed* (*bb*) to wild-type (bb^m: magnified *bobbed*) was due to a progressive increase in rRNa cistrons in the X chromosome to reach in bb^m a value approximating that of the wild-type locus (bb^+). Ritossa (1968) also observed that the accumulated rDNA, which in suitable conditions can be indefinitely maintained, can be lost in other conditions [this phenomenon has been called 'reduction' by Tartof (1973)].

Ritossa and collaborators (Ritossa, 1972, 1973; Boncinelli *et al.*, 1972; Ritossa, Scalenghe, Di Turi & Contini, 1974) have shown that rDNA magnification at the *bobbed* locus, which occurs both in the X and in the Y chromosome, is a step-wise process in which the newly acquired rDNA is often no more than twice the initial amount. Moreover, it seems to be associated with crossing-over at meiosis. Normally, rDNA regions at the *Drosophila* X and Y do not cross over because crossing-over is virtually absent in males and the cistron polarities are reversed. When an inversion of the nucleolus organizer is used, XY cross-overs are about 0.01%, but this value is increased to 0.3% in flies undergoing magnification. When magnification is completed, cross-overs are again reduced in number. These and other observations have led to a model which proposes that in phenotypically *bobbed* males extra copies of rDNA are formed, the number in each cell not exceeding that in the chromosomal rDNA block. In meiosis, but not in somatic cells, the rDNA copies can be circularized and integrated into the chromosomal rDNA locus of either X or Y, thereby increasing the cistron number. The model predicts that occasionally the extrachromosomal rDNA circle may insert simultaneously into both X and Y loci leading to the exchange of outside markers (increased recombination as observed in magnifying flies).

Another model proposes unequal sister chromatid exchange in mitotic cells (somatic crossing-over) as a mechanism for magnification (Tartof, 1974). In Tartof's experiments, severely *bobbed* males (X^{bb}/Y^{bb-}) were mated singly to females in which one X possessed a deleted organizer (bb^o) and the other X carried an inversion (*dl-49*) blocking meiotic recombination (X^{bb^o}/X*dl-49*). The female progeny which carried the X^{bb} from the male parent and the X^{bb^o} from the female parent (X^{bb}/X^{bb^o}) were studied. The magnification that occurred in the X^{bb} chromosome when present in the *bobbed* male parent was seen as a change from *bobbed* to wild-type. The high frequency of magnification, observed under conditions in which crossing-over was eliminated, strongly suggests that magnification is a premeiotic (mitotic) event. Tartof also observed that, in crosses between X^{bb+}/Y^{bb-} males and X^{bb^o}/X*dl-49* females, some of the X^{bb+}/X^{bb^o} progeny were *bobbed*, showing that the paternal X chromosome had undergone reduction. These and other observations led Tartof (1974) to propose that rDNA magnification–reduction results from unequal mitotic sister chromatid exchange. This would lead to the production of two new sister chromatids, one containing a greater and the other a lesser number of rDNA tandem repeats than originally contained in either parental chromatid. Obviously, Ritossa's and Tartof's theories are not mutually exclusive: the magnification–reduction phenomena may well reflect two distinctively separate occurrences of intrachromosomal exchange, one meiotic, the other mitotic (Tartof, 1974).

Another process of increase in the amount of rDNA per NO, which occurs during the development of *Drosophila melanogaster*, is 'rDNA compensation' (Tartof, 1971, 1973); it is not heritable, being limited to somatic cells. When a single X^{bb} or X^{bb+} chromosome is maintained opposite another chromosome that is deficient in rDNA (X^{bb+}/O males, $X^{bb+}/X^{bb''}$ females and X^{bb+}/Y^{bb-} males), there occurs an increase in the amount of rDNA per X chromosome NO. For instance, Tartof (1971) found that the reiteration of the ribosomal genes in an X/O fly, as determined by hybridization between rRNA and the DNA extracted from whole animals, was about 80% that of an X/X fly instead of the expected 50% (one NO instead of two). Spear & Gall (1973) have shown, however, that in X/O males no compensation occurs in brain and imaginal disk cells (50% reiteration), whereas full compensation (100% of the value of an X/X individual) occurs in salivary gland cells. The interest of these observations is twofold: (i) the differential replication of ribosomal genes – and possibly other DNA sequences (see below) – in somatic tissues might play a role in cell differentiation and function and (ii) in studies of magnification and compensation, hybridization data based on DNA extracted from whole animals might give an erroneous estimate of the actual level of reiteration of ribosomal cistrons (Harford, 1974). As pointed out by Harford (1974) and Williamson & Procunier (1975), several aspects of magnification and compensation are still unclear: among others, the question whether the NO in the Y chromosome responds to magnification and/or compensation in the same manner as the NO in the X chromosome in all genetic situations.

So far, magnification and compensation of ribosomal cistrons have been observed only in *Drosophila melanogaster.* Tests for rRNA–DNA hybridization on a whole-plant basis have failed to detect ribosomal cistron magnification or compensation in maize (Phillips, Weber, Kleese & Wang, 1974). Plants provide, however, evidence that, as in *Drosophila*, the number of ribosomal cistrons per nucleolus organizer is not constant. The best examples are offered by aneuploids in hyacinth (Timmis, Sinclair & Ingle, 1972) and in hexaploid wheat, *Triticum aestivum* (Mohan & Flavell, 1974). *Triticum aestivum* (2n = 42) possesses four different nucleolar chromosomes (1A, 1B, 5D, 6B) per haploid genome. Two of the four nucleolar organizers (chromosomes 1A and 5D) are relatively inactive and do not form nucleoli in the presence of the other two 'strong' nucleolar organizers (chromosomes 1B and 6B). Aneuploid stocks carrying additional 'strong' nucleolar organizers (e.g. tetrasomic 1B) have increased numbers of nucleoli, whereas stocks with additional 'weak' nucleolar organizers (e.g. tetrasomic 1A) do not show such an increase (Crosby, 1957; Longwell & Svihla, 1960). Molecular hybridization between rRNA and DNA extracted from aerial plant parts of twelve different genotypes – the euploid variety Chinese Spring and eleven of its aneuploid deriva-

TABLE 6.3. *Hybridization levels of rRNA as a percentage of the hybridization level of the euploid DNA in hexaploid wheat* (from Mohan & Flavell, 1974)

		Nucleolar organizers				
		'Strong'		'Weak'		
Genotype	Chromosome number	1B	6B	1A	5D	Hybridization level percent of control
Euploid	42	2	2	2	2	100 (control)
Tetrasomic 1A	44	2	2	4	2	101
Ditelosomic $1A^L$	40+2 telocentrics	2	2	0	2	58*
Tetrasomic 1B	44	4	2	2	2	128*
Ditelosomic $1B^L$	40+2 telocentrics	0	2	2	2	84†
Tetrasomic 5D	44	2	2	2	4	76*
Ditelosomic $5D^L$	40+2 telocentrics	2	2	2	0	105
Tetrasomic 6B	44	2	4	2	2	109
Tetrasomic 5A	44	2	2	2	2	109
Ditelosomic $5A^L$	40+2 telocentrics	2	2	2	2	109
Tetrasomic 6A	44	2	2	2	2	103
Ditelosomic $6A^L$	40+2 telocentrics	2	2	2	2	97

* Significantly different from euploid at 0.1% probability level.
† Significantly different from euploid at 1% probability level.

tives – has clearly shown that the number of ribosomal cistrons of each nucleolar organizer is not fixed (Table 6.3) but may be under genetic control (Mohan & Flavell, 1974). In the search for mechanisms responsible for the instability in the redundancy of ribosomal cistrons, tests on compensation and magnification in hexaploid wheat would be of especial interest.

DNA puffs in polytene chromosomes

The occurrence of extra DNA synthesis leading to the production of DNA puffs in polytene chromosomes of sciarid flies has been discussed in the preceding chapter. In sciarid flies, only RNA puffs are formed during most of the larval development; the DNA puffs appear exclusively in salivary gland cells during late larval and prepupal stages (in *Rynchosciara angelae*, only RNA puffs are found at all developmental stages in the polytene chromosomes of Malpighian tubules and intestine: Guevara & Basile, 1973). As first shown by Breuer & Pavan (1955) in *R. angelae*, extra DNA synthesis in a given puff may start simultaneously with intense RNA synthesis, whereas in another puff RNA synthesis may follow DNA synthesis (see also Pavan & Da Cunha, 1969). The extra synthesized DNA remains attached to the chromosome at its site of origin: e.g. salivary gland cells of *R. angelae* (Breuer & Pavan, 1955) and foot-pad cells of *Sarcophaga bullata*

(Whitten, 1969). On the other hand, in *Hybosciara fragilis*, the extra DNA is released into the nucleoplasm, and thereafter into the cytoplasm, in the form of DNA granules associated with RNA-rich material. Part of these granules are probably true micronucleoli because they originate at the nucleolar puff of the X chromosome (Da Cunha, Pavan, Morgante & Garrido, 1969; Da Cunha *et al.*, 1973).

There is thus far no proof that at least part of the micronucleoli of *Hybosciara* polytene chromosome cells contain rDNA. *In situ* hybridization experiments in polytene chromosome cells of *Rhynchosciara hollaenderi*, *Sciara coprophila* (Pardue *et al.*, 1970; Gerbi, 1971) and *Phaseolus coccineus* (Avanzi *et al.*, 1972) have shown that rDNA is present both in the chromosomal nucleolus organizing regions and in some of the micronucleoli scattered throughout the nucleus. The presence of rDNA-containing micronucleoli in these cells raises the question whether it reflects amplification above the rDNA level of diploid cells. Gerbi (1971) found that the *Xenopus* rRNA saturation levels for *Rynchosciara* DNA extracted from adult male carcasses, larval male carcasses and male salivary glands containing micronucleoli were virtually identical. To explain this finding, Gerbi assumed that the rDNA at the nucleolus organizing regions may be under-replicated in polytene chromosomes, in which case the formation of micronucleoli might occur to adjust the rDNA level back to the original diploid percentage of the genome. Gambarini & Meneghini (1972) have shown, however, that in salivary gland cells of *R. angelae* the number of ribosomal cistrons per haploid genome is about 100 and seems to be maintained virtually unchanged throughout the end of the fourth instar. In *Phaseolus coccineus*, it has been shown that the rRNA genes are under-represented in polytene chromosome cells (embryo suspensor) as compared to roots, ovular integuments and shoots (Lima-de-Faria *et al.*, 1975). Thus, the formation of micronucleoli which are released from the nucleolus organizing system of chromosome S_1 (Avanzi *et al.*, 1972) is such as not to compensate for the under-replication of rDNA which occurred during polytenization.

The *in situ* hybridization experiments of Pardue, Gerbi, Eckhardt & Gall (1970), Gerbi (1971) and Avanzi *et al.* (1972) have shown that DNA puffs in polytene chromosomes of *Rhynchosciara hollaenderi*, *Sciara coprophila* and *Phaseolus coccineus* do not contain rDNA; moreover, DNA puffs in *Rhynchosciara hollaenderi* do not contain a light satellite DNA which is present in centromeric heterochromatin and some telomeric regions (Eckhardt & Gall, 1971). These observations leave unanswered the question of the nature of the DNA sequences which are amplified at DNA puff formation. The studies of Balsamo *et al.* (1973*a*, *b*, *c*) have shown the existence in the genome of *R. angelae* of three DNA classes: (i) a slow-renaturing fraction, composed of 'unique' sequences and making about 86% of the genome; (ii) a fast-renaturing fraction (7.3% of the genome) composed of about 2×10^3 copies of sequences of 3.9×10^6 daltons and (iii) a fraction (6.7% of the genome) with an intermediate renaturation velocity composed of about 2.2×10^2 copies of sequences of 3.3×10^7 daltons. It is this fraction that is amplified at the time of giant DNA puff formation. Analysis of the DNA classes present in roots,

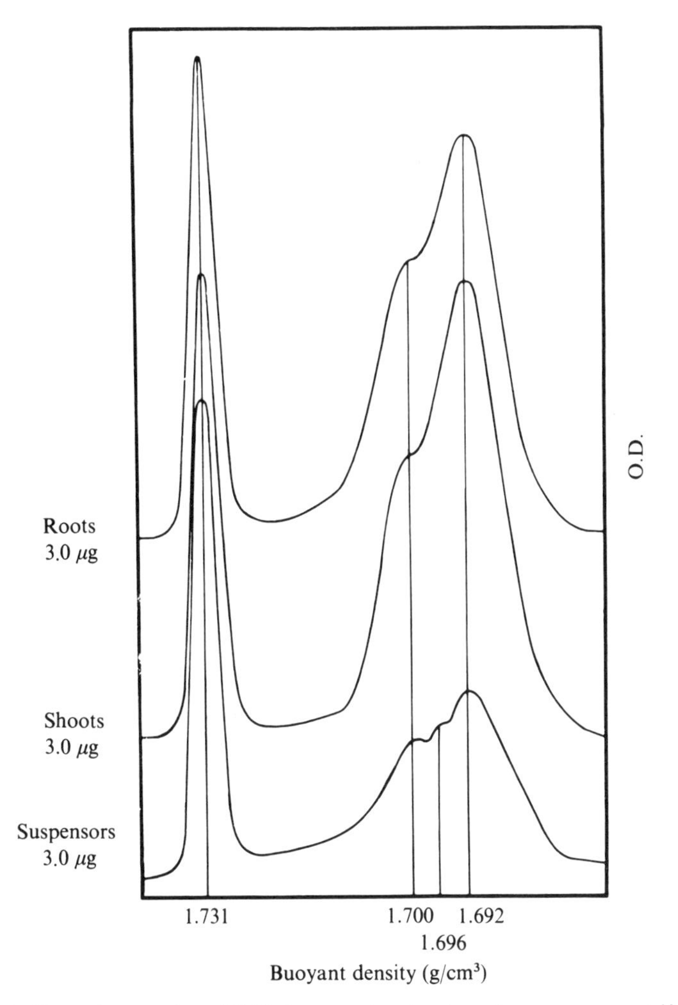

Fig. 6.4. *Phaseolus coccineus* DNA from roots, shoots and suspensors centrifuged in the analytical ultracentrifuge. Besides the DNA satellite (buoyant density of 1.700 g/cm³) in the three tissues, a second DNA satellite (buoyant density of 1.696 g/cm³) appears in the suspensors, which contain polytene chromosome cells (after Lima-de-Faria *et al.*, 1975).

shoots and suspensors of *Phaseolus coccineus* (Fig. 6.4) has shown that suspensors contain a satellite DNA (buoyant density: 1.696 g/cm³) which is not detected in the other two organs. It is tempting to speculate that this fraction, which makes about 13% of total suspensor DNA, represents the amplified DNA during formation of DNA puffs in polytene chromosomes (Lima-de-Faria *et al.*, 1975). Hopefully, future work will elucidate the nature and function of the DNA sequences which are amplified during DNA puff formation.

The initiation of DNA puff formation seems to be under hormonal control. Injection of the molting hormone ecdysone into *Sciara coprophila* larvae at 3–4 days after the third molt induces 24 hours later both DNA synthesis all along the chromosome and a DNA puffing pattern comparable to that of late fourth instar larvae; at two and three days after injection, the glands show DNA puff development normally observed in prepupae or very young pupae (Crouse 1968). Premature development of DNA puffs in larvae injected with ecdysone has also been observed in *Rhynchosciara hollaenderi* (Stocker & Pavan, 1974) and *R. americana* (= *R. angelae*) (Berendes & Lara, 1975). In *R. hollaenderi*, differences in amplification and puffing response between treated larvae and normally developing larvae have been observed. In *R. americana* indications have been found that (i) in the DNA puff regions, the rate of DNA chain elongation is higher than that during a normal replication cycle and (ii) the higher rate during amplification may depend upon *de novo* RNA synthesis. These observations are worth special consideration in studies aiming to clarify the mechanism of amplification in DNA puffs.

The flax genotrophs

Certain environmental conditions can induce genetic changes in particular varieties of flax, *Linum usitatissimum* (Durrant, 1962, 1971, 1972) and *Nicotiana rustica* (Perkins, Eglington & Jinks, 1971). The best examples of this environmental conditioning occur in a few varieties of flax, including Stormont Cirrus. When this variety is grown for the first five weeks in a heated greenhouse in particular combinations of fertilizer, soil pH, etc., and then transplanted in the field, it can be changed from its original form (Pl: plastic genotroph) to either a stable large form (L: large genotroph) or a stable small form (S: small genotroph) depending on the fertilizer applied (inducing environment). The L genotroph, about twice the size of Pl, and the S genotroph, about half the size of Pl, have remained stable over twelve generations in all environments. The genetic difference between L and S is primarily nuclear since there is no transmission through reciprocal grafts and reciprocal crosses give equilinear inheritance in the F1.

Plant weight in the flax genotrophs is correlated with nuclear DNA content. L has 16% more nuclear DNA as measured by Feulgen photometry than S and Pl has an intermediate amount. The change in the amount of nuclear DNA can be measured week by week during the first weeks of growth of the Pl plants in the inducing environment, high nitrogen and high phosphorus respectively (Fig. 6.5). The location of the DNA differences is most probably chromosomal because a similar difference between the genotrophs is observed when both 4C interphase nuclei and well-flattened metaphases are measured (Evans, Durant & Rees, 1966; Evans, 1968). *In vitro* rRNA–DNA hybridization experiments have shown that Pl and L have about the same number of ribosomal cistrons while S has 70% less cistrons than L and Pl (Timmis & Ingle, 1973; Cullis, 1975). Since this difference

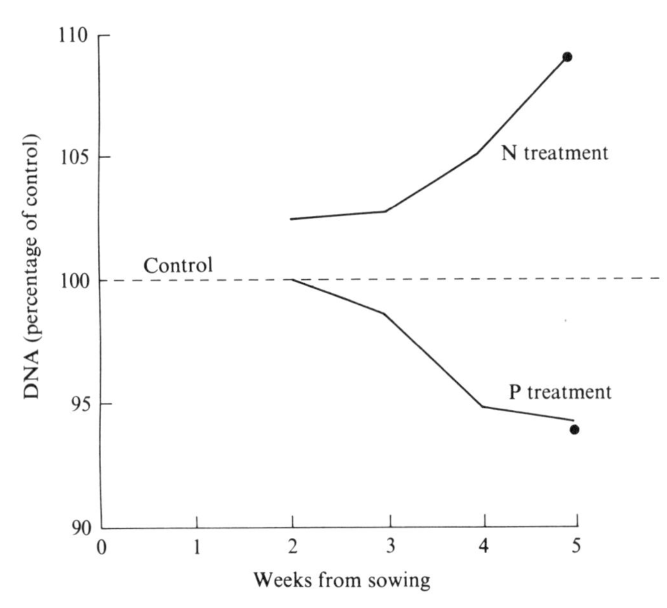

Fig. 6.5. Changes in the nuclear DNA content of plastic genotrophs of flax grown in an inducing environment (nitrogen fertilizer or phosphorus fertilizer) during the first five weeks of growth. The two points show the percent DNA values of large and small genotrophs. As judged by changes in nuclear DNA content, the induction is complete by the end of the first five weeks (after Evans *et al.*, 1966).

in rRNA genes only accounts for 0.23% of the difference between the L and S genotrophs, other sequences are also involved. The nature of the DNA associated with genotroph induction has been studied by analytical ultracentrifugation, thermal denaturation and renaturation kinetics (Ingle, Pearson & Sinclair, 1973; Cullis, 1973, 1975; Timmis & Ingle, 1974). These studies have shown that no gross differences can be detected between the DNAs of the L and S genotrophs and the DNA of the original variety Pl. However, the renaturation analyses of Cullis (1973, 1975) indicated that the DNA from L contained a class of moderately repeated DNA sequences which was not present in the DNA from either S or Pl; it also contained a lower proportion of highly repeated sequences than either S or Pl (S and Pl showed the same renaturation kinetics).

In L and S genotrophs, the amount of nuclear DNA can be reverted towards the intermediate value (the Pl value) by growing the plants at a lower temperature for the first 5 weeks after sowing. After three generations there is no measurable difference in the amount of DNA between L and S. The reversion can be stopped at any point by growing the plants of the next generation for 5 weeks from sowing in a heated greenhouse so that any number of plants can be obtained containing any desired amount of nuclear DNA up to the inducible limits (Durrant & Jones, 1971; Joarder, Al-Saheal, Begum & Durrant, 1974).

In conclusion, the flax genotrophs provide an unique example of how some genomes can respond to particular environmental conditions with stable heritable changes that can be made to revert at any time if proper conditions are applied.

Intraindividual variations in the replicative status of DNAs

This chapter would be incomplete if we did not discuss the results of some recent investigations which show that the ratio of one DNA fraction to another differs not only between tissues but also between nuclei within tissues. As discussed in Chapter 5, there is ample evidence that during chromosome endoreduplication the genome is fully replicated at each endoreduplication cycle; so, many tissues are found to contain cells with 2C, 4C, 8C, 16C and higher nuclear DNA contents. There are, however, excellent examples, e.g. in insect tissues (Fox, 1969, 1970), of non-doubling DNA distributions which indicate differential replication of portions of the genome (Table 6.4). Analysis of the genome on the basis of euchromatin–heterochromatin content has shown a differential replication of these two components within a tissue, the testis wall of *Dermestes maculatus*. In the majority of nuclei the replication of euchromatin and heterochromatin remain in step, but in a minority class of nuclei heterochromatin has undergone two or three fewer rounds of replication than euchromatin (Fox, 1971). More recently, compelling evidence for differential replication of DNA in the plant genome comes from the work of Pearson, Timmis & Ingle (1974) and Ingle & Timmis (1975). It has been shown that the percentage of satellite DNA, as determined by neutral CsCl equilibrium centrifugation, varies in different tissues of melon and cucumber (Tables 6.5 and 6.6). Satellite DNA formed a higher percentage of the genome in meristematic tissues (seed, root tip) than in mature, differentiated tissues (hypocotyl, leaf, cotyledon, fruit). The buoyant density of the satellite peak, and in some cases of the main band, varied significantly between seed and fruit tissue

TABLE 6.4. *Examples of doubling and non-doubling series in insect tissues* (from Fox, 1969, 1970)

		DNA values relative to 2C class (= 1)				
Species	Tissues	2C	4C	8C	16C	32C
Dermestes vulpinus	Testis wall	1	2.00	4.02	9.06	15.81
	Fat body	1	2.04	3.96	8.08	—
Locusta migratoria	Midgut diverticulum	1	1.68	2.76	5.12	9.26
	Malpighian tubule	1	2.02	3.63	6.60	10.30
	Fat body	1	1.95	3.37	5.75	10.44

TABLE 6.5. *Variation in the amount of satellite DNA present in melon tissue* (from Pearson *et al.*, 1974)

Variety	Satellite (% total DNA) Seed	Root tip	Flower	Hypo-cotyl	Leaf	Cotyl-edon	Fruit
Hero of Lockinge	26.1	25.8	21.5	19.7	15.6	14.4	17.1
Honeydew	29.4	—	—	22.4	—	17.2	—
Musk (b)*	29.5	—	—	—	18.9	17.6	17.1
Spanish Winter (c)	27.7	—	—	21.9	—	17.9	18.5
Spanish Winter (e)	24.3	—	—	—	16.2	19.2	16.7
Spanish Winter (g)	30.3	29.7	—	21.9	—	20.4	15.2
Spanish Winter (h)	24.8	—	—	—	16.3	14.1	20.9
Spanish Winter (j)	34.1	—	—	—	—	20.9	23.9
Spanish Winter (k)	36.3	—	—	—	15.1	19.3	—

* The letter in parentheses identifies an individual melon.

TABLE 6.6. *Distribution and buoyant densities of mainband and satellite DNAs from seed and fruit of melon and cucumber* (from Pearson *et al.*, 1974)

	Satellite DNA (% of total)		Buoyant density (g/cm³)			
			Mainband		Satellite	
Variety	Seed	Fruit	Seed	Fruit	Seed	Fruit
		Melon				
Hero of Lockinge	26.1	17.1	1.6924	1.6938	1.7058	1.7085
Honeydew (e)*	29.2	21.8	1.6919	1.6933	1.7062	1.7076
(f)	24.7	18.4	1.6922	1.6936	1.7056	1.7079
(g)	23.4	18.1	1.6922	1.6926	1.7056	1.7066
(h)	25.5	17.3	1.6928	1.6939	1.7063	1.7082
Ogen (a)	30.6	18.2	1.6927	1.6937	1.7056	1.7088
(c)	24.9	19.1	1.6929	1.6930	1.7063	1.7067
Musk (a)	24.2	16.3	1.6923	1.6927	1.7057	1.7070
(b)	29.5	17.1	1.6926	1.6940	1.7056	1.7071
Spanish Winter (c)	27.7	18.5	1.6917	1.6931	1.7063	1.7075
(e)	24.3	16.7	1.6931	1.6932	1.7068	1.7076
(g)	30.3	15.2	1.6923	1.6928	1.7064	1.7079
(h)	24.8	20.9	1.6918	1.6928	1.7057	1.7070
(j)	34.1	23.9	1.6921	1.6925	1.7056	1.7069
		Cucumber				
Kariha sat. 1	41	34	1.6939	1.6943	1.7011	1.7034
sat. 2					1.7060	1.7084

* The letter in parentheses identifies an individual melon.

(Table 6.6). These differences in the percentage of satellite DNA and in the buoyant densities of both satellite and main band DNAs indicate massive differential replication. The rRNA genes, however, appear to be normally replicated in the seed and fruit of melon. An analysis of the genome on the basis of euchromatin–heterochromatin content showed that the percentage of satellite DNA is correlated with a difference in heterochromatin content of nuclei from cucumber root tips and fruit. Moreover, the fruit was found to contain two populations of polyploid cells, one with under-replication and the other with over-replication of heterochromatin.

Two populations of nuclei, one the result of regular chromosome endoreduplication and the other containing over-replicated (amplified) heterochromatic DNA, have been found in the differentiating parenchyma of *in vitro* cultured protocorms of the orchid *Cymbidium*. In extreme cases the amplified heterochromatin reached the 1024C level in nuclei in which euchromatin was at the 32C level. Among the growth regulators tested, 2,4-dichlorophenoxyacetic acid (2,4-D) appears to be the most effective in inducing DNA amplification in cultured protocorms (Nagl, 1972*b*; Nagl & Rücker, 1972; Nagl, Hendon & Rücker, 1972). The nature of the amplified DNA in *Cymbidium* is not known. The DNA of *Cymbidium* comprises, in addition to main band (density: 1.694 g/cm^3), an A–T-rich satellite with density of 1.682 g/cm^3 (Capesius *et al.*, 1975). Staining of *Cymbidium* nuclei with different fluorochromes gives a very bright fluorescence in centromeric heterochromatin suggesting that it contains the A–T-rich satellite. The fluorescence analyses also show that, in nuclei with amplified heterochromatic DNA, the A–T-rich satellite is not amplified; rather, amplification involves a 'non-A–T-rich satellite' (Schweizer & Nagl, 1976).

The ability of 2,4-D to induce DNA amplification, as first observed in cultured *Cymbidium* protocorms, has also been demonstrated for another *in vitro* system, the *Nicotiana glauca* pith tissue (Parenti *et al.*, 1973). When primary explants of this tissue were cultured aseptically in the presence of 2,4-D a dense satellite DNA (1.722 g/cm^3) appeared at 24 hours; its amount steadily increased up to 72 hours and then decreased, disappearing at 120 hours. The transient appearance of the dense satellite correlated with the cytological events (Nuti-Ronchi, Bennici & Martini, 1973) suggests that DNA amplification may play some role in cell dedifferentiation. Whether or not DNA amplification in *Cymbidium* and *Nicotiana glauca* is a response to the culture conditions or is a normally occurring event (though not necessarily with the same degree of magnitude) remains to be elucidated.

In conclusion, the data discussed in this chapter strongly suggest that differential DNA replication plays a key role in differentiation and development in some biological systems.

7

Gene expression during differentiation and development

We have already seen (Chapter 5) that, in many types of somatic animal and plant cells, obvious deviations from the chromosome complement of the germ line may occur (binucleation and multinucleation, polyploidy, polyteny, polynemy, reducted chromosome number, aneuploidy, etc.). None of these obvious nuclear changes can, however, be causally related to somatic cell differentiation, though probably each one exerts its effect on the functional activity of cells and tissues. In plants, single somatic cells are able to develop into a complete plant regardless of their chromosome complement and, in amphibia, nuclei of differentiated cells can support normal development when transplanted into enucleated eggs (Chapter 9). The conclusion to be drawn from these experiments is that, apart from a few possible exceptions, the differences between somatic cells in an organism cannot be ascribed to differences in the information content of their DNA. Hence, cell differentiation must be viewed in terms of control of gene expression.

In this chapter, we will discuss some aspects of gene expression at the level of RNA synthesis (transcription). We will not consider message translation (protein synthesis) because gene expression at the translational level, we feel, falls outside the scope of this book. For a discussion on gene expression at the translational level during differentiation and development, reference is made to Papaconstantinou (1969), Sirlin (1972) and Gurdon (1974).

Cytoplasm and the initial steps of cell differentiation

It is a common, and probably general, phenomenon that the cytoplasm of animal eggs contains an heterogeneous distribution of materials, which are laid down during oogenesis and usually rearranged at fertilization. The localization of a particular type of cytoplasm (germinal cytoplasm) at the vegetal pole of the egg is responsible for the differentiation of the germ line in the embryo. The differentiation of the germ line is but one example of the close relationship existing between particular regions of a fertilized egg and certain types of differentiation during embryogenesis. A striking case of visualization of this relationship is found in ascidians of the genus *Styela*. Early embryos at the 64-cell stage have five regions of differently colored cytoplasm, each of which is destined to be confined to cells of a certain type such as coelomic mesoderm, notochord, mesoderm etc. In most animal embryos, however, evidence that certain regions of egg or early embryo cytoplasm are associated with certain types of differentiation has been obtained by experimental means, such as the killing or removal of some blastomeres or localized damage to some portions of the cytoplasm in uncleaved eggs (see Davidson, 1968; Gurdon, 1974).

Unequal cytoplasmic distribution most probably occurs in early plant embryos; but, unfortunately, those materials are not amenable to experiments of the type performed on animal eggs and embryos. An unequal cell division with profound developmental consequences is the first division of the zygote in angiosperms. Of the two daughter cells formed, one is destined to produce the embryo, the other is destined to form an unicellular or multicellular, sometimes highly differentiated, structure (suspensor) which has a crucial role in embryogenesis (Alpi, Tognoni & D'Amato, 1975). Another classic example of unequal cytoplasmic division in single plant cells is the division of the microspore in flowering plants; at the telophase of this division, the generative nucleus, which happens to be surrounded by a small amount of cytoplasm due to the asymmetry of the mitotic spindle, is already strikingly different from the vegetative nucleus. As first shown by Sax (1935), high and low temperatures can alter the polarity of the first microspore mitosis, which then occurs with a symmetrical spindle, and the daughter nuclei fail to differentiate. The cytoplasmic inequality of the generative and the vegetative cells in the pollen grain is both quantitative and qualitative, especially with regard to the amount and types of RNA present. The biochemical differences certainly reflect the contrasting functions of the two cells: the function of growth for pollen tube formation in the vegetative cell (abundant cytoplasm, expanded nucleus with a much enlarged nucleolus) and, in the generative cell, the function of further mitosis for sperm formation (scarse cytoplasm, very condensed nucleus in which the nucleolus is greatly reduced in size or has disappeared) (La Cour, 1949; Steffensen, 1966; Jalouzot, 1969).

Unequal cytoplasmic distribution is also of common occurrence in the mitotic divisions which commit to differentiation one of the two daughter cells in somatic tissues of plants and animals. Unequal cell division in plants is quite common (reviews by Bünning, 1957; Stange, 1965). An area of active unequal cell division lies at the margin of apical meristems where the cells which are cut off basipetally differentiate. Well-studied cases of unequal cell division are those of the differentiation to hair-initial and hairless cells in the rhizodermis of monocotyledons and the differentiation of stomatal cells in leaves of monocotyledons (Fig. 7.1). In *Phleum* roots, the epidermal cells which divide asymmetrically show a more extensive polarization of endoplasmic reticulum at their basal pole and the unequal-sized sister cells (hair-initial and hairless cells) show clear differences in the activity of several enzymes.

A very regular sequence of unequal divisions characterizes the development of the grasshopper neuroblasts. The neuroblast always divides to form a large daughter neuroblast and a small ganglion cell. The neuroblast divides again, repeating the sequence. By micro-manipulation it is possible to rotate the mitotic

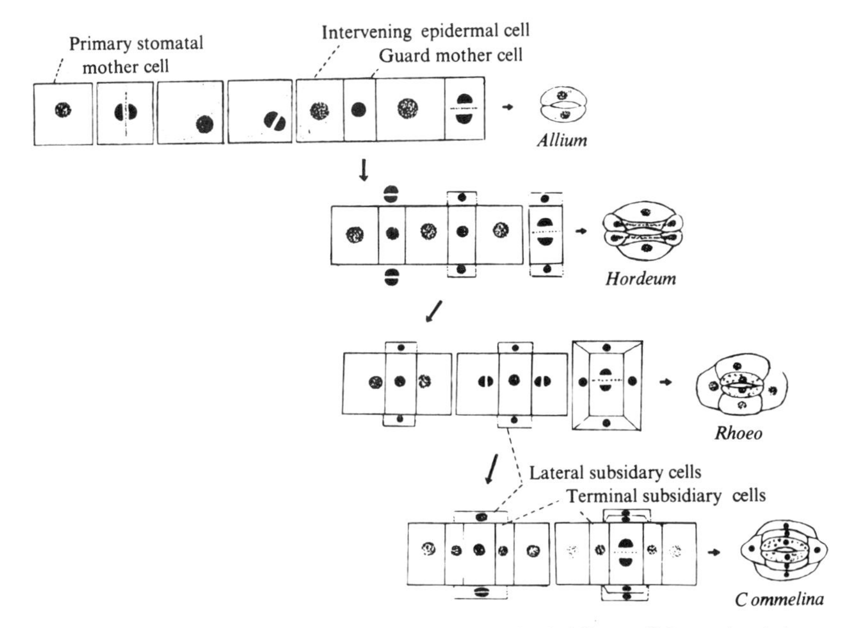

Fig. 7.1. Diagram showing the method of guard cell and subsidiary cell formation in leaves of *Allium*, *Hordeum*, *Rhoeo* and *Commelina*. Note unequal cell divisions (after Stebbins & Jain, 1960).

spindle by 180° so that those chromosomes which would have migrated to the ganglian daughter cell enter the neuroblast daughter cell and vice versa: the result is that the parts of the parent cell cytoplasm destined to form a ganglion or neuroblast cell do so (Carlson, 1952). This experiment again demonstrates cytoplasmic polarization in a cell whose division products have a divergent differentiation.

In summary, an heterogenous distribution in an egg or a somatic cell of materials produced under the direction of the nucleus is the preliminary step for unequal cytoplasmic distribution at mitosis which initiates cell differentiation. Once initiated, the process of cell differentiation proceeds to completion through further nuclear activities. The interplay of nucleus and cytoplasm in the process is so close that cytoplasmic differentiation is both the cause and the consequence of nuclear differentiation, that is, the state of repression or activation of gene functions (D'Amato, 1964*a*; Holliday & Pugh, 1975). Before exploring gene expression in the cell, it is adequate to discuss first the localization of gene function in the genome.

Localization of gene function

Knowledge of the chromosomal localization of particular genes and/or of the number of copies of a particular gene in a genome may be of great help in studies on gene activity during development, growth and differentiation. The development of techniques for RNA–DNA hybridization in standard cytological preparations (Gall & Pardue, 1969; John *et al.*, 1969) has made it possible to observe the chromosomal sites of production of specific classes of RNA (review by Wimber & Steffensen, 1973). So far, the cytological RNA–DNA hybridization technique has not succeeded in localizing single-copy genes, but information on such a class of genes can be gained from other procedures (see below).

18S and 28(25)S rRNA cistrons

The association of a specific chromosomal region, called the nucleolus organizer region (NO), with the formation of nucleoli was first demonstrated by Heitz (1931) and McClintock (1934). It was then observed by Lin (1955) and Longwell & Svihla (1960) that changes in nucleolar mass, nucleolar number and RNA content accompanied changes in the number of nucleolar organizers contained in the karyotypes of maize and wheat. Starting from the knowledge that the nucleolus is implicated in rRNA formation (Perry, 1962), Ritossa & Spiegelman (1965) studied the *in vitro* hybridization between tritiated 18S and 28S RNA and total DNA in stocks of *Drosophila melanogaster* (Urbana stock) carrying 1, 2, 3 or 4 doses of NO (in *D. melanogaster* the NO is located in the proximal portion of the X chromosome and in the short arm of the Y; so both females and males bear two organizers). They demonstrated that the DNA nucleotide sequences complementary to 18S and 28S RNA (ribosomal cistrons) are clustered in the nucleolus organizer and that each of the two types of ribosomal cistrons is present in an equal number of copies (*ca.* 130) in the NOs on the X and Y chromosomes. More recent experiments give a multiplicity of 200–220 rRNA cistrons per NO, in both the X and the Y chromosome (Tartof, 1973; Williamson & Procunier, 1975). Biochemical analyses and *in vitro* rRNA–DNA hybridization in another animal, *Xenopus laevis*, in which normal individuals and individuals homozygous and heterozygous for the deficiency of the NO were studied, clearly demonstrate that the ribosomal cistrons are clustered in or near the nucleolus organizer (Brown & Gurdon, 1964; Wallace & Birnstiel, 1966). In *Xenopus laevis*, the redundancy of the 18S and 28S rRNA genes is about 500 copies per haploid genome (Birnstiel, Chipchase & Speirs, 1971). Other studies have shown that the degree of reiteration of the ribosomal cistrons per haploid genome in eukaryotes ranges from less than a hundred to several thousand copies, depending on the species (Birnstiel *et al.*, 1971). High redundancy is common in plants (Ingle, Timmis & Sinclair, 1975). For example, in maize the number of ribosomal cistrons per diploid nucleus (one nucleolus organizer in the homologous chromosome pair 6) is 17000; this number is doubled in a stock carrying a duplication of the NO (Phillips, Kleese & Wang, 1971).

In the initial studies on *in situ* rRNA–DNA hybridization in somatic tissues, oogonia and oocytes of *Xenopus laevis* (Gall & Pardue, 1969; John *et al.*, 1969), it was shown that the rRNA genes were located, as expected, in association with the nucleoli and that the nuclear cap, a large accumulation of DNA which is seen at meiotic prophase I (Macgregor, 1968), concomitant with the amplification of ribosomal cistrons (Brown & Dawid, 1968; Gall, 1968) was made of rDNA. Pardue *et al.* (1970) then used the tritiated rRNA of *Xenopus* to localize the rDNA in the salivary gland chromosomes of three species of Diptera. In *Drosophila hydei*, rRNA hybridized to the intranucleolar DNA; in *Sciara coprophila*, the ribosomal cistrons were found at the nucleolus organizer on the centromere end of the X chromosome; in *Rhynchosciara hollaenderi*, they were found on one end of the X chromosome and one end of the C chromosome (a site not associated with the formation of a nucleolus). Moreover, both in *Rhynchosciara* and in *Sciara*, rRNA hybridization was detected in many of the micronucleoli scattered throughout the nucleus. Since in salivary gland cells of another species of *Drosophila* (*D. melanogaster*) the rDNA is found within the nucleolus (Wimber & Steffensen, 1973), it is probable that this localization reflects the distribution of intranucleolar DNA, which has been described in the polytene chromosome cells of species of *Drosophila* (Barr & Plaut, 1966; Rodman, 1969). Tritiated 18S and 28S RNA obtained from a kidney cell culture of *Xenopus laevis* has been used for *in situ* hybridization on lampbrush and mitotic chromosomes of the urodele, *Triturus marmoratus*. The rDNA has been localized in the NO which lies in the subterminal position on the long arm of chromosome X (Barsacchi-Pilone *et al.*, 1974).

The specific activity of the rRNA obtained from *in vitro* transcription of satellite DNA (rDNA) in *Xenopus laevis* and *X. mulleri*, and of the tritiated RNA extracted from a human tissue culture was sufficient to localize the rDNA in the mitotic chromosomes of these three organisms (Pardue, 1974; Henderson, Warburton & Atwood, 1972). Recently, cytological gene localization has been facilitated by the use of RNAs with a very high specific activity which are obtained by *in vitro* iodination (^{125}I) of RNA (Prensky, Steffensen & Hughes, 1973). In man, the rDNA is present in the satellite regions of chromosomes 13, 14, 15, 21 and 22 (Henderson *et al.*, 1972). Human chromosomes 13, 14, 21 and 22 are homologous to chromosomes 14, 15, 22 and 23 of the chimpanzee (*Pan troglodytes*); these four chromosomes and chromosome 17 bear rDNA in their satellite region as detected by *in situ* hybridization with ^{125}I-rRNA (Henderson, Warburton & Atwood, 1974).

In plants, cytological localization of the DNA complementary to 25S and 18S rRNA has been studied with the polytene chromosomes of the embryo suspensor of *Phaseolus coccineus* and *P. vulgaris* (Avanzi *et al.*, 1971, 1972; Brady & Clutter, 1972). In *P. coccineus* (Fig. 7.2), the rDNA has been localized in the nucleolus organizing region (satellite, nucleolar constriction and organizer) of the

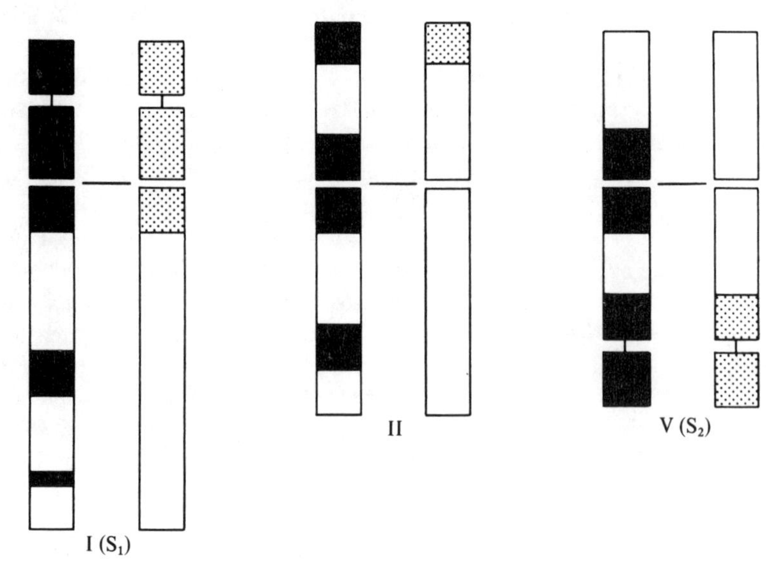

Fig. 7.2. Ideograms of one representative polytene chromosome of the pairs I (satellited chromosomes S_1), II and V (satellited chromosomes S_2) of *Phaseolus coccineus*. On the left: distribution of heterochromatic (black) and euchromatic (blank) segments (redrawn from Nagl, 1967*a*); on the right (dotted): chromosomal segments which show hybridization with tritiated 18S+25S rRNA (after Avanzi *et al.*, 1972).

nucleolar chromosomes I (S_1) and V (S_2) and in two other regions, the heterochromatic segment on the other side of the centromere in chromosome I and a terminal segment of chromosome II (rDNA is also present in the micronucleoli formed at the NO of chromosome I). These results raise the question whether other diploid organisms, either plant or animal, may have such a complex nucleolus organizing system as that of *P. coccineus*.

In situ rRNA–DNA hybridization has shown that the rRNA hybridizes over the nucleolar constriction in both *Xenopus laevis* and *Xenopus mulleri*, but these *in situ* experiments cannot exclude the possibility that some of the ribosomal genes are located within short regions immediately adjacent to the constriction (Pardue, 1974). An attempt to analyze the nucleolus-forming capacity of the nucleolar constriction and immediately adjacent regions was made by irradiating with an argon laser microbeam the nucleolus organizing chromosome of the salamander, *Taricha granulosa* (Fig. 7.3). Irradiation of the regions immediately adjacent to the constriction consistently resulted in the loss of nucleolus organizing capacity, whereas direct irradiation of the constriction did not affect the capacity to organize the nucleolus in 50% of the cases (Berns & Cheng, 1971). These observations, while not excluding the possibility of ribosomal cistrons being present in the nucleolar constriction, clearly demonstrate the importance of a region adjacent to the secondary constriction in nucleolus organization. In this connection, an

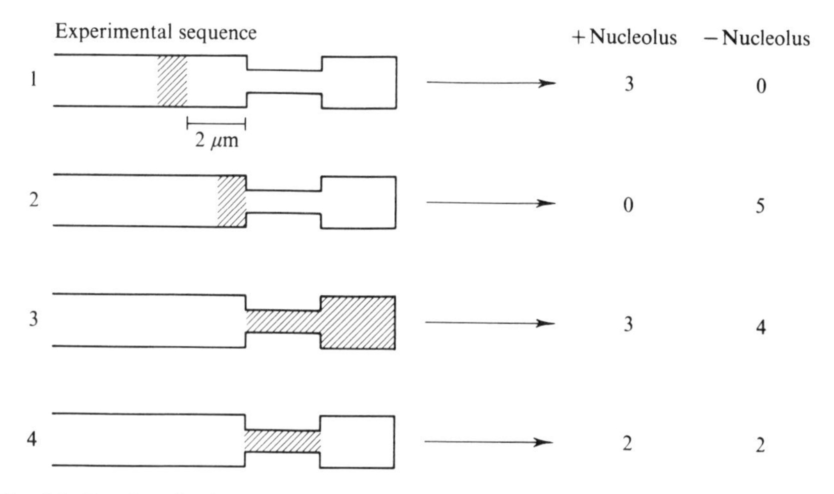

Fig. 7.3. Results of microbean irradiation of nucleolus organizing chromosomes in the salamander *Taricha granulosa*. The portion of the chromosome irradiated is indicated by shading. 1: irradiation of a region up to 2 μm down from the nucleolar constriction; 2: irradiation of the chromosome region immediately adjacent to the constriction; 3: irradiation of both the constriction and the distal satellited tip; 4: irradiation of the constriction alone. Cases of nucleolus formation (+) and no nucleolus formation (−) are indicated (after Berns & Cheng, 1971).

investigation of rRNA–DNA hybridization on the nucleolus organizing chromosomes of *Taricha* would be rewarding. It may be that the condition in *Taricha* is similar to that of maize in which the real nucleolus organizer is a region immediately adjacent to the secondary constriction (McClintock, 1934; Phillips *et al.*, 1971; Ramirez & Sinclair, 1975).

Before describing the cytolocalization of other types of cistrons we may summarize briefly the arrangement of ribosomal cistrons on the chromosomal DNA. Biochemical studies in different eukaryotic organisms have shown that the 18S and 28S RNA species are coded for by separate cistrons occurring in a 1:1 ratio. These cistrons are tandemly arranged (one 28S and one 18S cistron) and each cistron tandem is separated from the next by a segment of DNA ('spacer') which is transcribed only in part (Grierson & Loening, 1972). The rDNA units, each consisting of the ribosomal transcriptional unit together with intervening 'spacer' DNA, are repeated many times and clustered at the nucleolus organizer (Birnstiel *et al.*, 1971; Wensink & Brown, 1971; Scott & Ingle, 1973). The molecular models for rDNA derived from biochemical data were strikingly validated by the electron microscopic studies of Miller & Beatty (1969) on oocyte nucleoli. They showed that the nucleolar DNA strands contain serially repeated units, each unit consisting of a 'matrix segment' covered with RNA fibrils and apparently inactive 'spacer' DNA. For a review on visualization of RNA synthesis on chromosomes, see Miller & Hamkalo (1973).

The 28(25)S and 18S cistrons are transcribed as a unit to give a single RNA molecule, the ribosomal precursor RNA. During the processing of this precursor RNA to give 28(25)S and 18S RNA, some RNA, called the 'non conserved RNA fraction' or the 'excess RNA', is lost (Grierson, Rogers, Sartirana & Loening, 1970; Birnstiel *et al.*, 1971; Miassod, Cecchini, Becerra De Lares & Ricard, 1973).

5S RNA cistrons

It was thought for some time that the genes which code for the small molecular weight RNA of ribosomes (5S) could be located at the NO along with the 18S and 28(25)S RNA cistrons. It has been shown, however, that in such diverse organisms as *Xenopus* (Brown & Weber, 1968), *Drosophila* (Tartof & Perry, 1970) and man (Aloni, Halten & Attardi, 1971), the genes for 5S RNA are not contiguous with the other ribosomal genes. In *Drosophila melanogaster*, the haploid genome contains about 200 copies of the 5S RNA genes. As shown by *in situ* RNA–DNA hybridization, these genes are clustered in only one chromosome region, band 56 F in polytene chromosome 2 (Wimber & Steffensen, 1970, 1973). *In situ* RNA–DNA hybridization has also demonstrated the location of the 5S RNA genes in a single chromosomal site distinct from the NO in maize (Wimber *et al.*, 1974), *Triturus marmoratus* (Barsacchi-Pilone *et al.*, 1974), *Chironomus tentans* (Wieslander, Lambert & Egyhazi, 1975) and *Drosophila hydei* (Alonso & Berendes, 1975).

The single 'locus' distribution of 5S RNA genes found in these species strikingly contrasts with the situation found in other species. In *Xenopus laevis*, the number of copies of the 5S RNA cistrons in the haploid genome is about 24000 and in the 5S DNA the gene sequence that codes for 5S RNA alternates with a 'spacer' region (Brown & Sugimoto, 1973). The 5S DNA bands as a 'satellite' at lower buoyant density in CsCl gradients than the main band DNA; by *in vitro* transcription of this satellite in the presence of ^{3}H-uridine it is possible to obtain complementary RNA (cRNA) with specific activity higher than that of 5S RNA extracted from tissue culture cells of *Xenopus* fed with the same radioactive precursor. *In situ* RNA–DNA hydridization using these two types of RNA on *Xenopus* mitotic chromosomes has shown that 5S DNA is localized at or near the telomere region of the long arm of many, if not all, of the 18 chromosomes (Pardue, Brown & Birnstiel, 1973). In the chironomid fly, *Glyptotendipes barbipes*, salivary gland chromosomes were hybridized with ^{125}I-5S RNA. It was found that 5S DNA was present in one main locus (C-2) in chromosome IIIR and in several additional sites that appeared after a longer exposure to autoradiographic emulsion, probably because of the lower level of gene reiteration in these loci. None of the loci that hybridized with 5S RNA were labeled in *in situ* hybridization experiments with either 4S or 18S+28S RNA preparations (Wen, Leon & Hague, 1974). In man, the major 5S RNA locus is placed in the segment q32 to q44 on the long arm of chromosome 1, but 5S genes are present on two other chromosomes, probably chromosomes 9 and 16 (Steffensen & Duffey, 1974).

4S RNA (tRNA) cistrons

Ritossa, Atwood & Spiegelman (1966*a*) showed that the number of transfer RNA (4S RNA) cistrons in the haploid genome of *Drosophila melanogaster* is about 750, which leads to a 13-fold redundancy for each of the approximately 60 tRNA species. If the 13 copies of each tDNA sequence are clustered, they ought to be detectable on the *Drosophila* polytene chromosomes when their DNA is hybridized with ^{3}H-tRNA. Indeed, using this procedure, Steffensen & Wimber (1971) found about 70 sites of high annealing when scoring the X-chromosome and three-quarters of chromosome 2. As pointed out by Wimber & Steffensen (1973), the number of sites on the chromosomes that anneal to tRNA would be reduced if tRNA not contaminated with fragments of messenger RNA were used and the appropriate competition-type controls were introduced. That the results of *in situ* hybridization for tRNA and for other classes of RNA must be taken with some caution has been stressed by Grigliatti *et al.* (1974).

The redundancy of the tRNA cistrons in *Drosophila* is much lower than in *Xenopus laevis*, in which a haploid genome redundancy of 7800 to 9600 cistrons for total tRNA has been found; this means an average of 180 to 220 genes for each of the 43 to 44 families of tRNA sequences detected within unfractionated tRNA. In the tDNA of *Xenopus*, the sequences coding for one mature tRNA molecule alternate with 'spacer' DNA. Such an arrangement is serially repeated to form extensive isocoding tRNA gene clusters (Clarkson, Birnstiel & Serra, 1973). The chromosomal localization of these gene clusters is not yet known.

Other cistrons

In the sea urchin, *Psammechinus miliaris*, rapid nuclear and cellular divisions take place prior to the blastula stage, and a newly synthesized 9S RNA appears on small polysomes in the cytoplasm. This 9S RNA is the messenger for histone synthesis; the cistrons that code for this message are reiterated and clustered. There is some conservation of the nucleotide sequences of the histone cistrons because sea urchin 9S mRNA hybridizes to the DNA of different eukaryotic species, both animals and plants (Kedes & Birnstiel, 1971). Purified 9S mRNA from *P. miliaris* blastulae hybridizes to bands 39D and 39E and the entire chromosome region between them in polytene chromosomes of *Drosophila melanogaster*. Since the 9S message can be resolved on gels into five components, probably corresponding to the five main histone fractions, it seems likely that the expected five histone loci are all clustered in this chromosome region in *Drosophila* (Birnstiel, Weinberg & Pardue, 1973).

An interesting case of a 75S mRNA in the salivary gland cells of *Chironomus tentans* has been described. In the polytene chromosomes of this, as well as many other animal and plant species, RNA synthesis occurs at the level of regional swellings, the RNA puffs. Puffs are regarded as the morphological expressions of differential genome transcriptions (Beermann, 1962). A very large puff, the Balbiani ring (BR)2 on chromosome IV of *C. tentans* in the fourth instar larva, is filled with RNA containing large (450–500 Å) granules which are also found in the nucleoplasm and the cytoplasm. The BR2 can be isolated by microdissection and its RNA content analyzed separately from chromosomal and nuclear sap RNA. It is then shown that these granules are apparently formed in the BR2 from nascent RNA and preformed protein and are transferred to the cytoplasm through the pores in the nuclear envelope. A 75S RNA is present in the puff and in the nucleoplasm and cytoplasm, it is stable and has been shown to hybridize specifically with BR2 though some hybridization also occurs at BR1 on the same chromosome (IV). The hybridization is rapid enough to indicate that the DNA complementary to 75S RNA contains repeated nucleotide sequences (Lambert, 1972; Lambert *et al.*, 1972; Lambert & Edström, 1974; Edström & Tangway, 1974).

Although there is hope that cistrons present in only one or a few copies in the haploid genome may be localized in the future in mitotic chromosomes by RNA–DNA hybridization (Wimber & Steffensen, 1973), detection of 'single-copy' genes in the eukaryotic genome is now made possible by molecular hybridization experiments between DNA and a purified mRNA, as performed, for example, in the silkworm. In the silk glands of larvae at the fifth instar there is a large amount of the mRNA for silk fibroin. This fibroin messenger is 10000 nucleotides long (2.4×10^6 daltons) and hybridizes to purified DNA with a saturation level (0.002%), corresponding to 1, 2 or at most 3 genes per haploid genome (Suzuki, Gage & Brown, 1972).

Analysis of transcription at the chromosomal level

The activity of a gene is expressed by transcription, that is, the synthesis of an RNA molecule whose nucleotide sequence is complementary to the nucleotide sequence of one of the two DNA strands making that gene. RNA synthesis is most active in interphase nuclei, especially in some differentiated cells; at the microscopical level, it is generally studied by specific cytochemical methods and autoradiography following feeding with a radioactive RNA precursor. Transcription is strikingly visualized at the level of 'giant' chromosomes, that is, the lampbrush chromosomes (bivalents at the diplotene stage) of oocytes in urodeles, and the polytene chromosomes of Diptera and a few other organisms. In lampbrush chromosomes, a large number of presumably diverse genes are activated at a time; the many active chromomeres extend in form of loops whose DNA seemingly operates as one transcription unit. The loop is covered by a matrix made of ribonucleoproteins whose amount progressively increases from one end to the other of the loop, thus giving the loop an asymmetric ('polarized') appearance in the optical microscope (Callan & Lloyd, 1960). The electron microscope work of Miller *et al.* (1970) has beautifully visualized the attachment of RNA polymerase to the DNA axis of the loop and the unidirectionally increasing length of the

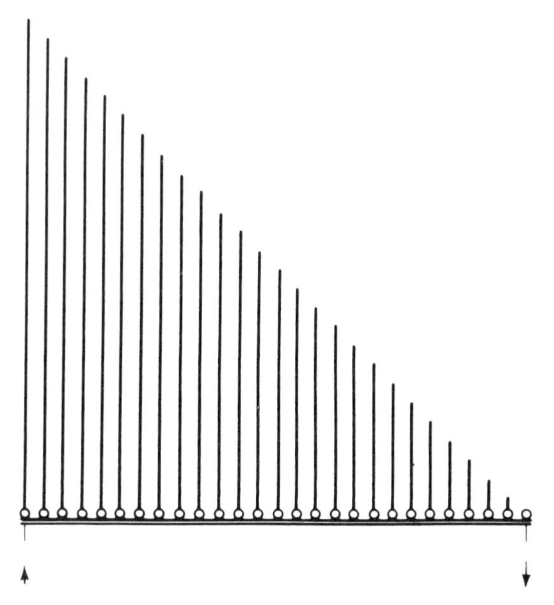

Fig. 7.4. Model of transcription derived from a lampbrush loop, entire DNA (thick line) constituting one unit of transcription. RNA is shown by a thin line and RNA polymerase by empty circles. Initiation and termination sites are shown by a downward and an upward arrow respectively (after Pelling, 1972).

nucleoprotein fibrils making the matrix (Fig. 7.4). In contrast with lampbrush chromosomes, polytene chromosomes have only a selected portion of genetic units which are most active at any given time. This is a favorable condition for studies on gene action in development and differentiation.

It is now accepted that the clear alternation of intensely staining bands (or chromomeres) and lightly stained interbands (or interchromomeres), as seen in dipteran polytene chromosomes, is a magnified view of the chromomere–interchromomere structure of mitotic chromosomes. On the basis of the rather extensive cytogenetic data concerning *Drosophila*, it can be stated that: (i) the number of 'essential' functional genetic units in a given chromosome segment is equal to that of the bands in that segment; (ii) only one essential genetic function can be associated with each band or band–interband complex (a band plus one of its adjacent interbands); (iii) the essential portion of each band–interband complex is restricted to a small portion at the band edge or, alternatively, only occupies the interband region (review by Beermann, 1972). A very short feeding of polytene chromosome cells with ^{3}H-uridine shows that the DNA of a great number of chromosome loci transcribes RNA, although incorporation of the radioactive RNA precursor is most active in puffed regions (references in Berendes, 1973). Puffing results from the uncoiling of the deoxyribonucleoprotein fiber; in a particular category of puffs, the Balbiani rings of the Chironomidae, the uncoiling is most conspicuous and the DNA is lifted from the chromosomal axis towards the surface of two hemispherical structures which may protrude tens of microns from the chromosomal axis. It is estimated that in *Chironomus* no more than 15% of the bands are active (puffed) at any given time in a polytene chromosome cell and in *Drosophila hydei* about 5% of all puffs are tissue-specific. In general, the amount of RNA synthesized in a given puff is closely related to its size, most probably due to the concentration of RNA polymerase molecules along the DNA fiber engaged in transcription (Pelling, 1972).

Since the now classical observations of Beermann (1952) on *Chironomus tentans*, Mechelke (1953) on *Acricotopus lucidus* and Breuer & Pavan (1955) on *Rhynchosciara angelae*, detailed investigations on puffing patterns in a number of species of Diptera have been published. Recently, the tissue and developmental specificity of puffing patterns was extensively reviewed by Ashburner (1970). It is now clear that: (i) different cell types (salivary glands, Malpighian tubules, etc.) show remarkable differences in their puff patterns; (ii) in a given cell type, temporal changes in the puff pattern occur, specific puffs appearing during certain periods of larval development; (iii) specific cell products (e.g. salivary proteins) can be demonstrated as being dependent on the presence of a particular puff. In 1963, Beermann demonstrated that in *Chironomus pallidivittatus* a relationship exists between a Balbiani ring located near the centromere of chromosome IV and the presence of granules in four special cells at the base of the salivary glands. In a related species, *Chironomus tentans*, neither the Balbiani ring nor the granules are present in the special salivary gland cells. Hybrids between these two species show that granule production behaves as a single gene, whose pattern of inheritance closely correlates with the location of the additional Balbiani ring on the IV chromosome (Fig. 7.5). In extending Beermann's observations, Grossbach (1969) showed that the salivary secretion of *C. pallidivittatus* and *C. tentans* differs in only one polypeptide, whose presence is related to the presence of the additional Balbiani ring in the IV chromosome of *C. pallidivittatus*. This special polypeptide is not synthesized in *C. tentans*. As discussed in the preceding section of this chapter, two of the three Balbiani rings that occur in the IV chromosome of *C. tentans*, BR1 and BR2, are involved in the production of RNA-containing granules in the salivary gland cells. BR2 has been shown to produce in large amounts a stable 75S mRNA. The question whether 75S RNA is polycistronic or monocistronic cannot yet be answered; the available data may indicate that the 75S RNA molecule contains more than one copy of the same cistron or, alternatively, two or more structurally related cistrons (Daneholt & Hosick, 1973). At any rate, it seems most probable that it contains the message for one or a few of the salivary gland polypeptides isolated by Grossbach (1969).

It has been pointed out by Ribbert (1972) that the functional metabolism of salivary gland cells is only vaguely related to developmental events during metamorphosis, though a close correlation in time between drastic changes in chromosomal activities and the molting process can be demonstrated (see below). In any

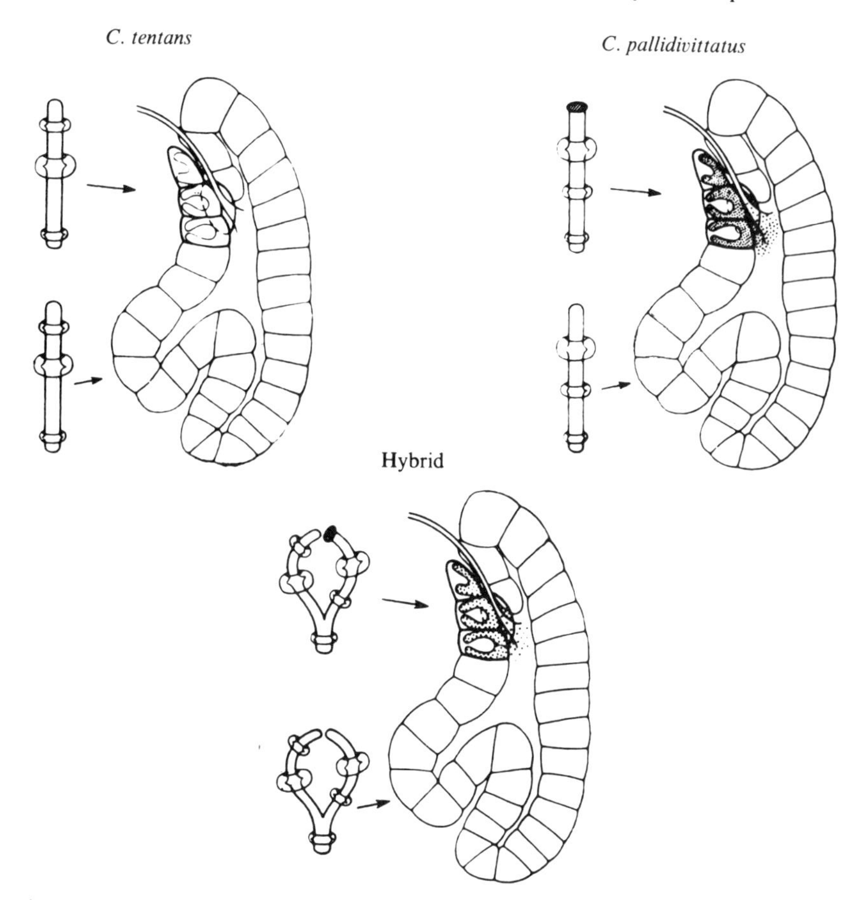

Fig. 7.5. The pattern of Balbiani rings in the two functionally different portions of the salivary gland in *Chironomus tentans*, *C. pallidivittatus* and their hybrid (after Beermann, 1963).

case, the salivary gland system is much more complex than some pupal polytene chromosome systems in which a fixed sequence of defined stages of cell differentiation during pupal development is demonstrated. The best-studied cases concern the morphogenesis of the bristles in *Calliphora erythrocephala* and of the footpads (pulvilli) in *Sarcophaga bullata* (references in Ribbert, 1972). The bristle-forming apparatus of *Calliphora* involves only two epidermal cells, of which the larger one produces the long tapering hair (trichogen cell) and the smaller (tormogen cell) produces a circular chitinous socket around the base of the bristle. During pupal development, these cells grow rapidly by chromosome polytenization to reach a nuclear volume some 3000 times that of the neighbouring diploid epidermal cell. The period of most active chromosome endoreduplication coincides with the

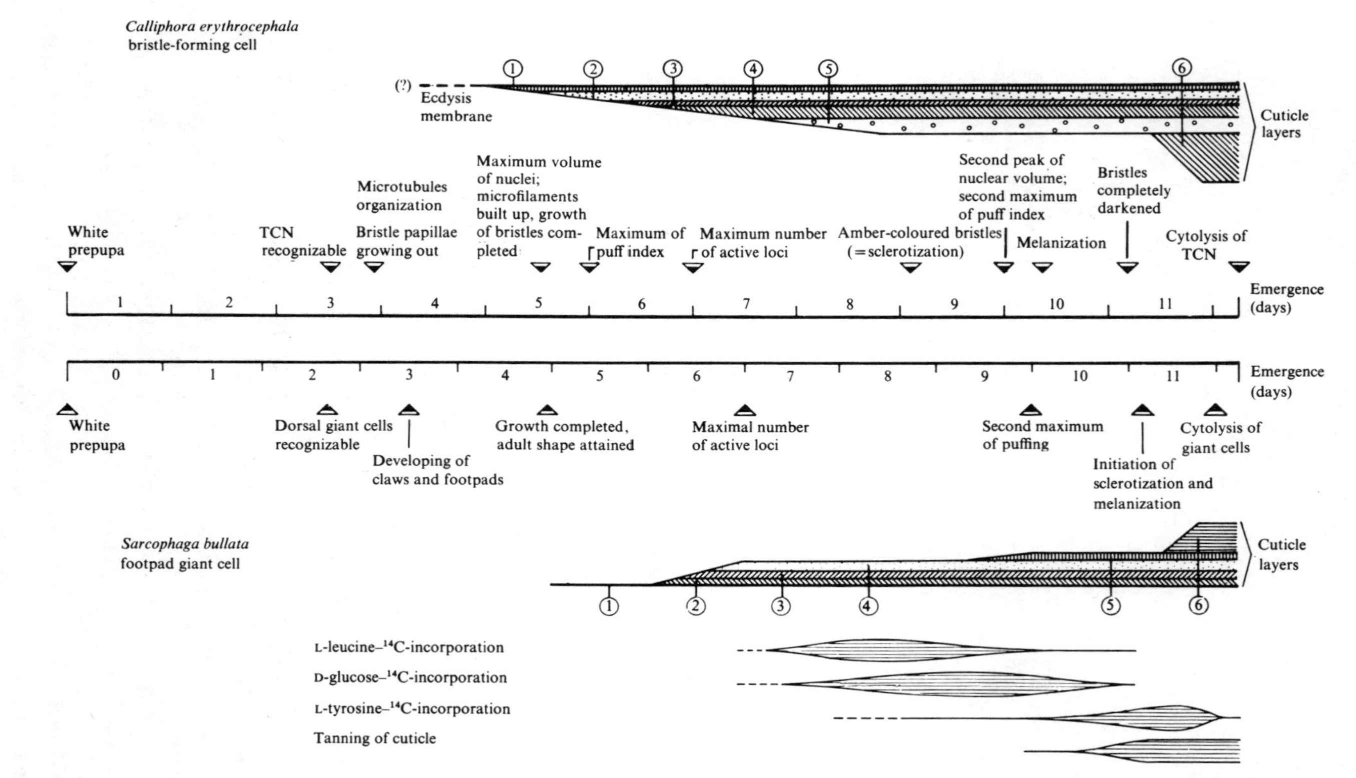

Fig. 7.6. Timetable of the morphogenetic sequences during pupal development in bristle-forming cells of *Calliphora erythrocephala* and in footpad cells of *Sarcophaga bullata* (after Ribbert, 1972).

growth of the bristle. At the end of the fourth day of pupal development, when the bristles have attained their definitive length, cuticle deposition begins. At least six cuticle layers are deposited in a strictly defined temporal sequence during the following days. At emergence, the cell content begins to degenerate and cytolysis is completed within the first two or three days of adult life. In *Sarcophaga bullata*, each footpad is produced by two epidermal cells, each of which is responsible for secreting the adult cuticle of each half footpad. The growth of the two giant cells that form the footpad is completed at the fifth day in pupal development, when the footpad attains its adult shape. Thereafter, cuticle secretion starts. The morphogenetic and biochemical capacities of the giant footpad cells in *Sarcophaga* closely resemble the processes of bristle formation (Fig. 7.6). The puffing patterns in the polytene chromosomes of trichogen cells in *Calliphora* and in giant footpad cells in *Sarcophaga* have been studied respectively by Ribbert (1967, 1972), Whitten (1969) and Bultmann & Clever (1969). In both systems, the strictly sequential changes in the puffing pattern run strictly parallel to the sequence of morphogenetic and biochemical events during pupal development. For further discussion, see Ribbert (1972).

Other examples of control of gene transcription

The studies on dipteran polytene chromosomes discussed in the preceding section illustrate one of the strongest cases in favor of transcriptional control in animals. It is the aim of this section to consider other cases of transcriptional control in animals and plants.

Gurdon (1974) has commented on the many investigations which show transcriptional control during early development in different animal species. Among these, the best-studied case is *Xenopus laevis*. Extensive biochemical studies, aided by the use of radioactive tracers, have established the relative rates of nucleic acid synthesis during *Xenopus* development from early oocyte to feeding tadpole (Fig. 7.7). It has been demonstrated that 28S and 18S (but not 5S) rRNA are always synthesized at the same relative rate (coordinately) during development and that 4S RNA, mRNA, 5S rRNA and 28S+18S rRNA are synthesized independently. As seen from Fig. 7.7, 5S RNA is produced in great excess of 28S+18S rRNA in the early stages of oogenesis. It has also been found by Ford & Southern (1973) that the 5S RNA synthesized in early oogenesis and the 5S RNA synthesized in development are coded for by different genes.

So far, due to methodological limitations, it has only been possible to measure collectively the synthesis of the several mRNAs present at different times of embryonic development in animals; so nothing is known on the course of synthesis of any particular mRNA during development. The situation is quite different in some specialized tissues which synthesize very large amounts of the mRNA coding for their cell-specific protein. For example, in chicken there is only one gene coding for albumin per haploid genome and the ovalbumin genes are present in equal

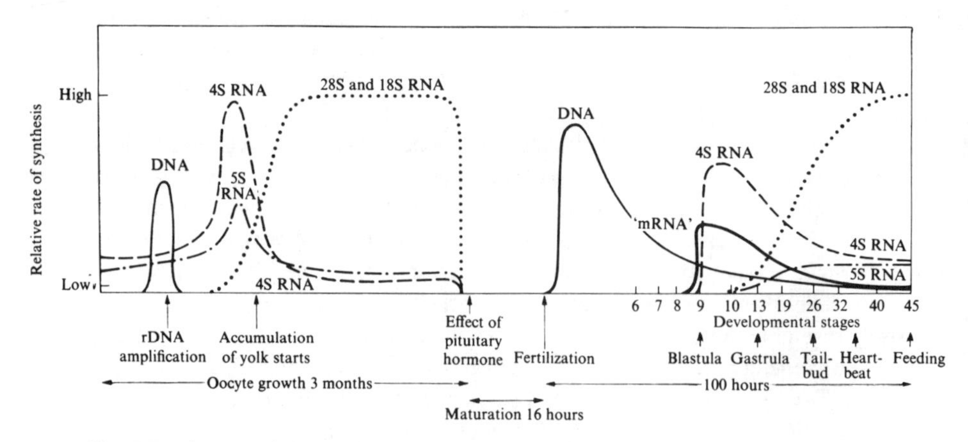

Fig. 7.7. Diagram of the relative rates of nucleic acid synthesis during amphibian development. Rates of synthesis are indicated very approximately on a low to high scale. The diagram shows that different classes of RNA are synthesized at non-coordinated rates during oogenesis and development (after Gurdon, 1974).

number in the liver and in the oviduct tissue – the tissue in which the genes are fully expressed (Sullivan *et al.*, 1973). In the oviduct cells, a close relationship exists between the amount of ovalbumin mRNA and the rate of ovalbumin synthesis (Schimke, Rhoads, Palacios & Sullivan, 1973). A similar condition is observed for the fibroin mRNA in the silkworm (1–3 gene copies per haploid genome) in which the amount of fibroin DNA is the same in the posterior gland, which is very active in fibroin synthesis, and in middle silk gland and other larval tissues, where fibroin is not detected (Suzuki *et al.*, 1972). Since it is generally assumed that genes coding for proteins are represented in one or a few copies, it might be speculated that the type of transcriptional control observed for the ovalbumin and fibroin genes (production of large amounts of mRNA without resorting to gene amplification) may apply to other structural genes in specialized cells.

Another clear case of transcriptional control is dosage compensation, observed for X-linked genes in species with XX/XY sex chromosomes. For instance, dosage compensation is easily observed in *Drosophila* polytene chromosomes (references in Ashburner, 1972). In *D. melanogaster* and *D. simulans* following puparium formation, the puffs at 2B5–6, 2B13–17 and 2EF are active for a longer time in male than in female larvae. Autoradiographic analyses following feeding with RNA precursors suggest that, in salivary glands at least, dosage compensation is due to hyperactivity of the single X chromosome in the male. Dosage compensation is also manifested at the level of rRNA synthesis. Brown & Gurdon (1964) were the first to show that *Xenopus* tadpoles of the wild type (2 nucleolus organizers) and tadpoles heterozygous for the anucleolate mutation (this mutation involves

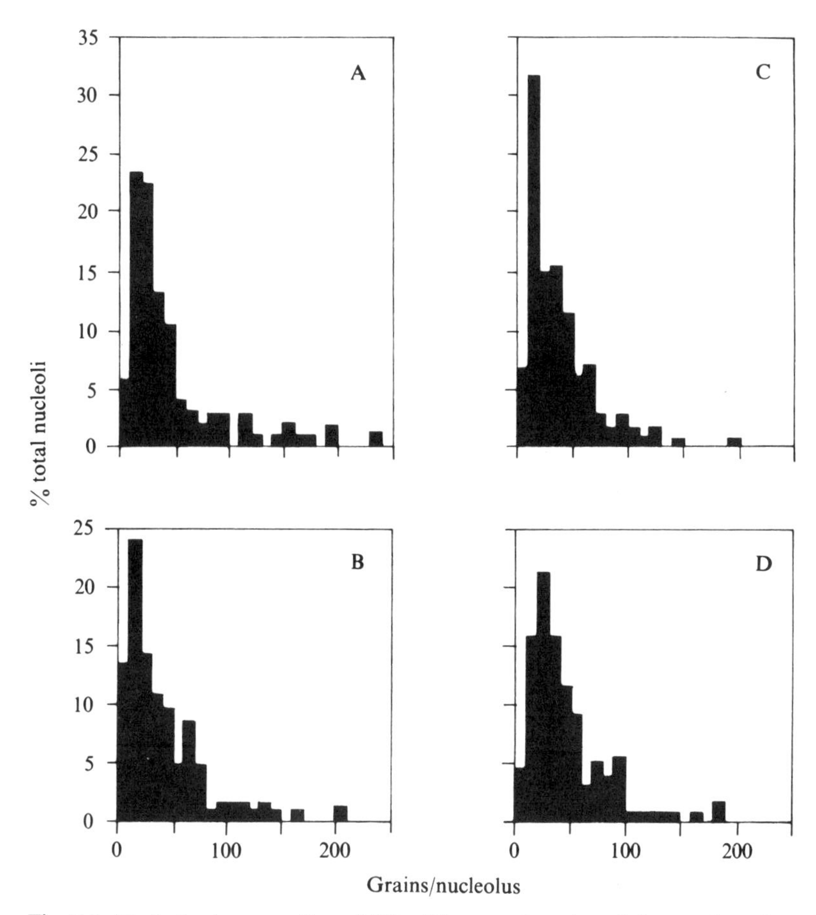

Fig. 7.8. Nucleolar incorporation of ^{3}H-uridine as a function of the number of nucleolus organizers in XX *Drosophila melanogaster* larvae bearing one (A), two (B), three (C) or four (D) organizers (after Krider & Plaut, 1972).

the total deletion of the ribosomal cistrons) synthesized rRNA at the same rate, although the heterozygotes had half the number of rRNA cistrons per diploid nucleus.

Evidence for regulation of rRNA synthesis is provided by studies on *Drosophila melanogaster.* In this species, the number of rRNA cistrons is not constant; individuals homozygous for the *bobbed* mutation, which is a partial deletion of the nucleolus organizing region (Ritossa *et al.*, 1966*b*), contain in their cells less than 130 rRNA cistrons, whereas phenotypically wild-type individuals have at least 130 cistrons (up to 600 in some genotypes) (Ritossa & Scala, 1969). It has been shown that the rate of rRNA synthesis is reduced in flies exhibiting the *bobbed* phenotype (these flies lay eggs at a reduced rate but of normal RNA content).

TABLE 7.1. *List of interspecific* Crepis *hybrids in which 'differential amphiplasty' ('disappearance' of the chromosome satellite) is known to occur* (from Navashin, 1934)

Hybrid	The satellite 'disappears' from the chromosome of:
C. capillaris × *C. alpina*	*C. alpina*
× *C. dioscoridis*	*C. dioscoridis*
× *C. neglecta*	*C. neglecta*
× *C. parviflora*	*C. capillaris*
× *C. tectorum*	*C. tectorum*
C. dioscoridis × *C. pulchra*	*C. pulchra*
C. leontodontoides × *C. capillaris*	*C. leontodontoides*
× *C. marschallii*	*C. leontodontoides*
× *C. parviflora*	*C. leontodontoides*
× *C. tectorum*	*C. leontodontoides*
C. palestina × *C. pulchra*	*C. pulchra*
C. parviflora × *C. tectorum*	*C. tectorum*
C. setosa × *C. tectorum*	*C. tectorum*

Approximately the same rate of rRNA synthesis is found instead in phenotypically wild flies independent of their rDNA contents (the number of rRNA cistrons is between 180 and 600). Females with such rDNA contents lay eggs of the same RNA content and at the same rate. Hence, not all rDNA in every cell is necessarily used at the same time (Mohan & Ritossa, 1970). Regulation of rRNA synthesis in *D. melanogaster* is also clearly seen in salivary gland cells. The rate of nucleolar RNA synthesis as determined by an analysis of ^{3}H-uridine incorporation has been found to be similar in female larvae bearing 1, 2, 3 or 4 nucleolus organizer regions (Fig. 7.8).

A classical case of regulation of nucleolar activity is 'differential amphiplasty', which was discovered and closely analyzed in the genus *Crepis* by Navashin (see Navashin, 1934). Many species of *Crepis* can be crossed to produce interspecific hybrids; in particular hybrid combinations, there invariably occurs the 'disappearance' of the satellite (probably due to condensation of the secondary constriction) in the nucleolus organizing chromosome of one definite species of the two entering the cross (Table 7.1). Differential amphiplasty is wholly independent of the direction in which the cross is made; this clearly excludes cytoplasmic effects and calls for the action of chromosomal genes. The process of differential amphiplasty, also called 'nucleolar suppression' (Keep, 1971), is reversible: the chromosomes suffering nucleolar suppression under the hybrid condition recover their normal shape (appearance of satellite) as soon as the chromosomes of the other species are eliminated. For instance, a pure *C. neglecta* plant, which segregated from a *C. neglecta-capillaris* hybrid, possessed normal satellited chromosomes, whereas in the hybrid plant no *C. neglecta* satellited chromosome could be seen (Navashin, 1934). Since nucleolar suppression is now known for many interspecific plant hybrids (Keep, 1971), there are ample opportunities for investigating the problem of nucleolar suppression with molecular biological techniques.

8

The regulation of gene activity

The cases of gene expression during differentiation and development discussed in the preceding chapter are best explained by assuming that (i) a given differentiated cell state is associated with the activity of a given set of genes, together with the total inactivity of those sets of genes which are associated with the differentiation of other cell types and (ii) the ordered sequence of gene activities during development reflects a temporal sequence of activations and repressions of genes. It is generally agreed that only a fraction of the total genetic information of a cell is expressed (that is, transcribes RNA) at any one time; regulatory mechanisms must, therefore, operate in activating and inactivating genes, or sets of genes, for transcription as a cell's needs change. In the bacterium *Escherichia coli*, the transcription of specific genes into RNA has been shown to be regulated by the binding and release of specific regulatory proteins at appropriate sites on the DNA (review by Lewin, 1974*a*). In animals and plants the genetic material is far more complex than in bacteria; the DNA is associated with proteins and a small amount of RNA to form chromatin, the essential component of chromosomes. So far, no specific regulatory proteins have been identified in eukaryotes, but several lines of evidence implicate chromosomal proteins as being the regulatory proteins. Models for the regulation of gene activity in eukaryotic cells have been proposed by several authors: e.g. Britten & Davidson (1969), Georgiev (1969*a*), Crick (1971), Paul (1972), Guillé & Quetier (1973), Holliday & Pugh (1975). In the following pages, the several aspects of repression and activation of genes in multicellular organisms will be discussed.

Heterochromatin

By studying the course of nuclear divisions in some plants, Heitz (1928, 1929, 1933) demonstrated that some chromosome segments or whole chromosomes do not decondense at telophase and remain condensed during interphase forming the 'chromocenters' of previous authors. Heitz called the chromosome material of such chromosomes or chromosome segments 'heterochromatin' and the chromosome material forming chromosomes or chromosome segments which undergo a typical cycle of condensation and decondensation during the mitotic cycle 'euchromatin'. Heitz believed heterochromatin to be genetically inert, a view strengthened by several observations in *Drosophila*; e.g. the demonstration

(Stern, 1929) that the Y chromosome (entirely heterochromatic) is essential for sperm motility, though without any effect on viability and sex phenotype of the male as is clearly seen in the XO flies. The original concept of total genetic inertness of heterochromatin was tempered as a result of further studies, but it can now be safely stated that major genes are absent or exceptional in heterochromatin. Controversial statements on the gene content and genetic activity of heterochromatin were partly due to the lack of an appropriate terminology. The confusion was clarified when Brown (1966) proposed the terms 'constitutive' and 'facultative' heterochromatin for two quite distinct types of condensed chromatin. Constitutive heterochromatin comprises segments of densely-staining material which occupy the same position in both members of a homologous chromosome pair (review by Yunis & Yasmineh, 1972); facultative heterochromatin is produced by inactivation (heterochromatization) of one of the two homologous chromosomes, as in the inactivation of one X chromosome in female mammals, or of the whole paternal chromosome set, as in the male mealy bug. In the last decade the studies on mammals and mealy bugs, together with studies employing such techniques as *in situ* nucleic acid hybridization, detection of bands in chromosomes by the use of specific staining and fluorescence methods, and the analysis of chromatin structure and activity, have much increased our knowledge of the nature of heterochromatin. For recent reviews on heterochromatin the reader is referred to Yunis & Yasmineh (1972), Shah, Lakhotia & Rao (1973), Vosa (1973) and Frenster (1974).

Constitutive heterochromatin

Constitutive heterochromatin, also called 'structural' heterochromatin, is believed to be permanently condensed chromatin devoid of structural genes and is believed to be mostly unable to support transcription *in vivo* (Frenster, 1974). Some rapidly renaturing DNA sequences in salivary gland nuclei of *Rhynchosciara angelae* have been found to synthesize a class of RNA which is limited to the nucleus not being transferred into the cytoplasm; there are indications that these sequences are localized in constitutive heterochromatin (Balsamo *et al.*, 1973*a*). Typically, constitutive heterochromatin is located near centromeres (pericentric heterochromatin) and nucleolus organizer regions (nucleolar heterochromatin); sometimes, it is also located as small segments along the length of chromosomes and/or at the chromosomes ends (telomeres). In some cases, whole chromosomes or parts of chromosomes are made of constitutive heterochromatin. *In situ* RNA–DNA and DNA–DNA hybridization experiments (see Chapter 2) have shown that constitutive heterochromatin contains highly repetitive DNA sequences. In a number of organisms, highly repetitive DNAs can be separated from the main DNA by ultracentrifugation (satellite DNAs). Hybridization *in situ* has shown that constitutive heterochromatin contains light (A–T rich) satellites in some species (Jones, 1970; Pardue & Gall, 1970; Eckardt & Gall, 1971; Gall *et al.*, 1971; Pardue, 1974) and heavy (G–C rich) satellites in other species, e.g. the centromeric heterochromatin in the salamander *Plethodon cinereus* (Macgregor & Kezer, 1971) and the heterochromatin in human chromosome 9 (Hsu, Arrighi & Saunders, 1972; Hearst *et al.*, 1974; Jones, Prosser, Corneo & Ginelli, 1973). Some highly repetitive DNAs have approximately the same buoyant density as main DNA. In these cases, they can be detected by centrifugation in Hg^{+} or Ag^{+}-Cs_2SO_4 (Corneo, Ginelli & Polli, 1970), by denaturation of DNA and separation of the rapidly renaturing material on hydroxyapatite (Laird, 1971) or by centrifugation of denatured and renatured DNA (Corneo *et al.*, 1970). Radioactive complementary RNA transcribed from rapidly reassociating fractions of denatured DNA has been used in hybridization experiments *in situ* on a number of animal species (Rae, 1970; Arrighi, Hsu, Saunders & Saunders, 1970; Botchan *et al.*, 1971; Hsu *et al.*, 1972; Hearst *et al.*, 1974). The amount of highly repetitive DNA which is separated from main DNA by the denaturation–renaturation method shows great variation among species; e.g. it is 2.5% of the genome in mouse and 22% in human (Hearst *et al.*, 1974). When highly repetitive DNA sequences are obtained, by further purification, as fairly pure satellite fraction, they may show specific localization as in the case of *Microtus agrestis* where the sex chromosomes are highly enriched for the satellite fraction (Arrighi *et al.*, 1970).

That constitutive heterochromatin is enriched for repetitive DNA sequences is also demonstrated by many comparative analyses of DNA isolated from the heterochromatic and the euchromatic fraction of nuclei in many species. On the basis of their observations on six species of mammals and two species of birds and of other observations, Comings & Mattoccia (1972) suggested that several different types of DNA may be localized to the constitutive heterochromatin of eukaryotes. They may be classified as follows:

(i) Repetitive satellite DNA: (a) A–T rich; (b) G–C rich; (c) same density as main band DNA;

(ii) Repetitive main band DNA;

(iii) Non-repetitive DNA: (a) G–C rich heavy shoulder DNA; (b) A–T rich main band DNA.

In summary, the sequences in DNA of constitutive heterochromatin appear to be very heterogeneous: very rapid renaturing sequences space sequences of far less redundant DNA, some of which are probably transcribed (e.g. Kram, Botchan & Hearst, 1972; Hearst *et al.*, 1974). *In situ* RNA–DNA hybridization in tissue sections prepared for electron microscopy shows that in mouse the redundant rDNA, which is normally transcribed, and the satellite DNA sequences are probably more intimately linked than hitherto assumed (Jacob, Gillies, Macleod & Jones, 1974).

In a number of species, mostly plants, constitutive heterochromatin can be detected by cold-treatment. When plants of *Trillium*, *Fritillaria* etc. are grown at low temperature, chromosomes show segments of reduced diameter and stainability (references in Vosa, 1973). This phenomenon, discovered by Darlington & La Cour (1938, 1940), was called by them 'nucleic acid starvation' and was attributed to inhibition of DNA synthesis in these chromosome segments. It has, however, been shown that cold-treated chromosomes contain the same amount of DNA as control chromosomes, supporting the concept that cold-induced segments are due to reduced spiralization (Woodard & Swift, 1964). Caspersson *et al.* (1969) and Vosa (1970) have studied the quinacrine fluorescence of cold-sensitive segments and shown that these segments may show enchanced or reduced fluorescence according to the species. It has also been shown that cold-sensitive segments can be detected by Giemsa staining after a denaturation-reannealing reaction (Vosa & Marchi, 1972). Thus, segments of constitutive heterochromatin can be detected by three methods (cold-treatment, quinacrine fluorescence, Giemsa staining) but quinacrine fluorescence is clearly specific. It has been shown that quinacrine binds to A–T rich sections of DNA (Weisblum & de Haseth, 1972)* and that the quinacrine-bright areas of the chromosomes of *Samoaia leonensis* can be labeled by ^{3}H-thymidine but not by ^{3}H-deoxycytidine, whereas the remaining chromosome areas are labeled by both precursors (Ellison & Barr, 1972). An elegant method based on immunofluorescence has recently allowed the localization of different bases in chromosomes. In man, the quinacrine fluorescent bands are identical with those shown by anti-A immunoglobulin (Dev *et al.*, 1972); conversely, anti-G immunoglobulin gives a reverse banding pattern (Schreck *et al.*, 1973).

The function of constitutive heterochromatin is not known. Several possible roles of constitutive heterochromatin have been proposed by Yunis & Yasmuneh (1972).

Facultative heterochromatin

Facultative heterochromatin is euchromatin in a condensed genetically inactive state. In fact, quinacrine staining and the various Giemsa techniques have failed to show any difference from normal euchromatin (see Vosa, 1973) and autoradiographic analyses have shown no or little synthesis of RNA in facultative heterochromatin (see Frenster, 1974).

Work on facultative heterochromatization began in 1921 – when Schrader described the unusual chromosome constitution of the male mealy bugs (coccids) – and was further carried out by the Schraders (review by Hughes-Schrader, 1948) and others (review by Brown & Nur, 1964). In the first cleavage divisions the chromosomes of all embryos appear to be euchromatic but, at the blastula stage in the male embryos, an entire haploid set of chromosomes becomes hetero-

* The pericentromeric chromatin of mouse chromosomes, which contains the A–T rich satellite DNA, shows reduced fluorescence when stained with quinacrine. The reason for the decreased quinacrine fluorescence of mouse satellite DNA has been explained by Weisblum (1974).

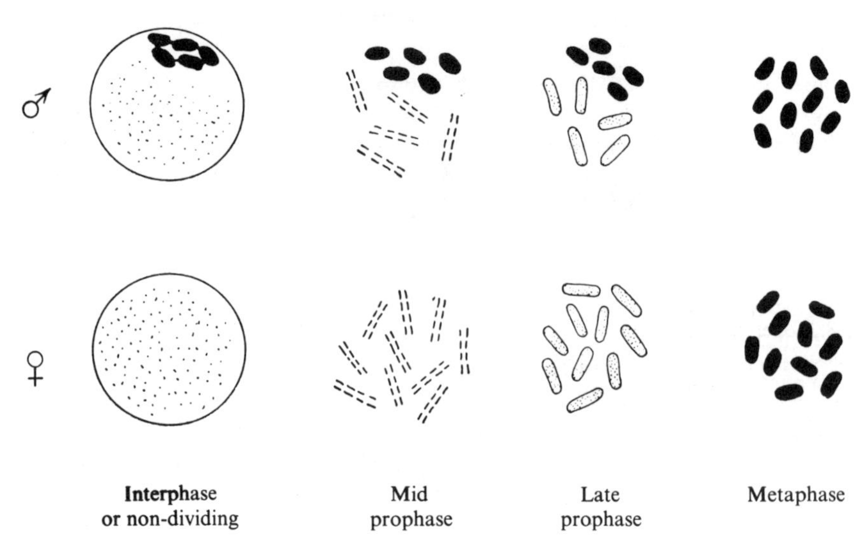

Fig. 8.1. Mitosis in male and female mealy bugs (2n = 10 is the commonest chromosome number). The chromosomes of the female remain euchromatic during development. Half the chromosomes of the male become heterochromatized at an early embryonic stage and remain so throughout development. The heterochromatic chromosomes are clumped at interphase and appear more condensed than the others during prophase. At metaphase all chromosomes of both sexes are equally condensed (after Brown & Nur, 1964).

chromatic and remains heterochromatic during further dcvelopment. In the female embryos, however, all chromosomes of the diploid complement remain euchromatic. At all mitoses in the male, the heterochromatic set divides synchronously with the euchromatic set (Fig. 8.1). When spermatogenesis occurs, the first division is equational; it is followed by a second division in which the two haploid chromosome sets segregate to opposite poles of the cell. The result of these two divisions are four nuclei; two euchromatic which develop into sperm and two heterochromatic which degenerate (Fig. 8.2). The Schraders suggested that the heterochromatic set in the males is of paternal origin and that the heterochromatic set is genetically inert, so that the male, though having a diploid chromosome complement, would be virtually haploid. These two hypotheses were confirmed by Brown & Nelson-Rees (1961) who studied the common mealy bug, *Planococcus citri.* They showed that after X-irradiation of fathers, chromosome aberrations appeared in the heterochromatic set of male embryos (since the chromosomes are holokinetic, fragments persist as heterochromatic fragments), whereas after irradiation of mothers, chromosome aberrations including fragments were found in the euchromatic set. Brown & Nelson-Rees also showed that after paternal irradiation dominant lethality is induced in daughters (progressive reduction in survival with increasing doages) but not in males (control survival even after 16000 rep). Obviously, dominant lethals are not expressed when they are in the

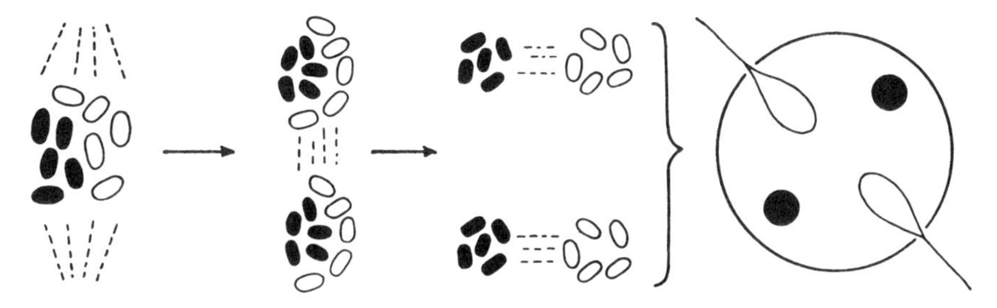

Fig. 8.2. Spermatogenesis in the mealy bug. The euchromatic and heterochromatic sets divide equationally at first division and segregate during the second. Only the euchromatic products form sperm; the heterochromatic products degenerate (after Brown & Nur, 1964).

heterochromatic set. If, however, very high radiation doses (70000 and 90000 rep) are used for irradiating males to be mated to stock females, the sons (X_1 males) are sterile (Nelson-Rees, 1962). This and other experimental data indicate that the paternal heterochromatic set is not totally genetically inert (Brown & Nur, 1964). It is now clear that the external morphology of males in *Planococcus* is determined by cells in which the paternal set is heterochromatic, but in some tissues cells have no heterochromatic set due to the reversal of heterochromatization during development (e.g. Malpighian tubules). It is these tissues that are most heavily damaged in third instar males developing after 60000 rep paternal irradiation (Nur, 1967).

An interesting phenomenon of heterochromatization reversal ('deheterochromatization') in some chromosomes of the complement has been shown to depend on particular ecological conditions in Japanese populations of the orchid *Spiranthes sinensis* (2n = 30) (Tanaka, 1969*a*, *b*). The standard karyotype consists of: twenty chromosomes (1–20 in Fig. 8.3), each with a euchromatic short arm and a heterochromatic long arm ending with a short euchromatic segment (h chromosomes), eight chromosomes with a euchromatic short arm and a partly heterochromatic long arm (he chromosomes), and two wholly euchromatic chromosomes (e chromosomes). In some clones, one to four he chromosomes may undergo reversal of heterochromatization (Fig. 8.3). An analysis of 487 clones collected from thirty-four localities covering almost all areas of the Japan archipelago has shown that the clones with standard karyotype (20h+8he+2e) predominate among cold-moor populations, whereas clones with two or more deheterochromatized chromosomes are most common among warm-moor and grassy land populations; moreover, increasing deheterochromatization is expressed in an increasing plant height (Table 8.1). Obviously, the climatic conditions of warm-moor and grassy land stations, by favoring deheterochromatization, activate genes or sets of genes thereby increasing overall growth.

We now turn to the clearest and best-known system of facultative heterochromatization, that of the X chromosomes in mammals. It was first observed by

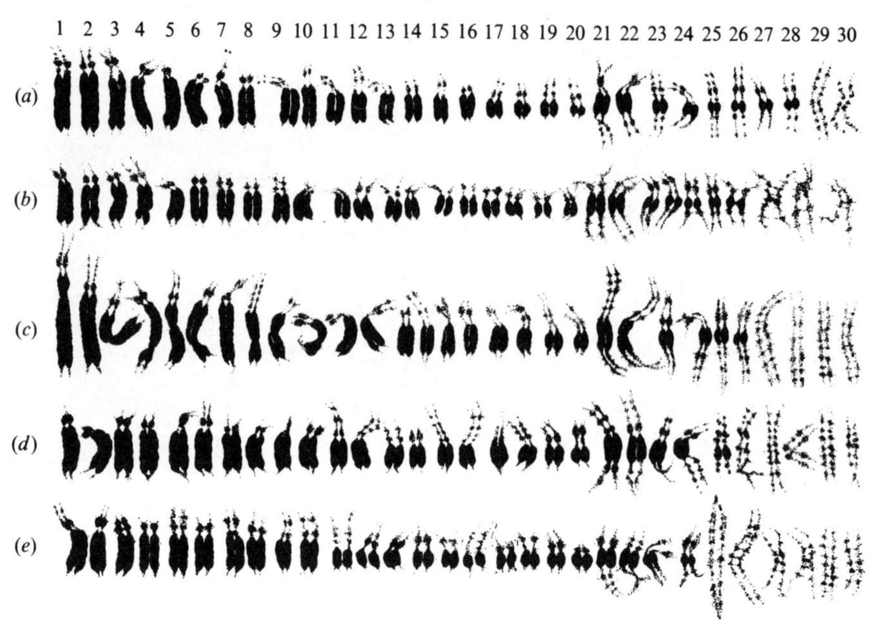

Fig. 8.3. Somatic prophase chromosomes of five clones of *Spiranthes sinensis* (2n = 30) showing typical differences in deheterochromatization. (*a*) 20h+8he+2e (standard karyotype); (*b*) 20h+7he+3e; (*c*) 20h+6he+4e; (*d*) 20h+5he+5e; (*e*) 20h+4he+6e (after Tanaka, 1969*a*).

TABLE 8.1. *Frequency distribution of clones of* Spiranthes sinensis *with different karyotypes among populations of cold-moor, warm-moor and grassy land in the Japan archipelago, and stem lengths (range) of plants of the different clones. For details see text and Fig. 8.3* (based on data from Tanaka, 1969*b*)

Karyotype of clone	Frequency (%) of clones among populations			Stem length (cm) range
	Cold-moor	Warm-moor	Grassy land	
20h+8he+2e	89.66	17.98	5.14	7–27
20h+7he+3e	2.30	3.37	9.65	15–32
20h+6he+4e	8.05	49.44	82.95	14–45
20h+5he+5e	0	11.24	2.25	30–60
20h+4he+6e	0	17.98	0	50–75

Barr & Bertram (1949) that in the cat the nuclei of neurons of females contained a heterochromatic element which did not occur in the males; this element was termed 'sex chromatin' (Barr, 1951) and is now also called 'Barr body'. Further work, reviewed by Mittwoch (1967), has shown that one of the two X chromosomes in mammalian females is heterochromatic. In humans, patients with Turner's syndrome (XO) lack sex chromatin, whereas sex chromatin is detected in patients with Klinefelter's syndrome (XXY); in patients with multiple X chromosomes, only one of the X is euchromatic, all the remainder becoming heterochromatic. Thus, the number of Barr bodies is one less than the number of X chromosomes.

In the mouse, X-linked coat color genes show a mosaic expression; in heterozygous females, carrying the wild-type gene on one X and a recessive mutant allele on the other X, the coat shows patches with wild-type phenotype and patches with mutant phenotype. Lyon (1961) interpreted this phenomenon as the result of a random heterochromatization (genetic inactivation) of one of the two X chromosomes: the mutant (recessive) phenotype was expressed in those cells, and their descendants, in which the X chromosome carrying the wild-type allele was inactivated. Lyon (1961) also proposed that X chromosome inactivation occurs at an early developmental stage and that once a given X has become heterochromatic in a cell it retains its state through numerous subsequent divisions. In the year 1961, further evidence in favor of the random X chromosome inactivation in the female mouse was presented by Russell. It was shown that variegated-type ('V-type') position effect* is manifested in females, but not in males, when some genes are translocated on one X chromosome. The best studied case was that of a reciprocal translocation between the X-chromosome and chromosome 8 which bears the brown locus. Normally, the wild-type allele (+) is completely dominant over the brown (*b*) allele, so that the heterozygote has wild-type phenotype. However, when the + allele is transposed to its new position near a piece of X chromosome, its action is impaired. Therefore, although the animal is heterozygous (+/*b*), certain regions of the body are brown. The random mingling of brown and wild-type regions produces what is called, in accordance with *Drosophila* terminology, a variegated phenotype. The restriction of the variegation effect to the females is easily explained when considering that the only X present in males is active; so, the wild-type allele is not affected in its expression when translocated on the X. When some translocations of autosomal segment on the X chromosome were studied (Russell, 1963; Russell & Montgomery, 1970) it was shown that the inactivating effect of the X appears to spread along the autosomal genes in the form of a gradient so that the more distant autosomal genes may fail to be

* Position effect variegation was studied in *Drosophila melanogaster* by Lewis (1950). He observed that heterozygotes in which the gene for eye colour was translocated close to heterochromatin showed a variegation effect in which some of the facets of the eye were white instead of red. Inactivation seemed to spread along the translocated segment proving most effective at the region closest to heterochromatin. The extensive literature on position effect (review by Baker, 1968) demonstrates the spreading effect of heterochromatin; it is the variability of the spreading effect that produces variegation.

inactivated. It has also been observed that the inactivation of autosomal genes depends on the translocation breakpoint, that is, on the position in the X to which the autosomal segment is attached. These results led Russell & Montgomery (1970) to suggest that single X chromosome inactivation may be initiated at one inactivation center from which it spreads to the entire X. For a detailed discussion of X chromosome inactivation in the mouse the reader may consult Eicher (1970).

As shown by numerous biochemical, cytological and genetic studies (review by Lyon, 1972), the random inactivation of one of the two X chromosomes in females is characteristic of eutherian mammals (placental mammals, including man, with the development of offspring to full term *in utero*). In some hybrid mammals, however, X chromosome inactivation may be non-random. This is the case of the hybrids between donkey and horse, that is, mules (male parent: donkey) and hinnies (male parent: horse). In early experiments with cultured leucocytes of female mules, Mukherjee & Sinha (1964) found that in half the metaphase cells one parental X was late labeling and in the remaining cells the other parental X labeled late. Other experiments, however, suggest a non-random inactivation. Since the X chromosomes of donkey and horse code for glucose-6-phosphate dehydrogenases (G6PD) of different electrophoretic mobilities, Hamerton *et al.* (1971) studied the expression of G6PD in female mules and hinnies by cloning their fibroblasts. Any particular cell line showed the presence of only one of the two types of G6PD, but more cells were found with the horse enzyme. Support for this preferential inactivation of the donkey X chromosomes was obtained from the observation that in cultured cells the late labeling X chromosome is more often that of the donkey than that of the horse. The preferential expression of the horse genes, however, is not absolute (see Lewin, 1974*b*).

The lack of suitable genetic markers active at early embryogenesis does not permit a really accurate determination of the time of X-inactivation, but useful information has been gained from analyses of sex chromatin and late replication (references in Lyon, 1972). Thus, in the pig embryo, sex chromatin appears at the stage of about fifty cells and in the rabbit embryo heterochromatization occurs when a few hundred cells are present, well before implantation. In murine rodents, X-inactivation begins at or soon after implantation, that is, at a time of a very large increase in the synthesis of RNAs.

In contrast with eutherian mammals, marsupial mammals are characterized by a direct inactivation of the paternal X. Euros, wallaroos and kangaroos are distinguishable by the possession of morphologically distinct X chromosomes and by the different electrophoretic mobilities of the G6PD enzymes specified by their X chromosomes. The female progeny of such hybrids show only the electrophoretic band corresponding to that of the mother (Richardson, Czuppon & Sharman, 1971). Incubation of the leucocytes of the hybrid progenies with ^{3}H-thymidine shows late replication of the paternal X in almost every case (Sharman, 1971).

The mechanism by which one of the X chromosomes of female mammalian cells is inactivated is not known. Several models have been proposed; among the recent ones, reference may be made to Eicher (1970), Cooper (1971), Lyon (1972), Brown & Chandra (1973), Ohno (1974) and Riggs (1975).

B chromosomes

The term 'B chromosomes' was introduced by Randolph (1928) to denote the chromosomes which occurred in maize in addition to the zygotic chromosome complement (2n = 20) and were not homologous to any chromosome of the complement (A chromosomes). B chromosomes are now known under a variety of different names, the most commonly used being 'accessory' and 'supernumerary' chromosomes. There are three excellent reviews of the subject (Battaglia, 1964; Müntzing, 1974; Jones, 1975). Jones lists 591 species of plants and 116 species of animals in which B chromosomes have been found.

It was first shown by Randolph (1941) and Darlington & Upcott (1941) that maize apparently can tolerate up to 25 B chromosomes with a slight loss in fertility as the major detectable effect. It has then been shown that, although phenotypic differences are not generally found between normal individuals and individuals carrying B chromosomes in populations of plants and animals, a variety of effects of B chromosomes can be demonstrated (Jones, 1975, Tables XIII and XIV). These include effects on growth and vigor, fertility, seed germination, flowering and A chromosome recombination. These manifold effects clearly disprove the long-held view that B chromosomes are genetically inert. This view was based on the common cytological observation that in the majority of species B chromosomes are heterochromatic (species with euchromatic B chromosomes are not rare among plants) and when analyzed by autoradiography show the late replication typical of heterochromatin. Information on the chemical composition of heterochromatin in B chromosomes is quite fragmentary. Gibson & Hewitt (1970) have shown that in the grasshopper *Myrmeleotettix maculatus* the DNA extracted from individuals with B chromosomes contains a highly repetitive DNA (light satellite) which is not present in the DNA extracted from individuals without Bs. By contrast, the heterochromatic Bs of maize (Chilton & McCarthy, 1973) and rye (Jones, 1975; Rimpau & Flavell, 1975) have been shown to have a DNA composition practically identical to that of the A chromosomes. This result indicates that the Bs of maize and rye possess a large amount of unique DNA sequences which are kept in a repressed state by heterochromatization. On the other hand, it has been observed on many occasions that B chromosomes in some species are not invariably heterochromatic.

As pointed out by Fernandes (1949) and Hewitt (1974), heterochromatization would be the most direct method for the incorporation in the genome of a species of an extra chromosome because its possible deleterious effects would be minimized or abolished by genetic inactivation. This raises the important question of the origin of some B chromosomes as a consequence of polysomy resulting from either mitotic or meiotic disturbances; for a detailed discussion on the origin of Bs, see Battaglia (1964) and Jones (1975). In natural populations of *Narcissus bulbocodium* in Portugal, Fernandes (1943, 1949) found heterochromatic B chromosomes which, although more condensed, corresponded in morphology to chromosomes of the diploid complement. He assumed that these B chromosomes were derived from A chromosomes by heterochromatization and showed that heterochromatization was due to a dominant gene (H) which was assumed 'to control the amount of active chromatin in the nucleus'. More recently, several observations point to the origin of B chromosomes from chromosomes of the regular complement in Orthoptera. In these animals, two major classes of B chromosomes are distinguished: small Bs which seem to have originated from small autosomes and large Bs which seem to have originated from X chromosomes (Hewitt, 1974). Often in grasshopper meiosis, extra chromosomes are heterochromatized; a very clear case is that of *Corthippus parallelus* (Hewitt & John, 1968). In one population the germ line has been found to be a mosaic of normal cells and cells polysomic for the medium sized M_4 chromosome. The M_4s in excess of two are heterochromatized at mitosis and meiosis when they do not form multiples with the normal M_4s. The heterochromatized M_4s are late labeling and have a much reduced incorporation of ^{3}H-uridine as compared with normal M_4s (Fox, Hewitt & Hall, 1974). Thus, heterochromatization not only inactivates extra chromosomes but also masks their homology. This observation raises the question whether the non-association of Bs to As at meiosis, which is a general phenomenon, may depend on heterochromatization or is due to differences in homology between Bs and chromosomes or chromosome segments of the complement.

An interesting property of the majority of B chromosomes is their accumulation in successive generations. In plants, the most common mechanisms of accumulation of Bs are mitotic nondisjunction at different ontogenetic stages (first or second pollen mitosis, first egg cell mitosis) or to preferential meiotic segregation in embryo sac mother cells. Premeiotic nondisjunction, which is rare in plants, is a common mechanism of Bs accumulation in animals, in which also cases of preferential segregation at meiosis of Bs have been reported (see Battaglia, 1964; Müntzing, 1974; Jones, 1975).

Histones

In the preceding section it has been seen that a condensed state of chromosomes or chromosome segments, as obtained by heterochromatization, inhibits gene transcription. The mechanism of heterochromatization is not known, but we know of a class of chromosomal proteins, the histones, which act as non-specific repressors of gene activity.

In 1950, Stedman & Stedman found that actively growing tissues contained less histone than nongrowing tissues and suggested that histones inhibit gene activity. The Stedmans' hypothesis has found ample confirmation but their method of histone extraction has since been shown to be inaccurate; it has been reported on many occasions that both genetically active and inactive chromatin have the same histone content (e.g. Frenster, 1965; Comings, 1967; McCarthy *et al.*, 1974; see, however, Bonner, 1975). For reviews of the role of histones in regulating gene transcription, reference is made to Georgiev (1969*b*) and Johnson *et al.* (1974).

The first demonstration that histones restrict the template capacity of DNA in RNA synthesis was provided by Huang & Bonner (1962). They used a cell-free system in which chromatin or DNA was used as template for the synthesis *in vitro* of RNA in the presence of RNA polymerase and RNA precursor molecules. Addition of histone to the cell-free system inhibited RNA synthesis, the inhibition being maximal at a DNA/histone ratio of 1:1, which is that normally found in the nucleus. Since the inhibition was reversed by addition of more DNA, it was clear that histone blocked transcription by binding to DNA. At about the same time Allfrey & Mirsky (1963) showed that removal of histone from isolated nuclei greatly increased the rate of RNA synthesis. The experiments of Huang & Bonner (1962) and Allfrey & Mirsky (1963) led to the conclusion that histones function as inhibitors of gene transcription. This conclusion has been strengthened by an extensive experimentation. It is now clear that the template activity of chromatin is exclusively a function of the total amount of histone bound to DNA and that no histone fraction is uniquely responsible for the repression effect. Since histones also influence the state of condensation of chromatin, it seems probable that histones repress gene activity by affecting both template activity and the state of chromatin condensation (Johnson *et al.*, 1974).

Although no differences in histone content are generally found between active and inactive chromatin, there is evidence in several systems that the histone fractions of active and inactive chromatin differ in degree of phosphorylation and acetylation (references in Johnson *et al.*, 1974). Phosphorylation and acetylation of histones have in some instances been found to precede increased RNA synthesis and putative gene activation. Thus, chemical modifications of histones may act as modulators of transcription. Since phosphorylation, acetylation and methylation of histones are mediated by specific enzymes, histones may be, for their more detailed and specific roles, dependent on these nonhistone enzyme proteins (Swift, 1974).

As to the amount of genome expressed in eukaryotic cells, information can be obtained from experiments in which labeled single-stranded fragments of unique DNA are prepared by annealing and hydroxyapatite fractionation and hybridized with RNA in gross excess for extended periods of time. A comparison between bacteria and eukaryotes (Table 8.2) shows that in *E. coli* the majority of the genome (80–100%) is expressed, whereas in eukaryotes the portion of the genome which is expressed is different in different species and in different tissues within the same

TABLE 8.2. *Estimates of the extent of transcription of various genomes.* Condensed from McCarthy *et al.* (1974) where references are reported

Species and material	Genome size (daltons)	Percentage transcribed*
Escherichia coli	2.8×10^{9}	40–50
Drosophila melanogaster	120×10^{9}	—
Cultured cells	—	15
Embryos	—	15
Adults	—	10
Xenopus laevis	2×10^{12}	—
Oocytes	—	0.6
Chicken	1×10^{12}	—
Red cells	—	2
Liver	—	15
Mouse	3×10^{12}	—
Liver, kidney, spleen	—	4–5
Embryo	—	8
Brain	—	11
Human	3×10^{12}	—
Liver, kidney, spleen	—	4–5
Brain	—	22

* Assuming that only one DNA strand is transcribed, these values would be multiplied by 2.

species, ranging from 1.2% in *Xenopus* oocytes to 44% in human brain. It must be pointed out, however, that these estimates are minimal ones due to technical reasons, e.g. the concentration of rarer RNA species and the long incubation times used (McCarthy *et al.*, 1974).

Nonhistone chromosomal proteins

The transcriptional repression imposed on chromatin by histones implies that the differences in gene activity in different cells or tissues in an organism and the temporal sequences of entry into action of genes during differentiation and development must depend on a specific derepression (activation) of genes or sets of genes. In the search for molecules capable of regulating specific genes in eukaryotes much attention has been paid to nonhistone chromosomal proteins since Frenster's (1965) initial observation that euchromatin is enriched with non-histone proteins (reviews by Johnson *et al.*, 1974; Stein, Spelsberg & Kleinsmith, 1974; Wang & Nyberg, 1974). Unlike histones, proteins with properties similar to those of nonhistone proteins of eukaryotes are found in protokaryotes; the extensively studied *lac* repressor of *E. coli*, which operates negative control, has properties closely resembling those of some nonhistone proteins of eukaryotes (Beyreuther, Adler, Geisler & Klemm, 1973). Like histones, nonhistone proteins are synthesized outside the nucleus in the cytoplasm and are transported into the nucleus. Whereas histones are associated with DNA immediately after synthesis, only some of the nonhistone proteins become associated with DNA soon after synthesis, others become associated with DNA after various periods of time. As a group, the nonhistone proteins show a faster rate of turnover than histones. Lifetimes of individual nonhistone proteins range from very short to long, so that nonhistone proteins – at least, in part – are in a state of dynamic flux. There are several lines of evidence in favour of the view that some of the nonhistone proteins are involved in the regulation of gene activity. In addition to the very extensive literature on clear differences in nonhistone protein spectra between different species and between different tissues within the same species, also in relation to developmental or differentiative stages, there is experimental proof that some of the nonhistone proteins are involved in the regulation of specific gene expression.

The first evidence that nonhistone proteins determine organ-specific transcription of chromatin was presented by Paul & Gilmour (1968), who reconstituted chromatin from dissociated chromatin components in 2 M NaCl by gradient dialysis (gradual dilution to lower ionic strength). In a control experiment, they reconstituted rabbit thymus and bone marrow chromatin from their dissociated chromatin components and used these reconstituted chromatins as templates for RNA synthesis in the presence of RNA polymerase. The transcribed RNAs, when tested for their ability to bind to DNA from rabbit thymus or bone marrow, behaved as normally synthesized thymus and bone marrow RNAs. Then Paul & Gilmour pooled DNA and histones dissociated from the two chromatins and reconstituted chromatin by adding nonhistone proteins from either thymus or bone marrow. In the first case, reconstituted chromatin made RNA which behaved as thymus RNA, in the second, the reconstituted chromatin made RNA which behaved as bone marrow RNA. By an improvement in technique resulting in the production of reconstituted chromatin similar to native chromatin (as judged by the hybridization of the transcribed RNA to DNA), results confirming for other systems the original observations of Paul & Gilmour have been obtained (references in Stein *et al.*, 1974; Wang & Nyberg, 1974). Moreover, it has been shown that the specificity of RNA synthesis *in vitro* can be changed: e.g. from the pattern characteristic of rat liver to that characteristic of rat thymus by exchanging rat liver nonhistone proteins with rat thymus nonhistone proteins in the reconstitution of chromatin (Spelsberg & Hnilica, 1970). Fig. 8.4 illustrates the results of an experiment in which unlabeled RNA, formed from reconstituted chromatin and native chromatin templates, was made to compete against labeled RNA made from native rat liver chromatin template.

To demonstrate that nonhistone proteins regulate the activity of individual genes it is necessary to test for the synthesis of specific messengers. In the last few years

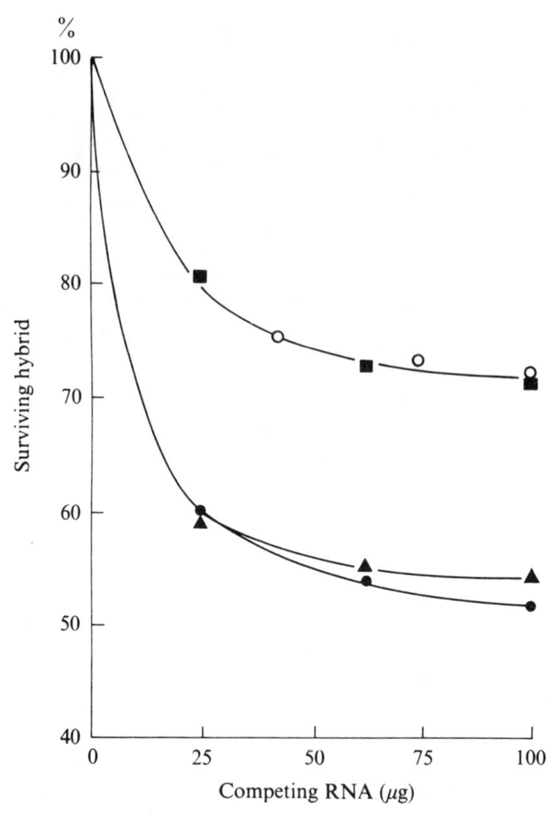

Fig. 8.4. Competitive hybridization of unlabeled RNA formed *in vitro* from the following templates: native rat liver chromatin (solid circles); hybrid chromatin composed of histones from rat liver and DNA with associated non-histone proteins from rat thymus chromatin (empty circles); native rat thymus chromatin (squares) and hybrid chromatin composed of histones from rat thymus and DNA with associated nonhistone proteins from rat liver chromatin (triangles). Chromatin reconstitution was performed by dialysis at pH 6.0. The unlabeled RNA was made to compete against ^{3}H-labeled RNA formed *in vitro* from native rat liver chromatin. The 100% hybridization represents 410 c.p.m. with 4 μg of DNA/filter (after Spelsberg & Hnilica, 1970).

it has become possible to synthesize *in vitro* radioactive copies of a given gene by placing its mRNA in the presence of RNA-dependent DNA polymerase ('reverse transcriptase') and radioactive DNA precursors. In this system, the mRNA serves as a template for the assembly of a complementary strand of DNA. From the hybrid RNA–DNA the radioactive complementary DNA (the synthesized gene) can be dissociated by hydrolysis and used for hybridization with a solution of RNAs: it will then hybridize only with its complementary RNA. This technique has been recently employed by Gilmour & Paul (1973) in a series of chromatin reconstruction experiments. In control experiments, chromatin was isolated from fetal mouse liver, which normally synthesizes the protein of hemoglobin, globin (the number of genes for globin in the haploid genome is one or two and not more than five), and from mouse brain, which does not synthesize globin. Each chromatin preparation was transcribed *in vitro* and the RNAs synthesized were tested for hybridization to the radioactive globin gene copies. Gilmour & Paul (1973) showed that fetal liver chromatin synthesized RNA coding for globin, but brain chromatin did not. They then dissociated the chromatins and reconstituted them in the presence of different types of nonhistone proteins and demonstrated that only chromatin reconstituted in the presence of nonhistone proteins isolated from fetal liver synthesized the globin messenger. That transcription of a given message depends on the presence of specific nonhistone proteins has also been shown for histone mRNAs; in these experiments too, use has been made of reverse transcriptase to synthesize DNA complementary to histone mRNAs (Stein *et al.*, 1975). After showing that chromatin from S phase cells transcribed histone mRNA and that chromatin from G_1 cells did not, Stein *et al.* demonstrated that chromatin reconstituted with S-phase nonhistone proteins transcribed histone mRNAs to the same extent as native S-phase chromatin. In contrast, chromatin reconstituted with G_1-phase nonhistone proteins did not transcribe any histone mRNA.

Direct evidence for the involvement of nonhistone proteins in the regulation of specific genes also comes from studies on gene activation induced by some hormones.

Hormones and the cell nucleus

The concept that hormones control gene activity was first proposed by Karlson (1965). He observed that 5–30 minutes after the injection of the steroid molting hormone, ecdysone, puffing was induced in dipteran polytene chromosomes. Although puffs or puff-like structures may develop in the absence of RNA synthesis, the majority of puffs can be taken as an expression of localized chromosomal activity. A variety of agents are known to cause agent-specific puffing responses. Among these, the most useful for experimental analysis are ecdysone (both α-ecdysone and β-ecdysone) and temperature shocks (reviews by Ashburner, 1972; Berendes, 1972). *In vivo* and *in vitro* administration of ecdysone to intact *Drosophila* intermolt larvae or isolated tissues induces changes in the puffing pattern which are essentially the same as those occurring in normal development during the 6-hour period prior to puparium. In isolated salivary glands of *Drosophila hydei* treated with ecdysone, the hormone-specific change in the puffing pattern is followed, 5–6 hours later, by the secretion of the mucopolysaccharide typical of the salivary gland (this is used by the larva to attach the puparium to a substrate). Inhibition of RNA synthesis or protein synthesis by actinomycin D or cycloheximide inhibits synthesis of the mucopolysaccharide, thus showing that the formation of this product is dependent on both *de novo* synthesis of RNA and *de novo* synthesis of protein. As shown by comparative analyses on the puffing reaction in different polytene-chromosome tissues in the same larva injected with ecdysone, the hormone simultaneously affects the activity of several chromosome loci.

Since ecdysone and ecdysone analogs induce similar puffs to those occurring at some time in the normal development in a given tissue, these agents are useful in the search for the mechanism of hormone action. One of the primary changes induced at the puffing site by the hormone (or by heat shock: this also induces puffs similar to those normally occurring in the tissue under investigation) is a net increase in protein dye binding due to an accumulation of nonhistone proteins (Holt, 1971). Helmsing & Berendes (1971) and Helmsing (1972) have analysed the nonhistone proteins isolated from nuclei of *D. hydei* with and without new puffs induced by ecdysone or heat shock (the new puffs induced by ecdysone occur at entirely different loci to those induced by heat shock). They have found a polypeptide fraction with a molecular weight of approximately 42000 in nuclei displaying ecdysone puffs and a fraction with a molecular weight of about 23000 in nuclei displaying temperature-induced puffs and in chromatin derived from these nuclei. Both fractions were absent from the controls. It was also established, by the use of cycloheximide, that the nonhistone fraction invading the nucleus concomitantly with the appearance of new puffs did not result from synthesis *de novo*, but was derived from a cytoplasmic pool from which it migrated into the nucleus. Since ecdysone appears to bind to proteins at the cytoplasmic as well as the chromatin level (Emmerich, 1972), it seems probable that this hormone behaves as other steroid hormones (see below).

There is much evidence that the action of steroid hormones is mediated at least in part at the transcriptional level (reviews by Tomkins & Martin, 1970; Turkington, 1971; Stein *et al.*, 1974; Wang & Nyberg, 1974). Steroid hormones enter the target cells and bind to a cytoplasmic receptor molecule to form a hormone-receptor complex which is transferred to the nucleus and becomes associated with chromatin (Fig. 8.5): this leads ultimately to activation of specific genes. This model seems to hold for a wide variety of steroid hormones including estrogen, progesterone, aldosterone, hydrocortisone and androgens.

The changes in gene expression induced by exogenous hormones in immature female chicks, in which it is possible to reproduce the differentiation in the oviduct mucosa that occurs during normal sexual maturation of the hen, will serve as an example. When 5-day-old female chicks are daily injected with 5 mg diethylstilbestrol (DES), the immature chick single-layered mucosa differentiates into three distinct epithelial cell types (tubular gland cells, ciliated cells and goblet cells) and ovalbumin is synthesized in the tubular gland cells. When chicks treated with DES for 12 days are daily injected with 5 mg progesterone, the goblet cells are induced to synthesize the egg-white protein avidin (Kohler, Grimley & O'Malley, 1968). In a series of studies, O'Malley and associates (see O'Malley

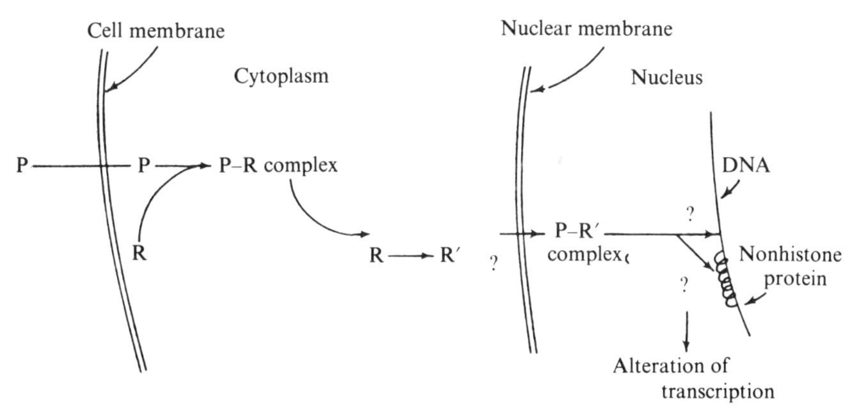

Fig. 8.5. Model of the intracellular fate of progesterone in the oviduct target cell. Progesterone (P) enters the cell and binds receptor protein (R) that is located in the cytoplasm. The P–R complex is transported to the nucleus, during which time the receptor undergoes some modification (R′). The P–R′ complex then binds to the chromatin and as a result the transcriptional process becomes altered (after Stein *et al.*, 1974).

et al., 1972) have identified a progesterone-binding protein in the cytoplasm of estrogen-treated chicks. After entering the target cells, progesterone binds to this receptor molecule and the steroid-receptor complex is then transported into the nucleus where it becomes associated with the target chromatin: the ultimate result is the induction of synthesis of avidin. Cell-free recombination studies using cytoplasm and nuclei have shown that: (i) the oviduct receptor protein is essential for the nuclear binding of progesterone, (ii) oviduct nuclei, but not spleen, lung, intestine or liver nuclei, can bind the progesterone-receptor complex and (iii) the nuclear-bound progesterone can be extracted still associated with the receptor. Moreover, the progesterone-receptor complex can be bound to purified chromatin *in vitro*: it binds much better to target cell chromatin than to other chromatins. These observations have led O'Malley *et al.* (1972) to postulate that the target cell genome may contain specific 'acceptor sites' for the progesterone-receptor complex. To examine directly the role of chromosomal proteins in the binding of the steroid hormone by target tissue chromatin, the dissociation and reconstitution of chromatin as described previously has been utilized. It has been shown that the binding of the progesterone-receptor complex to the reconstituted oviduct chromatin from which the nonhistone chromosomal proteins have been removed is markedly reduced as compared to the binding to reconstituted oviduct chromatin containing the nonhistone chromosomal proteins (O'Malley *et al.*, 1972). Thus, the nonhistone chromosomal proteins, but not the histones, appear to be involved in the specific binding of the progesterone-receptor complex.

Knowledge of the interaction between plant growth substances and the chromatin is still scanty; some effects of plant growth substances on the cell nucleus are discussed in Chapter 9. Although not much work has yet been carried out on the mode of action of the natural plant hormone, auxin, there is some evidence that stimulation of transcription by indole acetic acid is mediated by a 'nuclear' protein (Teissere *et al.*, 1975).

In summary, the observations discussed in this and the preceding section strongly support the notion that nonhistone chromosomal proteins act as regulator molecules in specific gene expression. Although much interest centers on non-histone–DNA interactions, the value of further studies on histone–nonhistone interactions and on chemical modifications of histones should not be minimized.

9

Regeneration and totipotency of cells and nuclei

General

Animals and plants are able to replace structures or organs which have been lost by accident or mutilation or spontaneously in the normal course of development. Examples of structures or organs which are periodically or continuously replaced during development are the feathers in birds, the fur and the cornified skin layer in mammals, the exoskeleton in arthropods, the epidermal scales of reptiles, the antlers of deer, the leaves and the bark of trees. The process of tissue or organ replacement which occurs as part of normal development is called physiological or repetitive regeneration, in distinction to reparative or restorative regeneration which follows injury. Since the latter type of regeneration is easily obtained experimentally, by wounding or mutilation, much knowledge has accumulated on reparative regeneration. Tissue and organ replacement is, however, only an aspect of the ability for regeneration in cells and organisms. In some invertebrates, such as the coelenterates (e.g. *Hydra*), the planarians (e.g. *Dugesia*) and the starfish, a small fragment of the body can restore a complete organism; in plants, a single cell *in vivo* (that is, within the plant body) or *in vitro* (somatic cells, pollen grains) can regenerate a complete plant which can be grown to maturity. Both *in vivo* and *in vitro*, plant regeneration from single cells may follow two distinct developmental patterns: embryogenesis, much in the fashion of the normal embryonic development, or (most commonly) formation of adventitious shoots which after rooting (either spontaneous or induced) develop into complete plants (reviews by Torrey, 1966; Vasil & Vasil, 1972; Reinert, 1973; Murashige, 1974; D'Amato, 1977). The demonstration that in some plant species plant regeneration can occur via embryogenesis and via adventitious shoot formation raises the important question of the genetic control mechanisms underlying the regenerative (either embryogenic or organogenic) capacity of the cell.

No case of cell totipotency of the sort observed in plants has been reported in any multicellular animal. Nuclear transplantation experiments in amphibian eggs clearly demonstrate, however, the totipotency of somatic cell nuclei (Gurdon, 1974).

Taken in the broadest sense, regeneration is the manifestation of a property which extends from the unicellular to the most complex multicellular organisms; namely, the capacity for growth and development. In those multicellular organisms in which complete individuals can be restored from single cells or multicellular buds (plants, coelenterates, tunicates) or by a spontaneous process of transverse fission (flatworms, annelids), regeneration is synonymous with vegetative reproduction (propagation). Among animals, the lower organisms have a greater regeneration potential than do the higher organisms; but in many instances closely related forms differ widely in their capacity for regeneration. This shows that, although the reduction or loss of regenerative potentiality generally depends on the degree of cell specialization, subtle control mechanisms operate in regeneration. A comprehensive treatment of regeneration in animals and plants can be found in recent books: Kiortsis & Trampusch (1965), Hay (1966), Steward (1968), Goss (1969).

Regeneration in coelenterates and planarians

Among coelenterates, the hydroid polyps, or Hydrozoa (species of *Hydra*, *Tubularia*, *Cordylophora* and some other genera), have been favourite materials for studies on regeneration since the classical experiments of Trembley (1744) on *Hydra* (see Tardent, 1963). The hydroid polyp, which consists of a tube-like hydrocaulus, a gastric apparatus and several tentacles, has a diploblastic organization (ectoderm or epidermis and endoderm or gastrodermis with an interposed noncellular layer of mesoglea which acts as a skeleton) enclosing a single body cavity which communicates with the exterior by one opening only (mouth). There is no central nervous system, nor any specialized organs for respiration or excretion. Of the 100000 or so cells that make up a hydra (Goss, 1969; Gierer *et al.*, 1972), there are only eight different cell types (Burnett, 1968) which serve numerous functions (i.e. limited degree of functional differentiation). A cell type with a cytoplasm unusually rich in RNA, the interstitial cells, situated among the cells of the ectodermal layer, are the most ubiquitous and seem to serve the function of stem cells. In an individual hydra, which under optimal life conditions can survive for decades reproducing asexually, cells are continuously lost at its distal (nematocysts) and proximal (glandular cells of the pedal disk) body extremities; this cell loss is continuously compensated for by the proliferation of embryonic cells, the interstitial cells in the opinion of many authors. Most probably because of the limited degree of functional differentiation of their cells, hydroid polyps are endowed with an extraordinary regenerative capacity; thus, a hydrocaulus of *Tubularia larinx*, with a total length of 4 cm, when subjected to repeated amputations, is capable of regenerating its terminal hydrant (digestive apparatus and tentacles) as many as 15 times within a period of 74 days (see Tardent, 1963). It has also been found that endoderm is able to regenerate the ectoderm and consequently reconstitute a complete hydra (Normandin, 1960; Haynes & Burnett, 1963), while the isolated ectoderm is able to reconstitute a complete polyp both in *Hydra* and other species of coelenterates (Lowell & Burnett, 1969, and here reported references). More recently, Gierer *et al.* (1972) have succeeded in regenerating a complete *Hydra attenuata* from disaggregated and reaggregated

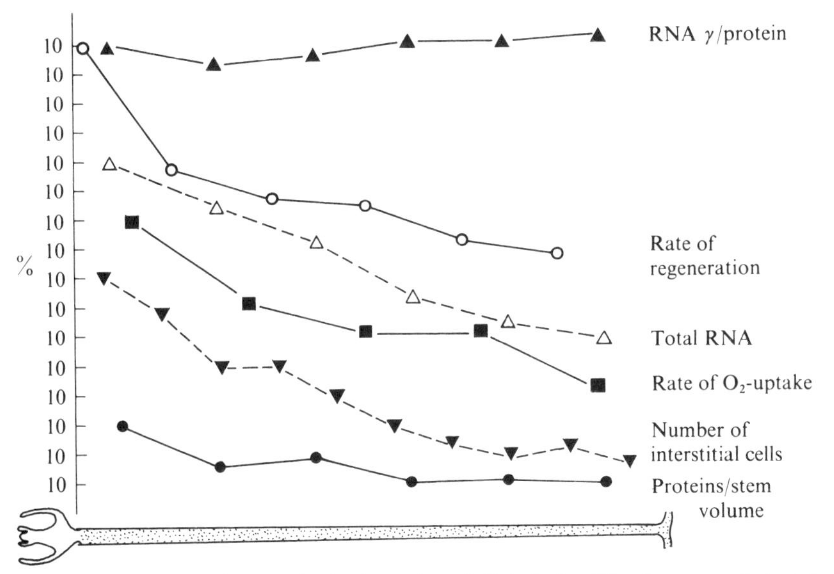

Fig. 9.1. The axial gradient in the hydrocaulus of *Tubularia*. All values recorded at the different axial levels are expressed as percentage of that recorded at the most distal fragment taken as 100% (after Tardent, 1963).

cells. They have found that the absolute minimum number of cells required for regeneration is 20000, which amounts to one fifth of the total number of cells making the mature polyp.

When a hydroid polyp regenerates, a blastema, an aggregate of undifferentiated cells, is not formed as it is in higher animals. It reconstitutes itself by morphallaxis (regeneration via blastema production is termed epimorphosis). When a hydroid is mutilated, there occurs a rapid healing of the wound, which is followed by an active migration of cells of different types in the primordial area where cell division and differentiation lead to reconstitution of the missing parts. Among the migrating cells, interstitial cells are very numerous. Evidence of their migration has been obtained from combined grafting and irradiation techniques; these show the ability of an irradiated distal half of a hydra to resume capacity for regeneration when grafted on to an unirradiated distal half. There are several lines of evidence in favour of the importance of interstitial cells in regeneration, namely: (i) irradiated polyps fail to regenerate because of the radiation sensitivity of the interstitial cells (this feature emphasizes their embryonic nature); (ii) the differences in regenerative capacity among different portions of the body strongly correlate with the incidence of interstitial cells (Fig. 9.1); (iii) in regenerating polyps, the number of interstitial cells decreases in the region adjacent to the developing primordium, suggesting that they have migrated into the primordial area; (iv) if a polyp is subjected to repeated regeneration, the stock of interstitial cells in the regenerant is gradually exhausted (references in Tardent, 1963). However important interstitial cells may be, they are not the only cells responsible for the production of a regenerate in Hydrozoa. In their study on regeneration in hydra from fragments of endodermis lacking interstitial cells, Haynes and Burnett (1963) observed that differentiated cells dedifferentiate to produce cell types which proliferate and then redifferentiate. These and other observations clearly demonstrate the importance of cell dedifferentiation in regeneration (Hay, 1968), but do not rule out the participation of interstitial cells in normal regeneration in coelenterates. Thus, both proliferation of embryonic cells and cell dedifferentiation seem to contribute to regenerate production in these animals (Tardent, 1963; Pedersen, 1972). Nothing is known of cell proliferation dynamics during regeneration in coelenterates. Hypostome regeneration in *Hydra* is accompanied by no increase in DNA synthesis (Clarkson, 1969); this might indicate that in this regenerating system cells divide from a G_2 population and therefore do not require DNA synthesis.

Other animals with an amazing power of regeneration are planarians (Brønsted, 1969). Great differences in regenerative capacity exist, however, between species and between biotypes within species. In some biotypes, sexual reproduction has been abolished and the animals propagate vegetatively by transverse division of their body (fission); many of these sterile vegetatively propagating planarians are polyploids (White, 1973). When a planarian is cut transversely, regeneration is initiated at both wound surfaces. The anterior end grows a new tail and the tail produces a new head. Complete animals can also be regenerated from tissue fragments. By the use of a simple nutrient medium, Montgomery & Coward (1974) have obtained complete cephalic regeneration from fragments of *Dugesia dorotocephala* as small as 0.08 mm^3. They have calculated that this volume is equivalent to 10000 cells and that the adult animal contains 200000 cells. Whether complete regeneration can be obtained, both in *Dugesia* and *Hydra*, from tissue fragments smaller than those used by Montgomery & Coward (1974) and by Gierer *et al.* (1972) remains to be investigated. It is probable that further manipulations and more refined cultural techniques may help in future to reduce further the minimum tissue size required for complete regeneration in these animals.

Regeneration in planarians involves the formation of a blastema, an accumulation of undifferentiated cells at the site of amputation. A crucial problem concerns the source of the blastema cells. Most authors subscribe to the original hypothesis of Wolff & Dubois (1948) that planarians possess in their parenchyma a stock of undifferentiated pluripotent free cells, usually called neoblasts, which migrate to the amputation site (Fig. 9.2), divide and form the blastema (references in Brønsted, 1969; Pedersen, 1972; Banchetti & Gremigni, 1973). Despite the practically unanimous consensus on the occurrence of neoblasts in planarians and the observations of many authors – a.o. Pedersen (1972) and Banchetti & Gremigni (1973) – that neoblasts are present in the regeneration blastema, the participation of neoblasts in the regeneration process is not definitely proved. Chandebois (1973*a*) found a very wide variation in nuclear DNA content, indicative of both

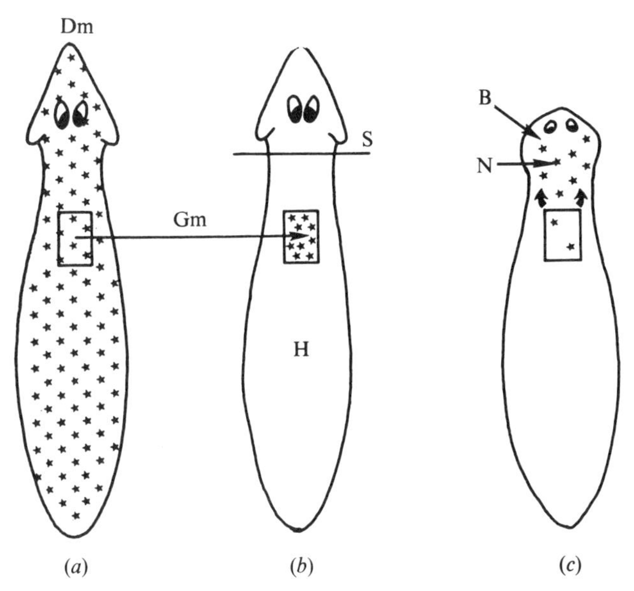

Fig. 9.2. Diagram of a grafting experiment in a Planaria. *a*; a fragment of tissue is taken from an animal which has incorporated ^{3}H-uridine (Dm); *b*: the fragment (Gm) is implanted in the equivalent position in a recipient animal (H) either unirradiated or irradiated *in toto*. 24 hours after a successful graft, the recipient animal is amputated behind its auricules (S); *c*: at regeneration, the blastema (B) contains labeled cells (N) (after Gabriel, 1970).

polyploidy and aneuploidy, in free cells of regenerating *Dugesia subtentaculata* which she interprets as neoblasts. If this interpretation were correct, it would be impossible to make polyploid and aneuploid cells responsible for regeneration, for they are also said to divide amitotically (Chandebois, 1973*b*). Chandebois has measured uniformly diploid DNA values in the interphase nuclei of syncytial elements which, in Chandebois' view, result from cell dedifferentiation and form the blastema, which also has a syncytial structure. The existence of syncytial elements in regenerating planarians is, however, denied by a number of workers (Banchetti & Gremigni, 1973, and here reported references).

That dedifferentiation of specialized cells might be a means of providing cells to the regeneration blastema of planarians in analogy to the situation in amphibian limb regeneration (see below) is suggested by several observations. In 1965, Woodruff & Burnett, working on *Dugesia tigrina*, confirmed the observation of previous authors that glandular cells of the gut in planarians dedifferentiate to provide cells for the regeneration blastema and, a few years later, Hay (1968) observed a gradient of tissue dedifferentiation extending proximally to the amputation site in the same planarian species. More recently, fine structure studies have provided further observations in favour of cell dedifferentiation during regeneration in *Dugesia* (Coward & Hay, 1972; Hay & Coward, 1975). Since cell dedifferentiation implies a remodelling of differentiated cells, lysosomal enzymes might be involved in the process. Assays on tissue homogenates and at the electron microscope level have demonstrated a marked increase in lysosomes and in the activity of the lysosomal enzyme acid phosphatase during cephalic regeneration in *Dugesia dorotocephala* (Coward, Bennett & Hazlehurst, 1974).

Although cell dedifferentiation might be an important factor in planarian regeneration, it does not necessarily exclude the participation of neoblasts and/or other types of pluripotent or totipotent cells in building the blastema. This is strongly supported by the recent observations of Banchetti & Gremigni (1973) and Gremigni & Puccinelli (1977). Of particular interest are the regeneration experiments performed on a biotype of *Dugesia lugubris* having a triploid ($2n = 3x = 12$) somatic cell line, a hexaploid ($2n = 6x = 24$) female germ line and a diploid ($2n = 2x = 8$) male germ line (in the developing embryo all cells are triploid). Sexually mature and immature individuals were amputated in the caudal region posterior to the testicles. Both the blastema reconstituting the caudal portion (A) and the blastema reconstituting the cephalic portion (B) were studied. The A blastema formed by mature individuals was found to consist of 95% triploid cells, 2% diploid cells and 3% cells with intermediate chromosome number; the B blastema formed by the mature individuals and the A and B blastema formed by the immature individuals were all found to consist of triploid cells only. According to Gremigni & Puccinelli (1977), the triploid cells in the blastema are derived from neoblasts, the diploid cells are derived from spermatogonia which have already completed elimination of one haploid set from originally triploid cells and the cells with intermediate chromosome number are derived from cells in the process of eliminating chromosomes to become spermatogonia. Thus, young germinal cells may participate in blastema formation in planarians.

Limb and tail regeneration in vertebrates

Larval and adult amphibians are able to reconstitute missing limb parts from the limb stump by means of a conical bud of undifferentiated blastema cells which accumulate under the wound epithelium. Following amputation, the cut surface of the limb is covered by an epidermis which derives from the epidermis of the stump as shown by an analysis of wound healing in regenerating newts whose epidermal cells had been labeled with tritiated thymidine (Hay & Fischmann, 1961). These experiments have also shown that no epidermal cell is incorporated in the blastema which starts to be formed after wound healing. Non-participation of epidermal cells in blastema formation is also demonstrated by the elegant experiments of Barr (1964). By grafting ectopically a denuded leg of the uninucleolate (one nucleolus, instead of two, per cell) mutant of *Xenopus laevis* on a host with normal nucleolar complement, a leg covered with a layer of the host's epidermis (binucleolate cells) was produced. When allowed to regenerate, the inner tissue of such grafts all contained cells with one nucleolus.

The source of the cells that participate in blastema formation has been a matter of dispute for decades. The long-held view of reserve cells distributed throughout the body was disproved in the early 1930s when it was shown that regeneration failed to occur from stumps X-irradiated in the region adjacent to the amputation site (Butler, 1933). These and other similar experiments proved that the cells of the blastema are of local origin. The demonstration, first offered by Thornton (1938), of muscle fiber dedifferentiation during forelimb regeneration in larval *Ambystoma*, opened the way to further investigations on the possible role of tissue dedifferentiation in the production of the undifferentiated cells of the blastema. In the last decade or so, much information has been accumulated on tissue dedifferentiation in the amphibian limb (Hay, 1962, 1968; Trampusch & Harrebomée, 1965, 1969; Trampusch, 1972). Dedifferentiation of muscle traumatized by the amputation is most dramatic. The multinucleate muscle fibers break up into individual mononucleate cells whose cytoplasm loses myofibrils, becomes intensely basophilic and develops a relatively abundant endoplasmic reticulum; thus, the cells become indistinguishable from blastemal cells. Muscle dedifferentiation is clearly seen during limb regeneration in both adult and larval animals. Another tissue whose dedifferentiation is easily observed in amputated limbs is cartilage. In the larval salamander, whose skeleton is not yet ossified, a large fraction of the cartilagineous skeleton in the amputated limb stump undergoes dissolution of the intercellular substance and the released cartilage cells (chondrocytes) begin to proliferate assuming the typical morphology of blastema cells. It seems also that Schwann cells and a few other cell types participate in blastema production. On the other hand, in both the larva and the adult, vessels seem to grow out directly from the corresponding tissue in the stump.

Dedifferentiation of different tissues into a single homogeneous cell type, the blastema cell, raises the question of cell transformation, or metaplasia; that is, whether or not cells of one line, such as muscle, can give rise to a different line, such as cartilage, when the regenerate differentiates from the blastema. Metaplasia is well documented in the case of regeneration in hydra from an endodermal fragment (Haynes & Burnett, 1963) and in lens regeneration in urodeles which will be discussed in another section of this chapter. Some recent grafting experiments seem to support the notion that metaplasia occurs in limb regeneration. When tritiated thymidine-labeled intestinal blastemal cells were implanted into limb blastemata of adult newts, the regenerate was found to contain labeled cartilage cells (Oberpriller, 1967). Using larval axolotl (*Ambystoma mexicanum*) Trampusch & Harrenomée (1965, 1969) have succeeded in inducing a supernumerary limb in the flank of the animal by grafting a single component of a regenerating limb stump (skeleton, muscle) and covering them with flaps of wounded limb-skin, thus simulating the situation at an amputation site. The graft was seen to dedifferentiate and form a blastema which developed into a supernumerary extremity. In addition to demonstrating cell dedifferentiation in the nerve sheath, the muscle, and the bone, these experiments have shown that the blastema derived from a single component of the limb stump is made of cells capable of differentiating into the different tissues of a regenerate. It would seem that metaplasia might be much more common than assumed so far.

Autoradiographic and cytophotometric methods have been used to study DNA, RNA and protein synthesis and changes in histone content during limb regeneration. Using tritiated thymidine, Hay & Fischmann (1961) showed that the dedifferentiating muscle fibers in regenerating adult newt limb undergo DNA synthesis to become incorporated as fibroblast-like cells in the blastema. More recently, Jeanny (1973) and Jeanny & Gontcharoff (1974) have used autoradiographic and cytophotometric techniques to study in a temporal sequence the changes which occur during the dedifferentiation of cartilage and muscle in the regenerating limb of the urodele *Desmognathus fuscus*. They have shown a reduction in fast-green stainable histone in the early phases of dedifferentiation and two peaks of nuclear DNA synthesis. The first peak occurs at the very beginning of dedifferentiation, 2 and 3 days after amputation, and in no case leads to 4C DNA content, whereas the second DNA synthesis occurs in the cells which have already accumulated in the blastema (10–15 days from amputation) and leads to the 4C (G_2) phase which immediately precedes mitosis. Jeanny & Gontcharoff (1974) have suggested that the early DNA synthesis might be a 'repair replication', but the possibility cannot be excluded that it may represent an amplification of particular DNA sequences in preparation for dedifferentiation. As to RNA, studies on the regenerating adult newt limb have shown that 28S, 18S, 5S, 4S and high molecular weight RNA have been found to increase over the period of dedifferentiation, accumulation and proliferation of blastema cells to a maximum during formation of the anlagen of the various limb parts. From this time on, synthesis of these RNA fractions, with the exception of high molecular weight RNA, decreases substantially. On the other hand, heterogeneous RNA appears to be synthesized at a constant rate up to anlagen formation (Morzlock & Stocum, 1971). Since an adequate nerve supply is crucial to morphogenesis and growth of the amphibian limb regeneration blastema (see Goss, 1969), the effect of denervation on RNA synthesis has been investigated. It has been found that denervation reduces the synthesis of ribosomal and transfer RNA by approximately 40–50% at any time during regeneration. It is not known, however, whether or not the synthesis of certain messenger RNAs is differentially inhibited during the regeneration process (Morzlock & Stocum, 1972).

Tail regeneration in lizards and salamanders has not been studied as intensively as limb regeneration in amphibia. Many species of lizard and salamander are not only able to regenerate a tail following amputation, but are also equipped with mechanisms for caudal autotomy. Autotomy of the tail occurs at preformed breakage planes across the caudal vertebrae; consequently, there occurs no tissue injury during caudal autotomy. The autotomized stump regenerates a new, though reduced, tail by means of a regeneration blastema. Not much is known about the origin of this blastema. Limiting the discussion to the striated muscle, it has been observed that it does not undergo, either in lizards (Simpson, 1970) or in the adult salamander *Desmognathus fuscus* (Mufti & Simpson, 1972), the classical dedifferentiation process and therefore does not release cells that could contribute to the blastema. Since the muscle of the regenerate is not derived from the dedifferentiated striated muscle fibers of the autotomy stump, it has been proposed that the regenerated muscle is derived either from connective tissue in the stump or from 'satellite cells' that have migrated out of the muscle and into the blastema. Satellite cells have been observed in the striated muscle of a wide range of vertebrates; e.g. in two species of lizards, in which some satellite cells have been found adjacent to the autotomy septum bordering the autotomy surface (Kahn & Simpson, 1974). Whether or not satellite cells of the lizard tail muscle participate in blastema formation must await further studies. In this context, it is worth noting that typical dedifferentiation is observed when the skeletal muscle is injured by amputation of the tail (Simpson, 1970).

Regeneration of mammalian liver

Following hepatectomy the liver remnant undergoes a process of compensatory hyperplasia whereby the liver reestablishes in a short time its initial mass. This process is clearly distinguished from typical regeneration because it does not replace the lobes which may have been removed and, therefore, does not reconstitute the original morphology (Harkness, 1957). Absence of metaplasia is the rule in the regeneration of liver and other glands in mammals: in fact, regeneration involves replacement of the same cell type by pre-existing cells without much dedifferentiation (Hay, 1968).

Liver response to hepatectomy is very rapid (Bucher, 1963; Alvarez, 1974). At 3 hours after removal of a third of the liver in male rats, both the diploid and the tetraploid parenchymal nuclei were found to have bound an acridine orange (AO) amount approximately 70% over the respective controls. This was followed by a decrease to normal levels at 6 hours and a subsequent depression to approximately 60% below the controls at 12 hours. Mean AO binding returned to slightly below control value by 24 hours (Fig. 9.3). The mean thermal stability of the chromatin of both diploid and tetraploid nuclei showed a decrease from control level at 3 hours post hepatectomy. This was followed by an increase to near controls for the diploids and below control for the tetraploids at 6 hours. Unlike the amount of AO bound, there was no apparent return to the control level at 12 and 24 hours (Fig. 9.4). Since the amount of DNA in liver nuclei during the time interval studied after operation did not change, the data on AO binding and thermal stability have been taken to reflect decondensation–condensation in liver chromatin which may be prerequisite for nucleic acid synthesis (Alvarez, 1974). An enhanced capacity to bind AO, which is usually paralleled by a decrease in the thermal stability of the chromatin, has been observed to precede RNA and DNA synthesis in other systems too. It has been suggested that 'activation' of

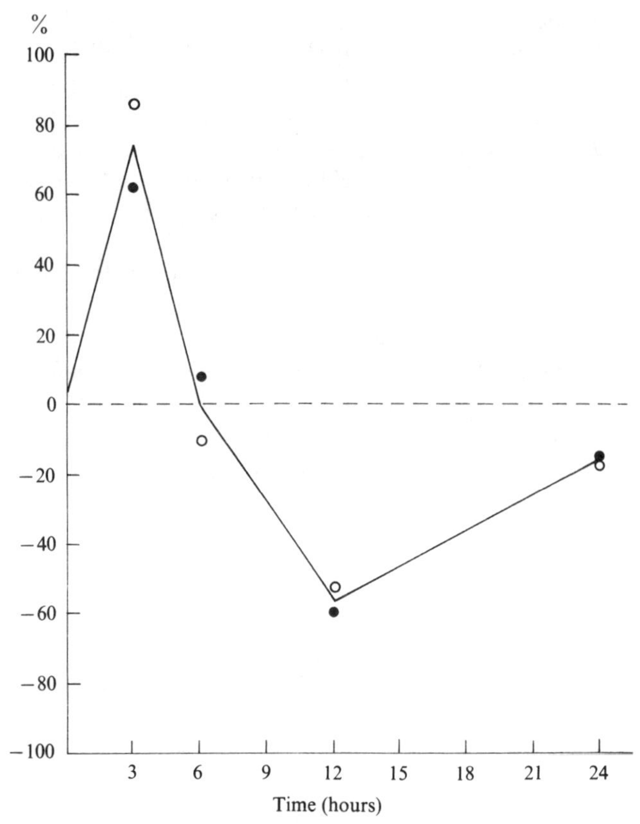

Fig. 9.3. Percentage change in mean acridine-orange nuclear fluorescence intensity at 530 nm (F_{530}) in the remnant lobes of rat liver as a function of time from partial hepatoctomy. Diploid and tetraploid cells are indicated by solid and empty circles respectively. The fluorescence intensity is taken to be proportional to the amount of bound dye. Note the similar changes in percent dye bound by diploid and tetraploid cells (after Alvarez, 1974).

cells for nucleic acid synthesis may be strictly dependent upon histone acetylation, which has been found to take place both in regenerating liver, where it occurs 1–3 hours after hepatectomy (Pogo, Pogo, Allfrey & Mirsky, 1968; Jones & Irvin, 1972) and in other systems (Darzynkievicz, Bolund & Ringertz, 1969; Berlowitz & Pallotta, 1972; Piesco & Alvarez, 1972). RNA synthesis in regenerating liver starts at approximately 6 hours after hepatectomy and is clearly manifested in a progressive increase in size of cells, nuclei and nucleoli which becomes maximal when the cells enter mitosis (Harkness, 1957). Depending on the age of the animal, the amount of liver removed and other factors, DNA synthesis in parenchymal cells starts at around 15–18 hours and reaches a peak at 22–26 hours after hepatectomy; the non-parenchymal cells, which are all diploid, lag approximately

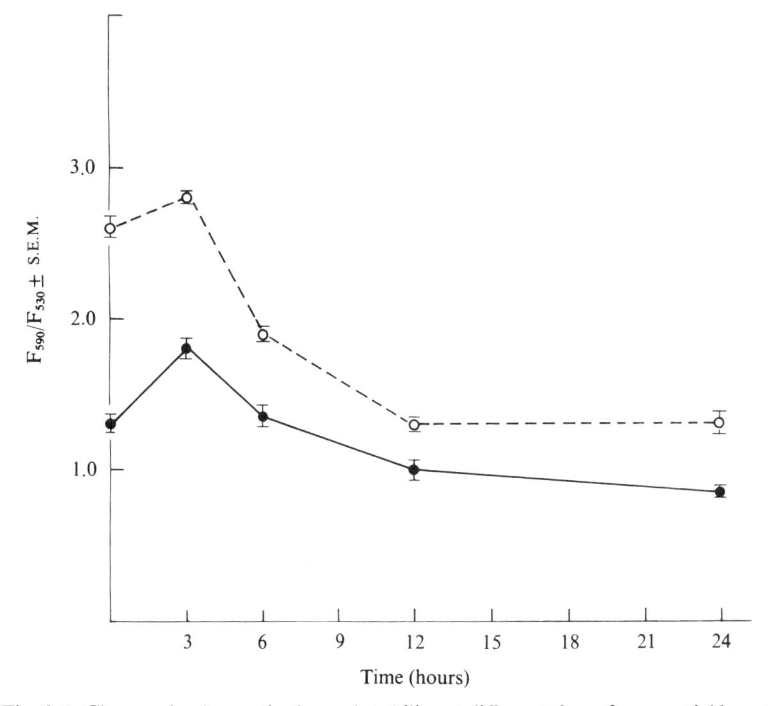

Fig. 9.4. Changes in chromatin thermal stability at different times from partial hepatoctomy in the rat. Diploid and tetraploid cells are indicated by solid and empty circles respectively. F_{590}/F_{530} is the ratio of fluorescence intensity of acridine–orange–stained nuclei at 590 and 530 nm. Isolated nuclei were heated to 90 °C prior to staining. Increasing numerical value on the ordinate reflects increasing denaturation (after Alvarez, 1974).

24 hours behind the parenchymal cells in synthesizing DNA (Bucher, 1963; Tidwell, Allfrey & Mirsky, 1968; Fukui, 1971). The peak in mitotic frequency in the liver remnant is observed at around 28 hours (Fig. 9.5).

Following the onset of mitosis the number of binucleate cells in the liver falls rapidly concomitant with a shift to a higher level of polyploidy, mostly due to fusion of mitotic spindles in the binucleate cell (references in Bucher, 1963). That binucleate cells are the source of tetraploid, or polyploid, cells is well documented by cytological analyses on liver regeneration in animals of different ages. When rats younger than 20 days are partially hepatectomized (their livers contain only 1–2% binucleate cells), tetraploid cells are not found in the liver remnant 4 days after operation (Nadal & Zajdela, 1966*a*). If, however, partial hepatectomy is made at an age when the incidence of binucleation in the liver is high, the liver remnant shows a dramatic decrease in the frequency of binucleate cells (Fig. 9.6) to which corresponds an increased frequency of tetraploid cells. Phenomena comparable to those described for the rat occur in the mouse liver following partial hepatectomy (Nadall & Zajdela, 1966*a*, *b*; Wheatley, 1972).

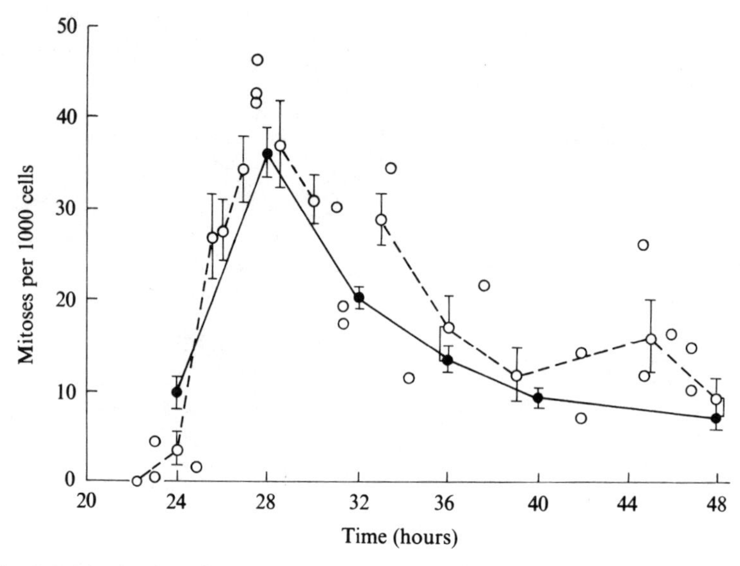

Fig. 9.5. Distribution of mitotic indices in the liver remnant at different times after partial hepatoctomy in rats. Data of two laboratories are shown by solid circles and empty circles (after Bucher, 1963).

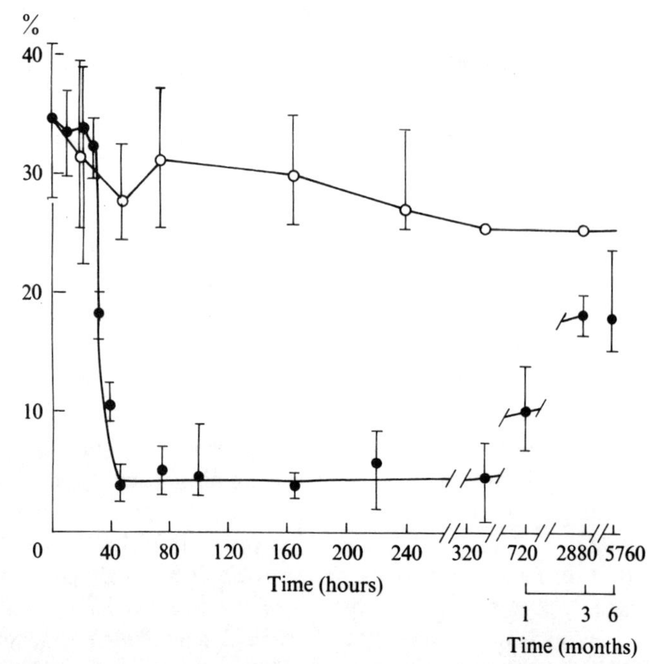

Fig. 9.6. Changes in percentage of binucleate hepatocytes after 70% partial hepatectomy in the rat. Partially hepatectomized and sham-operated rats are indicated by solid and empty circles respectively (after Wheatley, 1972).

Wolffian lens regeneration

In some species of urodeles, complete removal of the lens (lentectomy) is followed by formation of a new lens from a cell population derived from the dorsal part of the iris epithelium (Fig. 9.7). The phenomenon was first reported by Colucci (1885) and, in ignorance of Colucci's paper, by Wolff (1895), who published a series of papers concerning the subject. Since these early observations, Wolffian lens regeneration has been studied by a great number of workers, so that it represents the best-known process of regeneration in vertebrates. Wolffian lens regeneration has been shown to occur in the adult and larval stages in fourteen species of Salamandridae. Restriction of this type of regeneration to a small taxonomic group within the order Urodeles contrasts with the occurrence of limb and tail regeneration throughout this order. This suggests that the genetic factors controlling Wolffian lens regeneration are distinct from those controlling limb and tail regeneration (Yamada, 1967). That the capacity for lens regeneration is an intrinsic property of the dorsal iris and not the result of the intraocular environment has been demonstrated by grafting experiments. If the dorsal iris of a regenerating species (e.g. *Triturus*) is grafted into the optic chamber of another species of salamander which does not regenerate (e.g. *Ambystoma*), a lens is regenerated from the graft, whereas no lens regeneration occurs in the reciprocal graft. Grafting experiments have also shown the essential role of neural retina in Wolffian lens regeneration *in situ*. Quite recently, the influence of the neural retina has been demonstrated in the elegant experiments of Yamada, Reese & McDevitt (1973) and Yamada & McDevitt (1974) on lens tissue regeneration from iris epithelium *in vitro*. When the normal dorsal or ventral iris epithelium of adult *Triturus viridescens* was isolated and cultured in the presence of frog retinal complex, newt lens tissue was produced in almost all cultures. These lens tissues were positive for immunofluorescence for lens fiber-specific gamma crystallins as well as for total lens protein. Thus, Yamada's *in vitro* experiments have demonstrated unequivocally that Wolffian lens regeneration is due to the transformation of differentiated iris epithelial cells into another type of differentiated tissue, the lens tissue (metaplasia). On the other hand, several lines of indirect evidence, reviewed in Yamanda (1967, 1972), already concurred in supporting the notion of iris cell metaplasia during Wolffian lens regeneration.

The temporal sequence of events occurring in iris epithelial cells that differentiate into the primary lens fiber cells after lentectomy is now worked out (Table 9.1). The first prominent change observed after lentectomy in the adult newt (*Triturus viridescens*) is nucleolar activation. At the electron microscope level, the iris cell nucleoli observed 2 days after lentectomy already show an increase in the amount of the granular component and the fibrogranular region, and a close association with part of the chromatin. These changes in the nucleolar ultrastructure are further enhanced at 4 days after lens removal, when the size of the nucleoli is increased and their number is doubled (Dumont, Yamada & Cone, 1970).

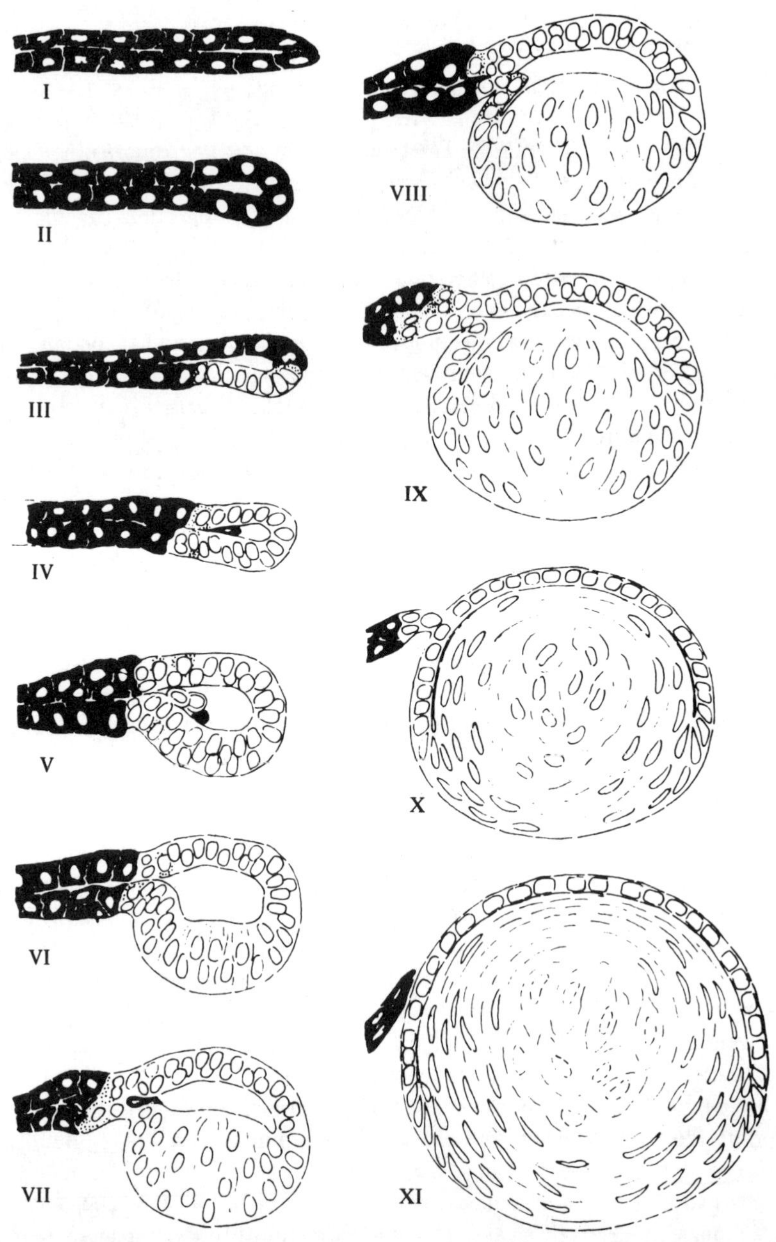

Fig. 9.7. Diagrams showing morphological stages of Wolffian lens regeneration in adult *Triturus viridescens*. Each figure represents a section through the mid-dorsal pupillary margin of the iris oriented perpendicular to the main body axis. The pigmented iris cells are indicated by black cytoplasm; depigmented regenerate cells by white cytoplasm. Incomplete depigmentation is shown by dots (after Yamada, 1967).

TABLE 9.1. *Program of cellular events in the transformation of iris epithelial cells during lens regeneration in salamanders* (from Yamada, 1972)

Days after lentectomy	Cell cycle	Cellular events
1–4	$G_0 \rightarrow G_1$	Nucleolar activation
3–5	S	First induced DNA replication
4–7	M	First induced mitosis
6–11	—	Cytoplasmic shedding including depigmentation
11–13	S, M	Final DNA replication followed by final mitosis
12–15	T*	Positive reaction for total crystallins
14–17	T	Start of lens fiber differentiation, Positive reaction for gamma as well as total crystallins
25–35	T	Degeneration of cell nucleus

* T: terminal phase.

TABLE 9.2. *The number of rRNA cistrons in normal dorsal iris and in iris at 7 days after lentectomy in adult newt,* Triturus viridescens (reprinted with permission from Collins, 1972. Copyright by the American Chemical Society)

	Normal iris	7-day iris	Liver
Total DNA (μg) used in the experiment	75	90	100
Total cpm in hybrid	590	1170	785
% rRNA cistrons	0.1	0.16	0.1
Number of rRNA cistrons	19900	31800	19900

The nucleolar changes are the visible manifestations of two biochemical processes which occur in the early phase of iris cell transformation, rRNA synthesis and amplification of ribosomal cistrons (rDNA). A significant increase in rRNA (18S and 28S) in iris cells occurs during 2 to 10 days after lens removal and a high level of rRNA is retained until the completion of lens regeneration (Reese, Puccia & Yamada, 1969). DNA–rRNA hybridization experiments have shown that a 1.6-fold increase in the number of rRNA genes occurs in cells of the dorsal iris within 7 days after lens removal (Collins, 1972). This is an amplification of 12000 units of these already redundant genes (Table 9.2). Since the activation of rRNA synthesis is the earliest biochemical change yet detected in lens regeneration, it is tempting to assume that this activation is the stimulus for re-entry of the resting (G_0) iris epithelial cells into the cell cycle (Yamada, 1972). DNA replication in iris epithelial cells after lentectomy starts on the third day; that is, one day later than the activation of rRNA synthesis. Subsequently, a significant increase in cell labeling occurs (Yamada & Roesel, 1969). The first mitotic cycle occurs between 3 and 7 days after lentectomy, when the iris cells are still heavily pigmented, and is followed by one or more cell divisions before the iris cells are converted into lens cells (Eisenberg & Yamada, 1966; Yamada & Roesel, 1969; Reyer, 1971).

Conspicuous cytoplasmic changes in iris epithelial cells occur at days 6–11 from lentectomy; they have been called 'cytoplasmic shedding' by Yamada (1972). The process involves the extrusion of the pigment granules (melanosomes) from the activated iris cells (depigmentation), which is associated with extensive alterations of the cell surface and loss of portions of cytoplasm. When most of the melanosomes are lost, the nucleus of the iris epithelial cells is left with a thin layer of cytoplasm surrounding it (Dumont & Yamada, 1972). When cytoplasmic shedding is completed (complete dedifferentiation), the incorporation of RNA precursors in the cell is accelerated and there are indications that the completion of cytoplasmic shedding is accompanied by alteration in the pattern of synthesis of nonribosomal RNA (Yamada, 1972). Recently, Zalik & Scott (1973) have found a surface RNA component, presumably associated with a carbohydrate in normal iris cells, and have suggested that this component is involved in keeping the iris cells in a differentiated state. They have also suggested that early disappearance from the cell periphery at the onset of depigmentation may be a signal to the cell to dedifferentiate. The population of cells which have completed cytoplasmic shedding is assembled into a vesicle at the dorsal margin of the iris epithelium and starts to accumulate lens-specific antigens and to manifest lens-specific morphogenesis (Yamada, 1972). Whether or not the program of cellular events in the transformation of iris epithelial cells may apply to other systems of cellular metaplasia cannot yet be decided.

Nuclear transplantation in animal eggs

Transplantation of a somatic cell nucleus into the enucleated egg of an animal provides a direct test of whether the somatic nucleus is able to substitute for the zygote nucleus in supporting normal development. The first real success in transplanting a living nucleus in animal eggs was achieved by Briggs & King (1952) in the frog *Rana pipiens*. Since then, nuclear transplantation has been successfully applied to many amphibian species (Table 9.3), to *Drosophila* and, with only limited success, to some other insect species (references in Signoret, 1973; Gurdon, 1974). Among the amphibians studied so far, the frog *Xenopus laevis* has provided the most convincing evidence for the genetic similarity of all somatic cell nuclei. In the experiments of Briggs and King (1952), the egg nucleus was removed from an unfertilized egg with a needle and an embryonic nucleus (from the blastula) was injected into the enucleated egg. In subsequent work on *Rana pipiens* and other amphibian species, successful transplantation experiments have been achieved not only with nuclei from embryonic stages (from blastula to tailbud), but also with larval germinal cells and different kinds of tissue from larvae or adults

TABLE 9.3. *Examples of development of normal tadpoles or adults from enucleated eggs injected with nuclei of different origin in Amphibia*

Species	Source of nuclei	Most advanced developmental stage attained	Reference
Rana pipiens	Blastula, gastrula	Tadpole	Briggs & King (1952, 1957)
	Primordial germinal cells	Tadpole	Smith (1965)
	Endoderm from late gastrula or tailbud stage	Tadpole	Hennen (1970)
	Lens epithelium from adult	Tadpole	Muggleton-Harris & Pezzella (1972)
Xenopus laevis	Intestinal epithelium from tadpole	Adult	Gurdon (1962)
	Skin from tadpole or adult	Adult	Gurdon & Laskey (1970), Gurdon (1974)
	Lung, skin, kidney from adult	Tadpole	Laskey & Gurdon (1970)
Ambystoma mexicanum	Blastula	Tadpole	Signoret, Briggs & Humphrey (1962)
Pleurodeles waltlii	Blastula, gastrula	Tadpole	Signoret & Picheral 1962
	Neurula and tailbud stage	Adult	Picheral (1962)
	Blastula of F1 hybrid *P. waltlii*×*P. poireti*	Adult	Gallien (1970)
Pleurodeles poireti	Blastula of F1 hybrid *P. waltlii*×*P. poireti*	Adult	Gallien (1970)

(Table 9.3). Nuclei from more advanced developmental stages and from more differentiated cells always promote a quantitatively and qualitatively less normal nuclear-transplant embryo development than nuclei from early developmental stages or undifferentiated cells (Fig. 9.8). As seen from Table 9.3, normal larvae or adults have been obtained from enucleated cells injected with nuclei of adult tissues in *Rana pipiens* and *Xenopus laevis*. In *Xenopus*, satisfactory results are obtained when adult tissues are explanted *in vitro* and the donor nuclei are taken from the cultured cells. On the other hand, very poor results are obtained if nuclei from adult cells are transplanted immediately after cell dissociation (Gurdon & Laskey, 1970; Laskey & Gurdon, 1970). Both in *Xenopus* and other amphibians, results are much improved if use is made of serial nuclear transfers (details in Gurdon, 1974). A very important characteristic of the intestine and skin cells used by Gurdon and his associates is their possession of a genetic nuclear marker. In these nuclear transfer experiments, donor nuclei were taken from cells of the

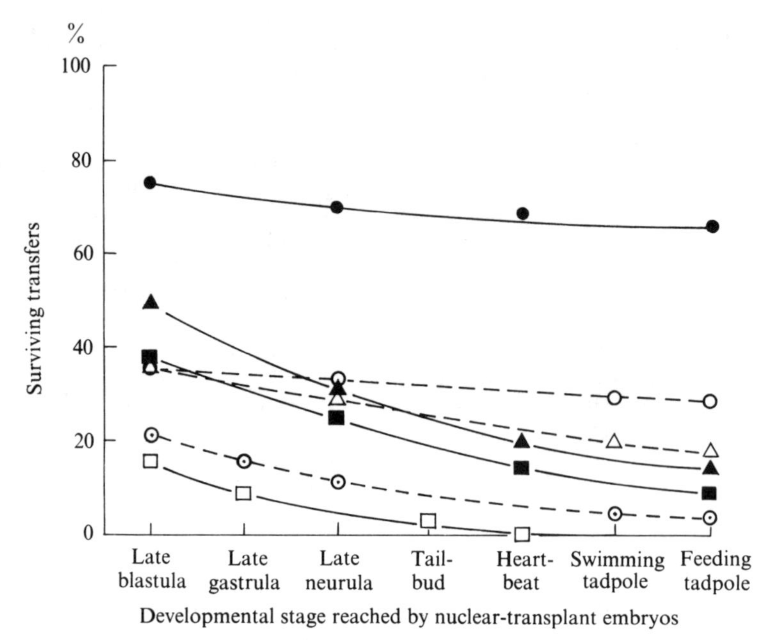

Fig. 9.8. Survival of nuclear-transplant embryos. The proportion of nuclear-transplant eggs which develop normally declines as donor nuclei are taken from progressively more advanced stages. The results shown are for endoderm nuclei and are based on data from different laboratories (after Gurdon, 1974). ●——● *Rana* blastula (a), ▲——▲ *Rana* neurula (a), ■——■ *Rana* tail-bud (a), ○ - - - - ○ *Xenopus* blastula (b), △ - - - - △ *Xenopus* neurula (b), □——□ *Rana* tail-bud (c), ⊙ - - - - ⊙ *Xenopus* swimming tadpole (b).

uninucleolate (1-*nu*) strain (one nucleolus per cell) and were transplanted into eggs of wild-type animals (two nucleoli per cell) whose nucleus was killed by UV irradiation. This made it possible to prove that every nuclear-transplant embryo was wholly of donor nucleus origin. The general conclusion to be drawn from the experiments discussed above is that at least some nuclei of differentiated tissues in amphibians are totipotent, carrying an entirely normal set of genes.

The decrease in the developmental capacity of transplanted nuclei with the increasing age of donor is expressed by a progressive increase, in frequency and degree, of developmental abnormalities of nuclear-transplant embryos. It is now agreed that the great majority, if not all, of such developmental abnormalities are due to chromosome abnormalities, which are themselves a consequence of nuclear transplantation. There is a close relationship between abnormal development and abnormal chromosome constitution in nuclear-transplant embryos (Gurdon, 1974). Certainly, chromosome breakage occurs after nuclear transplantation, but the factors responsible for the origin of chromosomal changes, not only at the first but also at later divisions of the transplanted nuclei, are not known. It has been shown by Hennen (1970) that the capacity to promote postgastrula development

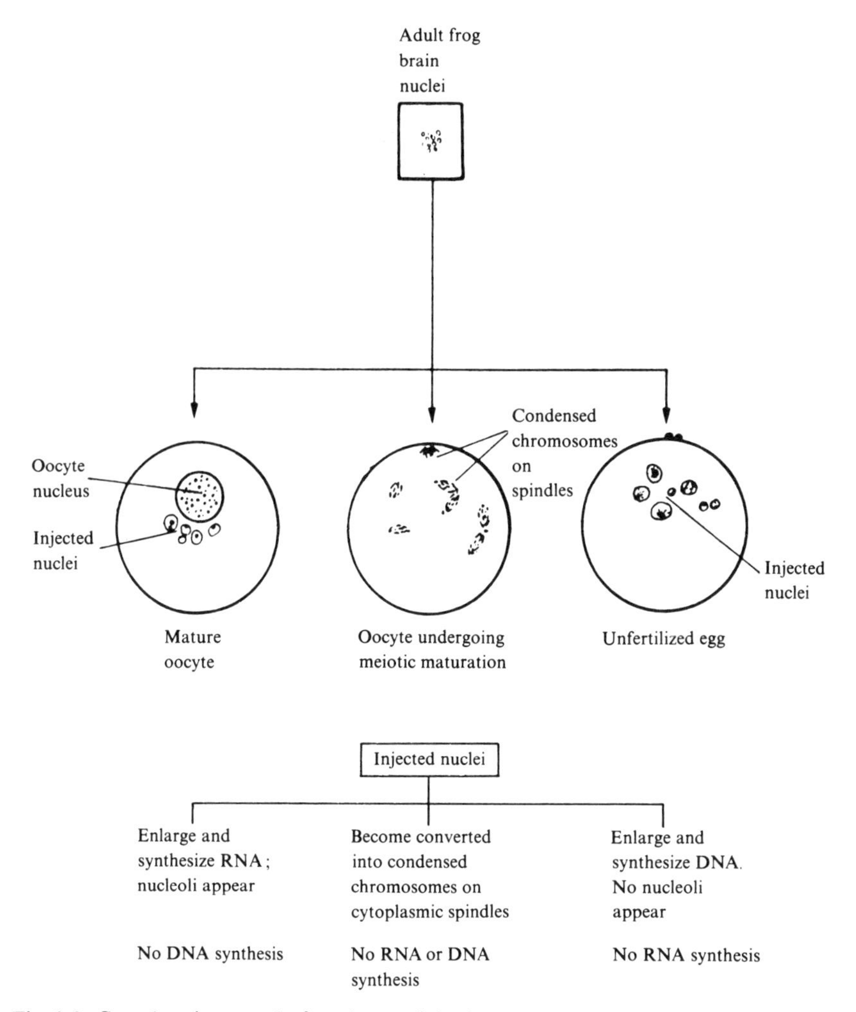

Fig. 9.9. Cytoplasmic control of nuclear activity in *Xenopus*. Adult brain nuclei, which do not normally synthesize DNA or divide, are made to change activity by exposure to different kinds of oocyte or egg cytoplasm (after Gurdon, 1974).

in *Rana pipiens* is significantly improved when the donor cells are exposed to the polyamine spermine prior to and during nuclear transplantation. This improvement is correlated with a reduction in the incidence of chromosomal abnormalities. The mechanism of the stabilizing effect of spermine on chromosome structure in nuclear transplants is not known.

Nuclear transplantation in *Xenopus* has provided a means of studying the effects of oocyte or egg cytoplasm on nuclear activity. Thus, multiple adult frog brain nuclei, which synthesize DNA extremely rarely, were injected into three different types of cell (Fig. 9.9): (i) mature oocytes, which synthesize RNA, but no DNA; (ii) oocytes undergoing meiotic maturation, whose chromosomes are condensed; and (iii) unfertilized eggs, whose nucleus synthesized DNA, but not RNA, immediately after activation or nuclear injection. In each case the adult brain nuclei were induced within a few hours to change activity or function so as to conform to that characteristic of the host cell (Graham, Arms & Gurdon, 1966; Gurdon, 1968).

Regeneration in plants

The amazing capacity for regeneration in plants offers a good opportunity for discussing cell totipotency both *in vivo* and *in vitro*. The discussion which follows will be limited to seed plants (Spermatophyta), because they represent the most highly differentiated phylum of plants and because they have been made the subject of extensive investigations. The recent rapid progress in research on organogenesis and embryogenesis *in vitro* (reviews by Vasil & Vasil, 1972; Reinert, 1973; Murashige, 1974) has provided new evidence on plant cell totipotency, in addition to several valuable observations made on plants *in vivo*. On the other hand, it must be realized that totipotency of a given cell type as expressed *in vitro* is merely a reflection of an inherent property of that cell type which, in the situation *in vivo*, must obey the several restrictions imposed by the pattern of development and growth of an organized system.

Plants from single cells *in vivo*

The most common expression of totipotency of somatic plant cells *in vivo* is the production of an embryo from single cells belonging to the nucellus or integument in the ovule (adventitious embryony). As seen in another section of this book (p. 75), adventitious embryony is a particular type of apomictic seed development (agamospermy) which has the typical attribute of vegetative (clonal) propagation. The cell destined to produce an adventitious embryo first enlarges and becomes densely cytoplasmic and then divides rapidly to form a proembryonic mass which continues development according to a typical embryogenic pattern. This mode of development differs from that of a zygotic embryo or, more generally, of embryos of embryo sac origin because in this case the first cell division is polarized and produces the initial cells of the embryo proper and the initial cell of the suspensor. Thus, the lack of a suspensor is a typical characteristic of adventitious embryos (cf. discussion in Haccius & Lakshmann, 1969). The nucellar cells of *Nothoscordum fragrans*, which give rise to adventitious embryos, are initially connected to the neighbouring cells by many plasmodesmata (Halperin, 1970); the same situation is found in epidermal cells of seedling stem segments of *Ranunculus sceleratus* which produce adventitious embryos when transplanted *in vitro* (Konar, Thomas & Street, 1972). Such evidence does not support the view (Steward,

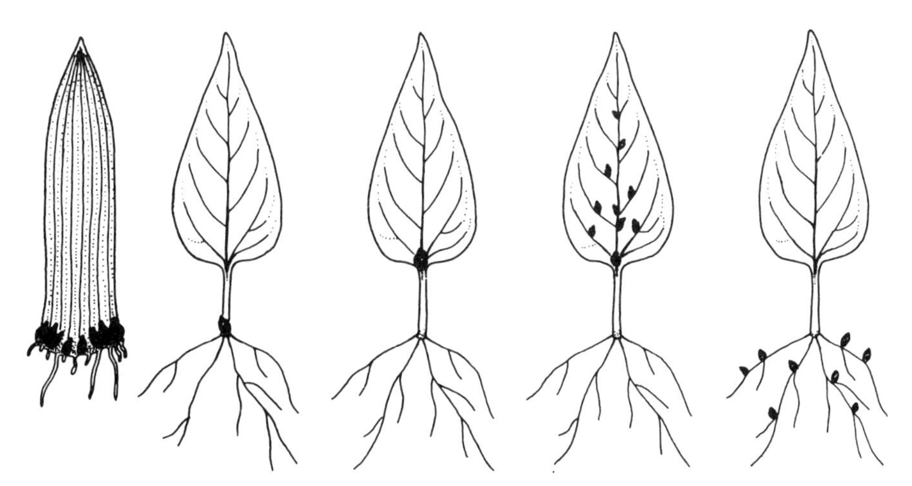

Fig. 9.10. Different types of location of adventitious bud formation on isolated leaves (after Broertjes *et al.*, 1968).

Blakely, Kent & Mapes, 1963) that the condition of the single isolated cell, ensuring physiological isolation, is necessary for the *in vitro* embryonic development of a somatic plant cell (see below).

Cell totipotency *in vivo* is also expressed under particular experimental conditions. Winkler (1916) was the first to show that stem decapitation (wound callus formation) in *Lycopersicum esculentum* and *Solanum nigrum* induces the formation of adventitious buds (and plants) with diploid, tetraploid and octoploid chromosome numbers. Having observed mitoses with the same ploidy levels in the stem following decapitation, Winkler proposed that the polyploid adventitious shoots originated from polyploid cells pre-existing in the stem. Curiously enough, subsequent workers, who studied the production of polyploid plants through the wound callus method, strove to try to show that polyploid nuclei originated by nuclear fusion in the wound tissue. Only in 1942, did Howard reestablish Winkler's notion, which is now generally accepted, that the polyploid adventitious shoots produced by the stem decapitation callus in species of Solanaceae and Cruciferae originate via mitosis stimulation of pre-existing polyploid (= endereduplicated) cells. Since both the diploid and the polyploid adventitious shoots obtained through stem decapitation or, in case of the potato tuber, through eye excision (Howard, 1961) are generally not mosaic, they must be traced back to a single initial cell (D'Amato, 1959).

Formation of adventitious buds, which eventually develop into a complete plant, can also be obtained on roots, bulbs, and leaves. Broertjes, Haccius & Weidlich (1968) list 350 angiosperm species, belonging to 50 families, in which detached leaves can produce adventitious buds when maintained under appropriate conditions (Fig. 9.10). In some of these species, it has been shown conclusively that the adventitious buds formed on detached leaves ultimately originate from only one epidermal cell (review by Broertjes, 1972). For example, in the African violet, *Saintpaulia*, the origin of adventitious buds from single epidermal cells was first shown histologically by Naylor & Johnson (1937) and convincingly proved by Sparrow, Sparrow & Schairer (1960), who obtained almost exclusively non-mosaic mutant plants after the irradiation of leaves. As shown by the extensive experiments of Broertjes on induced mutations in vegetatively propagated species, an origin from a single epidermal cell of adventitious buds regenerated from irradiated leaves is also demonstrated for *Achimenes*, *Streptocarpus*, and *Kalanchoë* (Broertjes, 1972). Since in *Saintpaulia*, *Achimenes*, *Streptocarpus* and *Kalanchoë* induced mosaic mutants are quite exceptional, it must be inferred that most, if not all, leaf epidermal cells in the adult leaves of these species must be in G_1 (Broertjes, 1972).

Plants from single cells *in vitro*

In 1939, Gautheret in Paris, Nobécourt in Grenoble and White in Princeton independently succeeded in obtaining the unlimited *in vitro* growth of plant tissues. Since then, *in vitro* cultures on either solid or liquid media have been obtained in a number of plant species. In seed plants, cultures have been established from such diverse materials as: ovaries, ovules and nucelli; excised embryos; seeds; differentiated portions of leaves, flowers, inflorescences; roots, stems, bulbs, tubers, rhizomes; tumor tissues; apical meristems; pith parenchyma; secondary phloem; pericarp and endocarp of fruits; endosperm; female gametophytes of gymnosperms; anthers; pollen grains, and wall deprived cells (protoplasts) (references in D'Amato, 1972*b*, 1977; Murashige, 1974).

By appropriate modification of the cultural medium, it is possible to regenerate plants *in vitro* from tissue and cell cultures. Knowledge of the nuclear cytology of plants regenerated *in vitro* is of special importance, not only for the problem of cell totipotency but also for the problem of genetic stability of cells cultured *in vitro* (reviews by D'Amato, 1975*a*, 1977). Plant regeneration *in vitro* may occur either through formation of an embryo which further grows to a plantlet or through formation of an adventitious bud giving rise to a shoot and eventually a plantlet (reviews by Vasil & Vasil, 1972; Reinert, 1973; Murashige, 1974). It is now certain that embryos *in vitro* originate from single cells. In carrot, the development of a single parenchymatic cell into an embryo has been documented by microcinematography (Backs-Hüsemann & Reinert. 1970). In tobacco and other species, isolated pollen grains in culture or pollen grains within cultured anthers have been shown to divide to form an embryo and plantlet. In general, plants regenerated *in vitro* from pollen grains are haploid, but in some species diploid, triploid, tetraploid and hexaploid embryos (plants) have also been obtained from pollen grains (Table 2 in D'Amato, 1977). The origin of diploids is not astonishing, since anthers may contain, albeit infrequently, unreduced pollen grains due to meiotic failure (e.g. restitution nucleus formation at the first meiotic division). Formation of triploid embryos is due to a peculiar nuclear process which has been elucidated in *Datura innoxia* (Sunderland, Collins & Dunwell, 1974): in the binucleate pollen grain, the generative nucleus, which has endoreduplicated (n diplochromosomes), and the vegetative nucleus (n chromosomes) divide on a common spindle to form two triploid daughter nuclei which continue division to form the embryo. Failure of the spindle mechanism or chromosome doubling in a triploid nucleus can easily explain the occasional development from pollen grains of hexaploid and triplo-hexaploid (mosaic) plants in *Datura innoxia* and *Petunia hybrida*. The mechanism of the formation of tetraploid embryos from pollen grains is not clear, but spindle (nuclear) fusion and/or chromosome endoreduplication probably operate in the transformation of an initially haploid, or diploid, pollen grain into a tetraploid embryonic structure.

In several species, *in vitro* cultures of pollen grains, or anthers, develop into an unorganized tissue mass, the callus, from which plants can be regenerated via induction of adventitious buds. Callus formation also occurs when somatic plant tissues are transplanted *in vitro*; the callus or cells isolated from the callus can be utilized to regenerate plants (regeneration in most species occurs via adventitious shoots). Only the shoot apex, a naturally proliferating organized structure, does not form callus *in vitro* provided that it is cultured under conditions which are not conducive to unorganized growth (callus). Under these conditions, the shoot apical cells maintain the original genetic constitution of the plant and can be used for its continued *in vitro* clonal propagation. Thus, it is now possible to maintain *in vitro* the genetic continuity of the meristematic cell line which every plant maintains *in vivo* throughout its life (D'Amato, 1977).

Regeneration of plants from tissue and cell cultures via adventitious bud formation traces back, at least generally, to a single cell. A single-cell origin has been inferred from chromosome counts in different meristems (root and shoot or leaf) in one and the same plant or from an analysis of microsporocytes in different flower twigs in a plant. Plant regeneration from *in vitro* cultures raises the question of the relationship between the nuclear conditions in the culture and the nuclear status of the regenerates (D'Amato, 1977). Apart from *Lilium longiflorum*, which keeps a relatively good stability at the diploid level in long-term cultures (Sheridan, 1975), the other species so far investigated experience nuclear instabilities of different degrees. Thus, in a callus from a somatic explant, in addition to diploid nuclei, nuclei with polyploid or aneuploid chromosome numbers or with chromosomal structural changes are found. Although polyploidy may be already present in the explant (endoreduplicated nuclei *in vivo*), the frequency and degree of polyploidy clearly increases as a consequence of the selective proliferation of polyploid cells made possible by some cultural regimes and the occurrence *in vitro* of spindle disturbances or chromosome endoreduplication. Aneuploidy is generally exceptional *in vivo*, but may be very frequent and involve a wide spread in chromosome numbers in *in vitro* cultures due to multipolar spindles, which also favour genome segregation (production of haploid and odd-ploid – e.g. triploid – chromosome numbers). Despite the great variation in the nuclear conditions in plant tissue and cell cultures of some truly diploid species (those in which 2n = 2x), mostly or only diploid plants have been regenerated *in vitro* (D'Amato, 1977, Table 1). It is not known whether the original karyotype is conserved in these cases; this seems to be so in *Lilium longiflorum* (Sheridan, 1975). Further karyotype and/or genetic analyses of diploid plants regenerated *in vitro* are needed to decide whether the meristematic cell line of the plant may be preserved even under conditions, such as callus induction and growth, which are conducive to nuclear changes. In plants of a polyploid nature, e.g. tobacco (2n = 4x = 48), plants regenerated *in vitro* directly from calli or from callus-derived cultures may be either euploid (48 or 96 chromosomes) or aneuploid. In any case, the aneuploidy includes chromosome numbers higher than 48, ranging from 50 to 71, from 82 to 90 and from 94 to 97. An interesting unexplained phenomenon is that the chromosome number of plants regenerated from *in vitro* cultures of tobacco is in all cases lower than the mode number (chromosome number most frequent in the culture) in all cases (D'Amato, 1977, Table 1).

In conclusion, plant cell totipotency is not restricted to diploid cells, but it is also expressed in conditions of haploidy, polyploidy and aneuploidy. The *in vitro* culture technique serves the purpose of triggering differentiated cells into a developmental, either organogenic or embryogenic, course. Certainly, cell dedifferentiation and diversion of cells from their differentiative course occur when plant tissues or cells are transplanted *in vitro*, but little is known of the basic mechanisms of these cellular events. For cell dedifferentiation (callus induction) to occur both an auxin (e.g. indole acetic acid (IAA), naphtalene acetic acid (NAA), 2,4-dichlorophenoxyacetic acid (2,4-D)) and a cytokinin are needed. During callus induction in pea epicotyl and tobacco stem pith, 2,4-D has been found to form a complex with lysine-rich and moderately lysine-rich histones thereby increasing the ratio of acidic proteins to basic proteins. Since lysine-rich histones inhibit DNA polymerase (Gurley, Irvin & Hoolbrook, 1964), it has been proposed that complex formation between 2,4-D and lysine-rich histones derepresses DNA polymerase (Yamada, Yasuda, Koge & Sekiya, 1971). As to cytokinins, the available evidence tends to support the view that during callus induction they act as modulators of protein synthesis in the cell (Skoog, 1971). During cell dedifferentiation there occurs an activation of rRNA and mRNA synthesis. Thus, the *in vitro* callus of tobacco stem pith contains a much greater amount of rRNA than control pith; moreover, this same callus, when transferred on a medium devoid of IAA and containing an increased concentration of kinetin, shows two additional RNA fractions, probably of messenger type (Yamada, Matsumoto & Takahashi, 1968). At the cytological level, activation of rRNA synthesis in callus cells is manifested by an enlargement of the nucleoli, their vacuolization and a clear segregation of the fibrillar and granular components in the nucleoli (Gifford & Nitsch, 1969; Vasil, 1971). Whether enhanced rRNA synthesis may be bound to an amplification of ribosomal cistrons (rDNA) is not known. There is, however, evidence of DNA amplification in plant tissues cultured *in vitro*. This aspect has already been discussed in the last section of Chapter 6.

Little is known of the changes in chromosomal proteins which may occur during cell dedifferentiation in plants. Carrot cell cultures do not show any change in histone after induction of embryogenesis; by contrast, limited but distinct changes in the chromosomal nonhistone proteins are observed (Gregor, Reinert & Matsumoto, 1974).

Although the temporal sequence of the cellular events which occur during plant cell dedifferentiation has not yet been worked out, it seems probable that it may follow the scheme encountered in dedifferentiating animal systems, e.g. iris cell dedifferentiation during Wolffian lens regeneration.

Plants from somatic cell hybrids

The presence of a pecto-cellulosic cell wall has hindered for a long time the development in plants of techniques such as cell fusion, nuclear transplantation, etc., so well developed in animal cells (reviews by Davidson, 1973; Gurdon, 1974). In recent times, methods for the mechanical or, more commonly, enzymatic removal of the wall and the *in vitro* culture of the naked cells (protoplasts) have been developed (review by Cocking, 1972). When cultured in appropriate media, protoplasts regenerate their wall and divide to form a callus from which plants can be regenerated. This result has now been achieved in several plant species (references in D'Amato, 1977).

Plants have been regenerated from both interspecific and intraspecific somatic cell hybrids. Carlson, Smith & Dearing (1972) were the first to produce by somatic cell hybridization ('parasexual hybridization') the amphidiploid *Nicotiana glauca*×*N. langsdorffii* (2n = 42). This amphidiploid had previously been obtained by several authors by crossing *N. glauca* (2n = 24) with *N. langsdorffii* (2n = 18) and doubling experimentally the chromosome number of the F1 hybrid (2n = 21 = 12+9). The amphidiploid thus obtained (2n = 42) is fertile and spontaneously produces tumors (genetic tumors); its tissues can be grown *in vitro* in the absence of auxin (auxin autotrophy). On the other hand, auxin is needed for the *in vitro* growth of tissues of the two parental species. This biological difference between the amphidiploid and its parental species was also expressed when protoplasts were cultured *in vitro*. In Nagata and Takebe's medium protoplasts of *N. glauca* and *N. langsdorffii* did not regenerate into a callus, whereas about 0.01% of the protoplasts of the amphidiploid divided to form a callus. This offered a selection method for preferentially recovering somatic hybrids. From populations of mixed protoplasts of *N. glauca* and *N. langsdorffii*, Carlson *et al.* (1972) isolated 33 regenerated calli which grew vigorously on a medium containing no added hormones. Three regenerated plants were shown to have 2n = 42 chromosomes and manifested biochemical and morphological characteristics identical to those of the sexual amphidiploid. More recently, plants have been regenerated from intraspecific cell hybrids. In tobacco, Melchers & Labib (1974) have succeeded in regenerating 48-chromosome plants from fused protoplasts of two haploid ($n = 2x = 24$) chlorophyll-deficient and light-sensitive strains bearing respectively the recessive genes *sublethal* (*s*) and *virescent* (*v*). These genes complement to normal leaf colour and resistance to high light intensity both in the sexual (F1: $s \times v$) and the somatic hybrid; both hybrids segregate in F2. Of the twenty-one somatic hybrids studied, ten had 48 chromosomes, seven had a chromosome number in the neighborhood of 70, one was tetraploid, one was triploid and the remaining two had 46 and 90 chromosomes respectively.

In the liverwort *Sphaerocarpos donnellii* ($n = 8$: 7A+X and 7A+Y in the female and the male gametophyte respectively), fusion has been obtained between protoplasts of a normal green, nicotinic acid-deficient female (nic_2) and a pale green, glucose-deficient male (pal_2). A somatic hybrid has been produced: the plant was autotrophic green and diploid (14A+XY) (Schieder, 1974). The use of genetic complementation for selecting somatic cell (plant) hybrids is a method of great potential value. It remains to be investigated whether genetic complementation may also occur at the interspecific level.

References and author index

The figure(s) in brackets at the end of each reference refer to the pages in which this reference appears.

Abel, W. O. (1967). Intergenische Rekombination und Chromosomenreduplication. *Ber. deutsch. bot. Ges.* **80**, 517–22. (58)

Adamson, D. (1962). Expansion and division in auxin treated plant cells. *Can. J. Bot.* **40**, 719–44. (114)

Agrell, I. (1957). The cellular division during the early embryonic development of the sea urchin embryo as influenced by temperature changes. *Kungl Fysiograf. Sällsk. Lund Förh.* **27**, 61–72. (37)

Agrell, I. (1964). Natural division synchrony and gradients in metazoan tissues. In *Synchrony in Cell Division and Growth*, ed. by E. Zeuthen, pp. 39–70. New York: Interscience Publishers. (35, 36, 37)

Alberts, B. M. (1970). Function of gene 32-protein, a new protein essential for the genetic recombination and replication of T_4 bacteriophage DNA. *Proc. Fed. Am. Soc. Exp. Biol.* **29**, 1154–63. (61)

Alfert, M. (1958). Variation in cytochemical properties of cell nuclei. *Exptl. Cell Res.* Suppl. **6**, 227–35. (32)

Alfert, M. & Geschwind, I. I. (1958). The development of polysomaty in rat liver. *Exptl. Cell Res.* **15**, 230–2. (116)

Allen, E. R. & Cave, M. D. (1968). Formation, transport and storage of ribonucleic acid-containing structures in oocytes of *Acheta domesticus* (Orthoptera). *Z. Zellforsch. Mikrosk. Anat.* **92**, 477–86. (149)

Allfrey, V. G. & Mirsky, A. E. (1963). Mechanisms of synthesis and control of protein and ribonucleic acid synthesis in the cell nucleus. *Cold Spring Harbor Symp. Quant. Biol.* **28**, 247–72. (195)

Aloni, Y., Halten, L. & Attardi, G. (1971). Studies of fractionated HeLa cell metaphase chromosomes. II. Chromosomal distribution of sites for transfer RNA and 5S RNA. *J. Mol. Biol.* **56**, 555–63. (172)

Alonso, C. & Berendes, H. D. (1975). The location of 5S (ribosomal) RNA genes in *Drosophila hydei*. *Chromosoma* **51**, 347–56. (172)

Alonso, P. & Perez-Silva, J. (1965). Giant chromosomes in Protozoa. *Nature* **205**, 313–14. (130)

Alpi, A., Tognoni, F. & D'Amato, F. (1975). Growth regulator levels in embryo and suspensor of *Phaseolus coccineus* at two stages of development. *Planta* **127**, 153–62. (166)

Alston, R. E. (1967). *Cellular Continuity and Development.* Glenview: Scott, Foresman and Co. (2)

Altenburg, E. & Browning, L. S. (1961). The relatively high frequency of whole-body mutations compared with fractionals induced by X-rays in *Drosophila* sperm. *Genetics* **46**, 203–12. (89)

Alvarez, M. R. (1974). Early nuclear cytochemical changes in regenerating mammalian liver. *Exptl. Cell Res.* **83**, 225–30. (211, 212, 213)

Ammermann, D. (1964). Riesenchromosomen in der Makronukleusanlage des Ciliaten *Stylonychia* spec. *Naturwiss.* **51**, 249. (130)

Ammermann, D. (1965). Cytologische und genetische Untersuchungen an dem Ciliaten *Stylonychia mytilus* Ehrenberg. *Arch. Protistenk.* **108**, 109–52. (130)

Ammermann, D. (1971). Morphology and development of the macronuclei of the ciliates *Stylonychia mytilus* and *Euplotes aediculatus. Chromosoma* **33**, 209–38. (130)

Ancora, G., Brunori, A. & Martini, G. (1972). Total protein to DNA ratio in 2C nuclei of dry seed meristems in relation to chromosome and chromatid aberrations induced by X-irradiation. *Caryologia* **25**, 373–81. (49)

Anderson, E. G., Longley, E., Li, C. H. & Retherford, K. L. (1949). Hereditary effect produced in maize by radiation from the Bikini atomic bomb. I. Studies on seedlings and pollen of the exposed generation. *Genetics* **34**, 639–46. (104)

Angold, R. E. (1968). The formation of the generative cell in the pollen grain of *Endymion non-scriptus* (L.). *J. Cell Sci.* **3**, 573–8. (12)

Antropova, E. N. & Bogdanov, Y. F. (1970). Cytophotometry of DNA and histone in meiosis of *Pyrrhocoris apterus. Exptl. Cell Res.* **60**, 40–4. (56)

Arrighi, F. E. & Hsu, T. C. (1971). Localization of heterochromatin in human chromosomes. *Cytogenetics* **10**, 81–6. (31)

Arrighi, F. E., Hsu, T. C., Saunders, P. & Saunders, G. F. (1970). Localization of repetitive DNA in the chromosomes of *Microtus agrestis* by means of *in situ* hybridization. *Chromosoma* **32**, 224–36. (24, 186)

Artman, M. & Roth, J. S. (1971). Chromosomal RNA: an artifact of preparation? *J. Mol. Biol.* **60**, 291–301. (25)

Ashburner, M. (1970). Function and structure of polytene chromosomes during insect development. *Adv. Insect Physiol.* **7**, 1–95. (130, 176)

Ashburner, M. (1972). Puffing patterns in *Drosophila melanogaster* and related species. In *Developmental Studies on Giant Chromosomes*, ed. by W. Beermann, pp. 101–51. Berlin: Springer-Verlag. (130, 180, 199)

Asseyeva, T. (1927). Bud mutations in the potato and their chimerical nature. *J. Genet.* **19**, 1–26. (97)

Astaurov, B. L. & Ostriakova-Varshaver, V. P. (1957). Complete heterospermic androgenesis in silkworms as a means for experimental analysis of the nucleus–cytoplasm problem. *J. Embryol. Exp. Morphol.* **5**, 449–62. (21)

Austin, C. R. (1960). Anomalies of fertilization leading to triploidy. *J. Cell. Comp. Physiol.* **56**, suppl. **1**, 1–15. (20, 21)

Austin, C. R. (1965). *Fertilization.* Englewood Cliffs: Prentice-Hall. (19, 20, 109, 110)

Austin, C. R. (1969). Variations and anomalies in fertilization. In *Fertilization* **2**, ed. by Ch. B. Metz and A. Monroy, pp. 437–66. New York: Academic Press. (21)

Austin, C. R. & Amoroso, E. C. (1959). The mammalian egg. *Endeavour* **18**, 130–43. (15)

Avanzi, M. G. (1950). Endomitosi e mitosi a diplocromosomi nello sviluppo delle cellule del tappeto di *Solanum tuberosum* L. *Caryologia* **2**, 205–22. (121)

Avanzi, S. (1972). Pattern of binding of tritiated actinomycin D to onion chromosomes in fixed material. *Accad. Naz. Lincei, Rendic. Cl. Sci. Fis. Mat. Nat. VIII* **52**, 215–19. (31)

Avanzi, S., Bruni, A. & Tagliasacchi, A. M. (1974). Synthesis of DNA and RNA in the quiescent center of the primary root of *Calystegia soldanella. Protoplasma* **80**, 393–400. (39, 40, 41, 45)

Avanzi, S., Brunori, A. & D'Amato, F. (1969). Sequential development of meristems in the embryo of *Triticum durum.* A DNA autoradiographic and cytophotometric analysis. *Develop. Biol.* **20**, 368–77. (50, 52, 53)

Avanzi, S., Brunori, A., D'Amato, F., Nuti-Ronchi, V. & Scarascia-Mugnozza, G. T. (1963). Occurrence of 2C(G_1) and 4C(G_2) nuclei in the radicle meristems of dry seeds in *Triticum durum.* Its implications in studies on chromosome breakage and on developmental processes. *Caryologia* **16**, 553–8. (49)

Avanzi, S., Buongiorno-Nardelli, M., Cionini, P. G. & D'Amato, F. (1971). Cytological localization of molecular hybrids between rRNA and DNA in the embryo suspensor cells of *Phaseolus coccineus. Accad. Naz. Lincei, Rend. Cl. Sci. Fis. Mat. Nat. VIII* **50**, 357–61. (169)

Avanzi, S., Cionini, P. G. & D'Amato, F. (1970). Cytochemical and autoradiographic analyses on the embryo suspensor cells of *Phaseolus coccineus. Caryologia* **23**, 605–38. (128)

Avanzi, S. & D'Amato, F. (1967). New evidence on the organization of the root apex in leptosporangiate ferns. *Caryologia* **20**, 257–64. (42, 43)

Avanzi, S. & D'Amato, F. (1970). Cytochemical and autoradiographic analyses on root primordia and root apices of *Marsilea strigosa.* (A new interpretation of the apical structure in Cryptogams). *Caryologia* **23**, 335–45. (42, 43, 44, 45, 46)

Avanzi, S., Durante, M., Cionini, P. G. & D'Amato, F. (1972). Cytological localization of ribosomal cistrons in polytene chromosomes of *Phaseolus coccineus. Chromosoma* **39**, 191–203. (128, 158, 169, 170)

Avanzi, S., Maggini, F. & Innocenti, A. M. (1973). Amplification of ribosomal cistrons during the maturation of metaxylem in the root of *Allium cepa. Protoplasma* **76**, 197–210. (153, 154)

Ayonoadu, U. W. & Rees, H. (1968). The regulation of mitosis by B-chromosomes in rye. *Exptl. Cell Res.* **52**, 284–90. (35)

Bachmann, K. & Cowden, R. R. (1965*a*). Quantitative cytophotometric studies on isolated liver cell nuclei of the bullfrog, *Rana catesbeiana. Chromosoma* **17**, 22–34. (118)

Backmann, K. & Cowden, R. R. (1965*b*). Quantitative cytophotometric studies on polyploid liver cell nuclei of frog and rat. *Chromosoma* **17**, 181–93. (118)

Bachmann, K. & Cowden, R. R. (1967). Quantitative cytochemical study of the 'polyploid' liver nuclei of the frog genus *Pseudacris. Trans. Amer. Microsc. Soc.* **80**, 454–63. (118)

Bachmann, K., Goin, O. B. & Goin, C. J. (1966). Hylid frogs: polyploid classes of DNA in liver nuclei. *Science* **154**, 650–1. (118)

Backs-Hüsemann, D. & Reinert, J. (1970). Embryobildung durch isolierte Einzelzellen aus Gewebekulturen von *Daucus carota. Protoplasma* **70**, 49–60. (224)

Baetcke, K. P., Sparrow, A. H., Naumann, C. H. & Schemmer, S. S. (1967). The relationship of DNA content to nuclear and chromosome volumes and to radiosensitivity (LD_{50}). *Proc. Nat. Acad. Sci.* **58**, 533–40. (135)

Baker, W. W. (1968). Position-effect variegation. *Adv. Genet.* **14**, 133–69. (191)

Balsamo, J., Hierro, J. M., Birnstiel, M. L. & Lara, J. S. (1973*a*). *Rhynchosciara angelae* salivary gland DNA: kinetic complexity and transcription of repetitive sequences. In *Genetic Expression and Its Regulation*, ed. by F. T. Kenney, B. A. Hamkalo, G. Favelukes and J. T. August, pp. 101–22. New York: Plenum Publ. Corp. (133, 158, 185)

Balsamo, J., Hierro, J. M. & Lara, F. J. S. (1973*b*). Transcription of repetitive DNA sequences in *Rhynchosciara* salivary glands. *Cell Differentiation* **2**, 119–30. (133, 158)

Balsamo, J., Hierro, J. M. & Lara, F. J. S. (1973*c*). Further studies on the characterization of repetitive *Rhynchosciara* DNA. *Cell Differentiation* **2**, 131–41. (158)

Banchetti, R. & Gremigni, V. (1973). Indirect evidence for neoblasts migration and for gametogonia dedifferentiation in exfissiparous specimens of *Dugesia gonocephala* s.l. *Accad. Naz. Lincei, Rendic. Cl. Sci. Fis. Mat. Nat. VIII* **55**, 107–15. (206, 207, 208)

Barlow, P. W. (1970). RNA synthesis in the root apex of *Zea mays. J. Exp. Bot.* **21**, 292–9. (41)

Barlow, P. W. (1972*a*). The ordered replication of chromosomal DNA: a review and a proposal for its control. *Cytobios* **6**, 55–80. (28, 30)

Barlow, P. W. (1972*b*). Differential cell division in human X chromosome mosaics. *Humangenetik* **14**, 122–7. (35)

Barlow, P. W. & Sherman, M. I. (1972). The biochemistry of differentiation of mouse trophoblast: studies on polyploidy. *J. Embryol. Exp. Morph.* **27**, 447–65. (128)

Barlow, P. W. & Sherman, M. I. (1974). Cytological studies on the organization of DNA in giant trophoblast nuclei of the mouse and the rat. *Chromosoma* **47**, 118–31. (128)

Barlow, P. W. & Vosa, C. G. (1969). The chromosomes of *Puschkinia libanotica* during mitosis. *Chromosoma* **27**, 436–47. (35)

Barr, H. J. (1964). The fate of epidermal cells during limb regeneration in larval *Xenopus. Anat. Rec.* **148**, 358. (208)

Barr, H. J. & Plaut, W. (1966). Comparative morphology of nucleolar DNA in *Drosophila. J. Cell Biol.* **31**, 17–22. (169)

Barr, M. L. (1951). The morphology of neuroglial nuclei in the cat according to sex. *Exptl. Cell Res.* **2**, 288–90. (191)

Barr, M. L. & Bertram, E. G. (1949). A morphological distinction between neurones of the male and female and the behaviour of the nucleolar satellite during accelerated nucleoprotein synthesis. *Nature* **163**, 676–7. (191)

Barry, E. G. (1969). The diffuse diplotene stage of meiotic prophase in *Neurospora. Chromosoma* **26**, 119–29. (56, 58)

Barsacchi, G. & Gall, J. G. (1972). Chromosomal localization of repetitive DNA in the newt, *Triturus. J. Cell Biol.* **54**, 580–91. (24)

Barsacchi-Pilone, G., Nardi, I., Batistoni, R., Andronico, F. & Beccari, E. (1974). Chromosome localization of the genes for 28S, 18S and 5S ribosomal RNA in *Triturus marmoratus* (Amphibia Urodela). *Chromosoma* **49**, 135–53. (169, 172)

Bateson, W. (1916). Root-cuttings, chimeras and 'sports'. *J. Genet.* **6**, 75–80. (96)

Battaglia, E. (1945). Fenomeni citologici nuovi nella embriogenesi (semigamia) e nella microsporogenesi (doppio nucleo di restituzione) in *Rudbeckia laciniata* L. *Nuovo Giorn. Bot. Ital.* **52**, 34–8. (75)

Battaglia, E. (1951*a*). The male and female gametophytes of angiosperms. An interpretation. *Phytomorphology* **1**, 87–116. (9–11, 73)

Battaglia, E. (1951*b*). Development of angiosperm embryo sacs with non-haploid eggs. *Amer. J. Bot.* **38**, 718–24. (70, 73)

Battaglia, E. (1955*a*). The concepts of spore, sporogenesis and apospory. *Phytomorphology* **5**, 173–7. (6)

Battaglia, E. (1955*b*). Unusual cytological features in the apomictic *Rudbeckia sullivantii* Boynton et Beadle. *Caryologia* **8**, 1–32. (75)

Battaglia, E. (1963). Apomixis. In: *Recent Advances in the Embryology of Angiosperms*, ed. by P. Maheshwari, pp. 221–64. Delhi: International Society of Plant Morphologists. (70, 71, 73, 74, 75)

Battaglia, E. (1964). Cytogenetics of B-chromosomes. *Caryologia* **17**, 245–99. (109, 193, 194)

Battaglia, E. & Boyes, J. W. (1955). Post-reductional meiosis: its mechanism and causes. *Caryologia* **8**, 87–134. (63, 64)

Bauer, H. (1933). Die wachsenden Oocytenkerne einiger Insekten in ihrem Verhalten zur Nuklearfärbung. *Z. Zellforsch. Mikroskop. Anat.* **18**, 254–98. (148)

Bauer, H. & Beermann, W. (1952). Der Chromosomencyclus der Orthocladinen (*Nematocera, Diptera*). *Z. Naturforsch.* **7b**, 557–63. (110)

Baur, E. (1909). Das Wesen und die Erblichkeitsverhältnisse der Var. *albo-marginata* hort. von *Pelargonium zonale. Z. Indukt. Abst. Vererbl.* **1**, 330–51. (94, 96)

Beadle, G. W. (1930). Genetical and cytological studies of mendelian asynapsis in *Zea mays. Cornell Agric. Exp. Sta. Mem.* **129**, 1–23. (65, 66)

Beams, H. N. & King, R. L. (1942). The origin of binucleate and large mononucleate cells in the liver of the rat. *Anat. Rec.* **83**, 281–97. (115)

Beasley, J. O. & Brown, M. S. (1942). Asynaptic *Gossypium* plants and their polyploids. *J. Agric. Res.* **65**, 421–7. (65, 66)

Beattie, W. G. & Skinner, D. M. (1972). The diversity of satellite DNAs of Crustacea. *Biochim. Biophys. Acta* **281**, 169–78. (24)

Beatty, R. A. (1967). Parthenogenesis in vertebrates. In *Fertilization* **1**, ed. by C. B. Metz and A. Monroy, pp. 413–40. New York: Academic Press. (77)

Beermann, W. (1952). Chromomerenkonstanz und spezifische Modifikationen der Chromosomenstruktur in der Entwicklung und Organdifferenzierung von *Chironomus tentans*. *Chromosoma* **5**, 139–98. (176)

Beermann, W. (1962). Riesenchromosomen. *Protoplasmatologia, VID*. Vienna: Springer-Verlag. (120, 130, 174)

Beermann, W. (1963). Cytological aspects of information transfer in cellular differentiation. *Amer. Zool.* **3**, 23–32. (176, 177)

Beermann, W. (1972). Chromomeres and genes. In *Developmental Studies on Giant Chromosomes*, ed. by W. Beermann, pp. 1–33. Berlin: Springer-Verlag. (130, 175)

Beetschen, J.-C. (1973). Réalisation experimentale de l'androgénèse animale. Perspectives récentes. *Soc. bot. Fr. Mémoires* 1973, *Coll. Morphologie*, 235–42. (21)

Belar, K. (1924). Die Cytologie der Merospermie bei freilebenden *Rhabditis*-Arten. *Z. Zell. Gewebel.* **1**, 1–21. (80)

Bell, P. R. (1960). Interaction of nucleus and cytoplasm during oogenesis in *Pteridium aquilinum*. *Proc. Roy. Soc. B* **153**, 421–32. (15)

Bello, L. J. (1968). Synthesis of DNA-like RNA in synchronized cultures of mammalian cells. *Biochim. Biophys. Acta* **157**, 8–15. (28)

Benazzi-Lentati, G. (1970). Gametogenesis and egg fertilization in planarians. *Int. Rev. Cytol.* **27**, 101–79. (20, 77, 79)

Bennett, M. D. (1970). Natural variation in nuclear character of meristems in *Vicia faba*. *Chromosoma* **29**, 317–35. (135)

Bennett, M. D. (1972). Nuclear DNA content and minimum generation time in herbaceous plants. *Proc. Roy. Soc. B* **181**, 109–35. (34)

Bennett, M. D. & Rees, H. (1969). Induced developmental variations in chromosomes of meristematic cells. *Chromosoma* **27**, 226–44. (135)

Berendes, H. D. (1972). The control of puffing in *Drosophila hydei*. In *Developmental Studies on Giant Chromosomes*, ed. by W. Beermann, pp. 181–207. Berlin: Springer-Verlag. (130, 199)

Berendes, H. D. (1973). Synthetic activity of polytene chromosomes. *Int. Rev. Cytol.* **35**, 61–116. (130, 175)

Berendes, H. D. & Keyl, H. G. (1967). Distribution of DNA in heterochromatin and euchromatin of polytene nuclei of *Drosophila hydei*. *Genetics* **57**, 1–13. (132)

Berendes, H. D. & Lara, F. J. S. (1975). RNA synthesis: a requirement for hormone induced DNA amplification in *Rhynchosciara americana*. *Chromosoma* **50**, 259–74. (160)

Bergann, E. (1967). The relative instability of chimeral clones, the basis for further breeding. In *Induced Mutations and Their Utilization*, ed. by H. Stubbe, pp. 287–300. Berlin: Akademie Verlag. (96, 97, 98, 99)

Berger, C. A. (1938). Multiplication and reduction of somatic chromosome groups as a regular developmental process in the mosquito, *Culex pipiens*. *Publ. Carnegie Inst.* **476**, 209–36. (137)

Berger, C. A. (1941*a*). A reinvestigation of polysomaty in *Spinacia*. *Bot. Gaz.* **102**, 759–69. (136)

Berger, C. A. (1941*b*). Multiple chromosome complexes in animals and polysomaty in plants. *Cold Spring Harbor Symp. Quant. Biol.* **9**, 19–22. (137)

Berger, C. A. (1946). Cytological effects of combined treatment with colchicine and naphtalene acetic acid. *Amer. J. Bot.* **33**, 817. (125)

Berger, C. A., Feeley, E. J. & Witkus, E. R. (1956). The cytology of *Xanthisma texanum*

D.C. IV. Megasporogenesis and embryo sac formation, pollen mitosis and embryo formation. *Bull. Torrey Bot. Club* **83**, 428–34. (109)

Berger, C. A., McMahon, R. M. & Witkus, E. R. (1955). The cytology of *Xanthisma texanum* D.C. III. Differential somatic reduction. *Bull. Torrey Bot. Club* **82**, 377–82. (109)

Berger, C. A. & Witkus, E. R. (1948). Cytological effects of alpha-naphtalene acetic acid. *J. Hered.* **39**, 117–20. (125)

Berger, C. A. & Witkus, E. R. (1954). The cytology of *Xanthisma texanum* D.C. I. Differences in the chromosome number of root and shoot. *Bull. Torrey Bot. Club* **81**, 489–91. (109)

Bergerard, J. (1962). Parthenogenesis in the Phasmidae. *Endeavour* **21**, 137–43. (77, 81)

Berlowitz, L. & Pallotta, D. (1972). Acetylation of nuclear protein in the heterochromatin and euchromatin of mealy bug. *Exptl. Cell Res.* **71**, 45–8. (212)

Berns, M. W. & Cheng, W. K. (1971). Are chromosome secondary constrictions nucleolar organizers? *Exptl. Cell Res.* **69**, 185–92. (170, 171)

Berrie, G. K. (1959). Mitosis in the microspore of *Encephalartos berteri* Carruth. *Nature* **183**, 834–5. (135)

Bertalanffy, F. D. (1964). Tritiated thymidine *versus* colchicine technique in the study of cell population cytodynamics. *Labor. Investig.* **13**, 871–86. (34)

Beyreuther, K., Adler, K., Geisler, N. & Klemm, A. (1973). The amino-acid sequence of lac repressor. *Proc. Nat. Acad. Sci.* **70**, 3576–80. (197)

Bier, K., Kunz, W. & Ribbert, D. (1967). Struktur und Funktion der Oocytenchromosomen und Nukleolen sowie der Extra-DNS während der Oogenese panoistischer und meroistischer Insekten. *Chromosoma* **23**, 214–54. (148)

Bird, A. P. & Birnstiel, M. L. (1971). A timing study of DNA amplification in *Xenopus laevis* oocytes. *Chromosoma* **35**, 300–9. (145, 146)

Bird, A. P., Rochaix, J.-D. & Bakken, A. H. (1973*a*). The mechanism of gene amplification in *Xenopus laevis* oocytes. In *Molecular Cytogenetics*, ed. by B. A. Hamkalo and J. Papaconstantinou, pp. 49–58. New York: Plenum Publ. Corp. (146, 147)

Bird, A., Rogers, E. & Birnstiel, M. (1973*b*). Is gene amplification RNA-directed? *Nature New Biol.* **242**, 226–30. (146)

Birnstiel, M. L., Chipchase, M. & Speirs, J. (1971). The ribosomal RNA cistrons. *Progr. Nucl. Acid Res. Mol. Biol.* **11**, 351–89. (168, 171, 172)

Birnstiel, M. L., Weinberg, E. S. & Pardue, M. L. (1973). Evolution of 9S mRNA sequences. In *Molecular Cytogenetics*, ed. by B. A. Hamkalo and J. Papaconstantinou, pp. 75–93. New York: Plenum Publ. Corp. (173)

Blackler, A. W. (1970). The integrity of the reproductive cell line in the Amphibian. In *Current Topics in Developmental Biology* **5**, ed. by A. Moscona and A. Monroy, pp. 71–87. New York: Academic Press. (109)

Blaser, H. W. & Einset, J. (1948). Leaf development in six periclinal chromosomal chimeras of apple varieties. *Amer. J. Bot.* **35**, 473–82. (100, 101)

Blaser, H. W. & Einset, J. (1950). Flower structure in periclinal chimeras of apple. *Amer. J. Bot.* **37**, 297–304. (100, 103)

Bloch, D. P., Macquigg, R. A., Brack, S. D. & Wu, J. R. (1967). The synthesis of deoxyribonucleic acid and histone in the onion root meristem. *J. Cell Biol.* **33**, 451–67. (32)

Blumenfeld, M. & Forrest, H. S. (1972). Differential under-replication of satellite DNA during *Drosophila* development. *Nature New Biol.* **239**, 170–2. (134)

Bogdanov, Y. F. & Antropova, E. N. (1971). Delayed termination of nuclear histone doubling after premeiotic DNA synthesis in *Triturus vulgaris* male meiosis. *Chromosoma* **35**, 353–73. (56)

Bogdanov, Y. F. & Jordansky, A. B. (1964). Autoradiographic investigations of nuclei of root meristem of germinating pea seeds with the employment of H^3-thymidine. *Zh. Obshch. Biol.* **25**, 357–63. (50)

Bogdanov, Y. F., Liapunova, N. A. & Sherudilo, A. I. (1967). Cell population in pea embryos and root tip meristem. Microphotometric and autoradiographic studies. *Tsitologia* **9**, 569–76. (50)

Bogdanov, Y. F., Liapunova, N. A., Sherudilo, A. I. & Antropova, E. N. (1968). Uncoupling of DNA and histone synthesis prior to prophase of meiosis in the cricket Grillus (*Acheta domestica*). *Exptl. Cell Res.* **52**, 59–70. (56)

Bogdanov, Y. F., Strokov, A. A. & Reznickova, S. A. (1973). Histone synthesis during meiotic prophase in *Lilium*. *Chromosoma* **43**, 237–45. (56)

Boncinelli, E., Graziani, F., Polito, L., Malva, C. & Ritossa, F. (1972). rDNA magnification at the bobbed locus of the Y chromosome in *Drosophila melanogaster*. *Cell Differentiation* **1**, 133–42. (155)

Bonner, J. (1975). Regulation of gene expression in higher organisms: how it all works. *Ciba Foundation Symp.* **28** (n.s.), 315–27. (25, 195)

Bonner, J., Dahmus, M. E., Fambrough, D., Huang, R. C., Marushige, K. & Tuan, D. Y. H. (1968). The biology of isolated chromatin. *Science* **159**, 47–56. (23, 24, 25)

Böök, J. A. (1964). Some mechanisms of chromosome variations and their relation to human malformations. *Eugenics Rev.* **56**, 151–7. (88)

Böök, J. A. & Santesson, B. (1960). Malformation syndrome in man associated with triploidy (69 chromosomes). *Lancet* **1**, 858–9. (87)

Bostock, C. J. & Prescott, D. M. (1971). Buoyant density of DNA synthesized at different stages of the S phase of mouse cells. *Exptl. Cell Res.* **64**, 267–74. (31)

Bostock, C. J., Prescott, D. M. & Hatch, F. T. (1972). Time of replication of the satellite and main band DNAs in cells of the Kangaroo rat (*Dipodomys ordii*). *Exptl. Cell Res.* **74**, 487–95. (31)

Botchan, M., Kram, R., Schmid, C. W. & Hearst, J. E. (1971). Isolation and chromosomal localization of highly repeated DNA sequences in *Drosophila melanogaster*. *Proc. Nat. Acad. Sci.* **68**, 1125–9. (24, 186)

Boveri, T. (1887). Über Differenzierung der Zellkerne während der Fürchung des Eies von *Ascaris megalocephala*. *Anat. Anz.* **2**, 688–93. (109)

Boveri, T. (1888). Über partielle Befruchtung. *Sitzber. Ges Morph. Physiol. München* **4**, 64–72. (21, 22)

Boveri, T. (1889). Die Vorgänge der Zellteilung und Befruchtung in ihrer Beziehung zur Vererbungsfrage. *Beitr. Anthropologie Urgeschicte Bayerns* **8**, 27–39. (21, 83)

Bower, F. O. (1930). *Size and Form in Land Plants*. London: Macmillan. (37)

Brabec, F. (1965). Propfung und Chimären unter besonderer Berucksichtigung der Entwicklungsphysiologischen Problematik. In *Encyclopedia of Plant Physiology* **15/2**, ed. by W. Ruhland, pp. 390–498. Berlin: Springer-Verlag. (92, 95)

Brachet, J. (1940). La localisation de l'acide thymonucléique pendant l'oogénèse et la maturation chez les amphibiens. *Arch. Biol.* **51**, 154–65. (143)

Brady, T. (1973). Feulgen cytophotometric determination of the DNA content of the embryo proper and suspensor cells of *Phaseolus coccineus*. *Cell Differentiation* **2**, 65–75. (128)

Brady, T. & Clutter, M. E. (1972). Cytolocalization of ribosomal cistrons in plant polytene chromosomes. *J. Cell Biol.* **53**, 827–32. (128, 169)

Brady, T. & Clutter, M. E. (1974). Structure and replication of *Phaseolus* polytene chromosomes. *Chromosoma* **45**, 63–79. (128)

Breuer, M. E. & Pavan, C. (1955). Behaviour of polytene chromosomes of *Rhynchosciara angelae* at different stages of larval development. *Chromosoma* **7**, 371–86. (134, 157, 176)

Brewbaker, J. L. (1959). Biology of the angiosperm pollen grain. *Indian J. Genet. Plant Breed.* **19**, 121–33. (12)

Brewbaker, J. L. & Emery, G. C. (1962). Pollen radiobotany. *Radiat. Bot.* **1**, 101–54. (12, 13)

Bridges, C. B. (1921). Triploid intersexes in *Drosophila melanogaster. Science* **54**, 252–4. (84)

Bridges, C. B. (1925). Haploidy in *Drosophila melanogaster. Proc. Nat. Acad. Sci.* **11**, 706–10. (85)

Briggs, R. & King, T. J. (1952). Transplantation of living nuclei from blastula cells into enucleated frogs' eggs. *Proc. Nat. Acad. Sci.* **38**, 455–63. (218, 219)

Briggs, R. & King, T. J. (1957). Changes in the nuclei of differentiating endoderm cells as revealed by nuclear transplantation. *J. Morph.* **100**, 269–302. (219)

Brink, R. A. (1962). Phase change in higher plants and somatic cell heredity. *Quart. Rev. Biol.* **37**, 1–22. (107)

Brink, R. A. & Cooper, D. C. (1947). The endosperm in seed development. *Bot. Rev.* **13**, 423–541. (119)

Brinkley, B. R. & Stubblefield, E. (1970). Ultrastructure and interactions of the kinetochore and centriole in mitosis and meiosis. *Adv. Cell Biol.* **1**, 119–85. (62)

Bristow, M. (1962). The controlled *in vitro* differentiation of callus derived from a fern, *Pteris cretica* L., into gametophytic or sporophytic tissues. *Develop. Biol.* **4**, 361–75. (18)

Britten, R. J. & Davidson, E. H. (1969). Gene regulation for higher cells: a theory. *Science* **165**, 349–58. (184)

Britten, R. J. & Kohne, D. E. (1968). Repeated sequences in DNA. *Science* **161**, 529–40. (23)

Broertjes, C. (1972). *Use in Plant Breeding of Acute, Chronic or Fractionated Doses of X-Rays or Fast Neutrons as Illustrated with Leaves of Saintpaulia.* Wageningen: Centre for Agricultural Publishing and Documentation. (224)

Broertjes, C., Haccius, B. & Weidlich, S. (1968). Adventitious bud formation on isolated leaves and its significance for mutation breeding. *Euphytica* **17**, 321–44. (223)

Brønsted, H. V. (1969). *Planarian regeneration.* Oxford: Pergamon Press. (206)

Brown, D. D. (1966). The nucleolus and synthesis of ribosomal RNA during oögenesis and embryogenesis of *Xenopus laevis. Nat. Cancer Inst. Mono.* **23**, 297–309. (145)

Brown, D. D. & Blackler, A. W. (1972). Gene amplification proceeds by a chromosome copy mechanism. *J. Mol. Biol.* **63**, 75–83. (145, 147)

Brown, D. D. & Dawid, I. (1968). Specific gene amplification in oocytes. *Science* **160**, 272–80. (144, 152, 169)

Brown, D. D. & Gurdon, B. (1964). Absence of ribosomal RNA synthesis in the anucleolate mutant of *Xenopus laevis. Proc. Nat. Acad. Sci.* **51**, 139–46. (168, 180)

Brown, D. D. & Littna, E. (1964). RNA synthesis during the development of *Xenopus laevis*, the South African Clawed Toad. *J. Mol. Biol.* **20**, 81–94. (145)

Brown, D. D. & Sugimoto, K. (1973). 5S DNAs of *Xenopus laevis* and *Xenopus mulleri*: evolution of a gene family. *J. Mol. Biol.* **78**, 397–415. (172)

Brown, D. D. & Weber, C. (1968). Gene linkage by RNA–DNA hybridization. I. Unique DNA sequences homologous to 4S RNA, 5S RNA and ribosomal RNA. *J. Mol. Biol.* **34**, 661–80. (172)

Brown, D. D., Wensink, D. & Jordan, E. (1972). A comparison of the ribosomal DNA's of *Xenopus laevis* and *Xenopus mulleri*: The evolution of tandem genes. *J. Mol. Biol.* **63**, 57–73. (145)

Brown, S. W. (1966). Heterochromatin. *Science* **151**, 417–25. (185)

Brown, S. W. & Chandra, H. S. (1973). Inactivation system of the mammalian X chromosome. *Proc. Nat. Acad. Sci.* **70**, 196–9. (193)

Brown, S. W. & Nelson-Rees, W. A. (1961). Radiation analysis of a lecanoid genetic system. *Genetics* **46**, 983–1007. (188)

Brown, S. W. & Nur, U. (1964). Heterochromatic chromosomes in the coccids. *Science* **145**, 130–6. (30, 187, 188, 189)

Brown, W. V. & Stack, S. M. (1968). Somatic pairing as a regular preliminary to meiosis. *Bull. Torrey Bot. Club* **95**, 369–78. (63)

Brunori, A. (1967). Relationship between DNA synthesis and water content during ripening of *Vicia faba* seed. *Caryologia* **20**, 333–8. (51)
Brunori, A. & Ancora, G. (1968). The DNA content of nuclei in the embryonic root apices of *Allium cepa* and their radiation response. *Caryologia* **21**, 261–9. (50)
Brunori, A., Avanzi, S. & D'Amato, F. (1966). Chromatid and chromosome aberrations in irradiated dry seeds of *Vicia faba. Mutation Res.* **3**, 305–13. (50)
Brunori, A. & D'Amato, F. (1967). The DNA content of nuclei in the embryo of dry seeds of *Pinus pinea* and *Lactuca sativa. Caryologia* **20**, 153–61. (49, 50, 51, 113)
Brunori, A., Georgieva, J. & D'Amato, F. (1970). Further observations on the X-irradiation response of G_1 cells in dry seeds. *Mutation Res.* **9**, 481–7. (50)
Bryant, T. R. (1969). DNA synthesis and cell division in germinating onion. I. Onset of DNA synthesis and mitosis. *Caryologia* **22**, 127–38. (50)
Bucher, N. L. R. (1963). Regeneration of mammalia liver. *Int. Rev. Cytol.* **15**, 245–300. (115, 211, 213, 214)
Buiatti, M. (1977). DNA amplification and tissue cultures. In *Applied and Fundamental Aspects of Plant Cell, Tissue and Organ Culture*, ed. by J. Reinert and Y. P. S. Bajaj, pp. 358–74. Berlin: Springer-Verlag. (143)
Buiatti, M., Tesi, R. & Molino, M. (1969). A developmental study of induced somatic mutations in *Gladiolus. Radiat. Bot.* **9**, 39–48. (106, 107)
Buller, A. H. R. (1941). The diploid cell and the diploidization process in plants and animals with special reference to the higher fungi. *Bot. Rev.* **7**, 335–431. (115)
Bullough, W. S. (1963). Analysis of the life cycle in mammalian cells. *Nature* **199**, 859–62. (111)
Bultmann, H. & Clever, U. (1969). Chromosomal control of footpad development in *Sarcophaga bullata*. I. The puffing pattern. *Chromosoma* **28**, 120–35. (179)
Bünning, E. (1957). Polarität und inäquale Teilung der pflanzlichen Protoplasten. *Protoplasmatologia VIII* (*9a*). Vienna: Springer-Verlag. (166)
Buongiorno-Nardelli, M., Amaldi, F. & Lava-Sanchez, P. A. (1972). Amplification as a rectification mechanism for the redundant RNA genes. *Nature New Biol.* **238**, 134–7. (144, 147)
Burnet, F. M. (1974). *Intrinsic Mutagenesis: A Genetic Approach to Ageing.* Lancaster: Medical and Technical Publ. Co. Ltd. (140, 142)
Burnett, A. L. (1968). The acquisition, maintenance and lability of the differentiated state in Hydra. In *The Stability of the Differentiated State*, ed. by H. Ursprung, pp. 109–35. Berlin: Springer-Verlag. (204)
Burns, J. A. (1972). Preleptotene chromosome contraction in *Nicotiana* species. *J. Heredity* **63**, 175–8. (63)
Butler, E. G. (1933). The effects of X-radiation on the regeneration of the forelimb of *Ambystoma* larvae. *J. Exp. Zool.* **65**, 271–313. (209)
Butler, W. B. & Mueller, G. C. (1973). Control of histone synthesis in HeLa cells. *Biochim. Biophys. Acta* **294**, 481–96. (32)
Buvat, R. & Genèves, L. (1951). Sur l'inexistence des initiales axiales dans la racine d'*Allium cepa* L. (*Liliacées*). *C.R. Acad. Sci.* **232**, 1579–81. (38)
Buvat, R. & Liard, O. (1951). Nouvelle constatation de l'inertie des soi-disant initiales axiales dans le méristème radiculaire de *Triticum vulgare. C.R. Acad. Sci.* **236**, 1193–5. (38)
Byrne, J. M. & Heimsch, C. (1970). The root apex of *Malva silvestris*. II. The quiescent center. *Amer. J. Bot.* **57**, 1179–84. (39, 41, 50)
Callan, H. G. & Lloyd, L. (1960). Lampbrush chromosomes of creasted newts, *Triturus cristatus* (Laurenti). *Phil. Trans. Roy. Soc. B.* **243**, 35–219. (58, 174)
Camenzind, R. (1966). Die Zytologie der bisexuellen und parthenogenetischen Fortpflanzung von *Heteropeza pygmaea* Winnertz, einer Gallmücke mit pädogenetischen Vermehrung. *Chromosoma* **18**, 123–52. (110)

Cameron, I. L. & Cleffmann, G. (1964). Initiation of mitosis in relation to the cell cycle following feeding of starved chickens. *J. Cell Biol.* **21**, 169–74. (49)
Capesius, I., Bierweiler, B., Bachmann, K., Rücker, W. & Nagl, W. (1975). An A+T-rich satellite DNA in a Monocotyledonous plant, *Cymbidium. Bioch. Biophys. Acta* **395**, 67–73. (164)
Carlson, J. G. (1952). Microdissection studies of the dividing neuroblast of the grasshopper, *Chortophaga viridifasciata* (De Geer). *Chromosoma* **5**, 199–220. (167)
Carlson, P. S., Smith, H. H. & Dearing, R. D. (1972). Parasexual interspecific plant hybridization. *Proc. Nat. Acad. Sci.* **69**, 2292–4. (227)
Carniel, K. (1963). Das Antherentapetum. Ein kritischer Überblick. *Öster. bot. Z.* **110**, 145–76. (120, 125)
Caspersson, T., Farber, S., Foley, G. E., Kudynowski, J., Modest, E. J., Simonsson, E., Wagh, U. & Zech, L. (1968). Chemical differentiation along metaphase chromosomes. *Exptl. Cell Res.* **49**, 219–22. (31)
Caspersson, T., Zech, L., Modest, E. J., Foley, G. E., Wagh, U. & Simonsson, E. (1969). Chemical differentiation with fluorescent alkylating agents in *Vicia faba* metaphase chromosomes. *Exptl. Cell Res.* **58**, 128–40. (187)
Cassagnau, P. (1968). Sur la structure des chromosomes salivaires de *Bilobella massoudi* Cassagnau (Collembola: Neanuridae). *Chromosoma* **24**, 42–58. (130)
Castro, D., Camara, A. & Malheiros, N. (1949). X-rays in the centromere problem of *Luzula purpurea* Link. *Genetica Iberica* **1**, 49–54. (63)
Catarino, F. M. (1965). Salt water, a growth inhibitor causing endopolyploidy. *Port. Acta Biol. (A)* **9**, 131–52. (126)
Catarino, F. M. (1968). *Endopoliploidia e Differenciacao.* Lisbon: Tipografia Matematica Lda. (120, 126)
Cave, M. D. (1972). Localization of ribosomal DNA within oocytes of the house cricket, *Acheta domesticus* (Orthoptera: Gryllidae). *J. Cell Biol.* **55**, 810–21. (149)
Cave, M. D. (1973). Synthesis and characterization of amplified DNA in oocytes of the house cricket, *Acheta domesticus* (Orthoptera: Gryllidae). *Chromosoma* **42**, 1–22. (149)
Cave, M. D. & Allen, E. R. (1969). Synthesis of nucleic acids associated with a DNA-containing body in oocytes of Acheta. *Exptl. Cell Res.* **58**, 201–12. (149)
Chandebois, R. (1973*a*). Teneur en ADN des éléments qui participent à la régénération chez la Planaire dulcicole, *Dugesia subtentaculata. C.R. Soc. Biol.* **167**, 286–91. (206)
Chandebois, R. (1973*b*). The syncytial and intercellular nature of dedifferentiated material in the parenchyma of regenerating and starved planarians. *Oncology* **27**, 356–84. (207)
Chiarugi, A. (1927). Il gametofito femminile delle Angiosperme nei suoi vari tipi di costruzione e di sviluppo. *Nuovo Giorn. Bot. Ital.* **34**, 5–133. (12)
Chiarugi, A. (1950). La poliploidia della generazione aploide femminile delle fanerogame. *Caryologia* **3**, 149–55. (12)
Chiarugi, A. (1952). Fondamento anatomico dell'accrescimento nei metafiti. *Accad. Naz. Lincei Quad. no. 28, Accrescimento negli organismi*, pp. 152–72. (37)
Chiarugi, A. (1954). La poliploidia somatica nelle piante. *Caryologia* **6**, Suppl. 488–520. (120, 123)
Chilton, M. D. & McCarthy, B. J. (1973). DNA from maize with and without B chromosomes: a comparative study. *Genetics* **74**, 605–14. (193)
Chittenden, R. J. (1927). Vegetative segregation. *Bibliographia Genetica* **3**, 358–442. (96)
Christensen, B. (1961). Studies on cyto-taxonomy and reproduction in the Enchytraeidae. With notes on parthenogenesis and polyploidy in the animal kingdom. *Hereditas* **47**, 387–450. (80)
Chu, E. H. Y., Thuline, H. C. & Norby, D. E. (1964). Triploid–diploid chimerism in a male tortoise-shell cat. *Cytogenetics* **3**, 1–18. (87, 88)

Chun, E. H. L. & Littlefield, J. W. (1963). The replication of the minor DNA component of mouse fibroblasts. *J. Mol. Biol.* **7**, 245–8. (31)

Cionini, P. G. & Avanzi, S. (1972). Pattern of binding of tritiated actinomycin D to *Phaseolus coccineus* polytene chromosomes. I. The nucleolus organizing chromosomes. *Exptl. Cell Res.* **75**, 154–8. (31, 128)

Clark, F. J. (1940). Cytogenetic studies of divergent meiotic spindle formation in *Zea mays*. *Amer. J. Bot.* **27**, 547–59. (65, 69, 70)

Clarkson, S. G. (1969). Nucleic acid and protein synthesis and pattern regulation in hydra. I. Regional pattern of synthesis and changes in synthesis during hypostome formation. *J. Embryol. Morphol.* **21**, 33–54. (206)

Clarkson, S. G., Birnstiel, M. L. & Serra, V. (1973). Reiterated transfer RNA genes of *Xenopus laevis*. *J. Mol. Biol.* **79**, 391–410. (173)

Clever, U. (1968). Regulation of chromosome function. *Ann. Rev. Genet.* **2**, 11–30. (130)

Clowes, F. A. L. (1956). Localization of nucleic acid synthesis in root meristems. *J. Exp. Bot.* **7**, 307–12. (38, 39)

Clowes, F. A. L. (1958). Development of quiescent centres in root meristems. *New Phytol.* **57**, 85–8. (41)

Clowes, F. A. L. (1959*a*). Apical meristems of roots. *Biol. Rev.* **34**, 501–29. (38, 39)

Clowes, F. A. L. (1959*b*). Reorganization of root apices after irradiation. *Ann. Bot.* **23**, 205–10. (39, 42)

Clowes, F. A. L. (1961). *Apical Meristems*. Oxford: Blackwell Scientific Publications. (38, 42, 46, 94, 96)

Clowes, F. A. L. (1964). The quiescent center in meristems and its behaviour after irradiation. *Brookhaven Symp. Biol.* **16**, 46–58. (39, 42)

Clowes, F. A. L. (1965). Meristems and the effect of radiation on cells. *Endeavour* **24**, 8–12. (39, 42)

Clowes, F. A. L. (1967*a*). The quiescent centre. *Phytomorphology* **17**, 132–40. (39)

Clowes, F. A. L. (1967*b*). The functioning of meristems. *Sci. Prog. Oxf.* **55**, 529–42. (39, 40, 41)

Clowes, F. A. L. (1968). The DNA content of the quiescent centre and root cap of *Zea mays*. *New Phytol.* **67**, 631–9. (40, 42)

Clowes, F. A. L. (1972*a*). Regulation of mitosis in roots by their caps. *Nature New Biol.* **235**, 143–4. (42)

Clowes, F. A. L. (1972*b*). The control of cell proliferation within root meristems. In *The Dynamics of Meristem Cell Populations*, ed. by M. W. Miller and C. C. Kuehnert, pp. 133–47. New York: Plenum Publ. Corp. (39, 40)

Clowes, F. A. L. & Hall, E. J. (1962). The quiescent centre in root meristems of *Vicia faba* and its behaviour after acute X-irradiation and chronic gamma irradiation. *Radiat. Bot.* **3**, 45–53 (42)

Clowes, F. A. L. & Juniper, B. E. (1964). The fine structure of the quiescent centre and neighbouring tissues in root meristems. *J. Exptl. Bot.* **15**, 622–30. (41)

Clowes, F. A. L. & Stewart, H. E. (1967). Recovery from dormancy in roots. *New Phytol.* **66**, 115–23. (42)

Cocking, E. C. (1972). Plant cell protoplasts. Isolation and development. *Ann. Rev. Plant Physiol.* **23**, 29–50. (227)

Coggins, L. W. & Gall, J. G. (1972). The timing of meiosis and DNA synthesis during early oogenesis in the toad, *Xenopus laevis*. *J. Cell Biol.* **52**, 69–76. (145, 146)

Collins, J. M. (1972). Amplification of ribosomal ribonucleic acid cistrons in the regenerating lens of *Triturus*. *Biochemistry* **11**, 1259–63. (153, 217)

Colucci, V. L. (1885). Sulla rigenerazione parziale dell'occhio nei Tritoni. Istogenesi e sviluppo. *Mem. R. Accad. Sci. Ist. Bologna V* **1**, 593–629. (215)

Comings, D. E. (1967). Histones of genetically active and inactive chromatin. *J. Cell Biol.* **35**, 699–708. (195)
Comings, D. E. (1972). Replicative heterogeneity of mammalian DNA. *Exptl. Cell Res.* **71**, 106–12. (31)
Comings, D. E. & Kakefuda, T. (1968). Initiation of deoxyribonucleic acid replication at the nuclear membrane in human cells. *J. Mol. Biol.* **33**, 225–9. (28)
Comings, D. E. & Mattoccia, E. (1972). DNA of mammalian and avian heterochromatin. *Exptl. Cell Res.* **71**, 113–31. (186)
Cooper, D. W. (1971). Directed genetic change model for X chromosome inactivation in eutherian mammals. *Nature* **230**, 292–4. (193)
Cooper, K. W. (1948). The evidence for long-range specific attraction forces during the somatic pairing of dipteran chromosomes. *J. Exp. Zool.* **108**, 327–35. (130)
Cooper, K. W. (1959). A bilateral gynandromorphic *Hypodynerus*, and a summary of cytologic origins of such mosaic hymenoptera. Biology of eumenine wasps. VI. *Bull. Florida State Mus. Biol. Sci.* **5**, 25–40. (83)
Corneo, G., Ginelli, E. & Polli, E. (1970). Repeated sequences in human DNA. *J. Mol. Biol.* **48**, 319–27. (186)
Corsi, G. & Avanzi, S. (1970). Cytochemical analyses on cellular differentiation in the root tip of *Allium cepa. Caryologia* **23**, 381–94. (125)
Coward, S. J., Bennett, C. E. & Hazlehurst, B. L. (1974). Lysosomes and lysosomal enzyme activity in regenerating planarian; evidence in support of dedifferentiation. *J. Exp. Zool.* **189**, 133–46. (208)
Coward, S. J. & Hay, E. D. (1972). Fine structure of cells involved in physiological and reconstitutive regeneration in planarians. *Anat. Rec.* **172**, 296. (207)
Cramer, P. J. S. (1954). Chimeras. *Bibliotheca Genetica* **16**, 193–381. (82)
Cremonini, R. (1974). Frequenza e localizzazione delle mitosi in primordi di radici laterali di *Marsilea strigosa. Giorn. Bot. Ital.* **108**, 155–9. (44, 45)
Crew, F. A. E. & Lamy, R. (1938). Mosaicism in *Drosophila pseudoobscura. J. Genet.* **37**, 211–28. (85)
Crick, F. H. C. (1971). General model for chromosomes of higher organisms. *Nature* **234**, 25–7. (184)
Crippa, M. & Tocchini-Valentini, G. P. (1971). Synthesis of amplified DNA that codes for ribosomal RNA. *Proc. Nat. Acad. Sci.* **68**, 2769–73. (146)
Crosby, A. R. (1957). Nucleolar activity of lagging chromosomes in wheat. *Amer. J. Botany* **44**, 813–22. (156)
Crouse, H. V. (1968). The role of ecdysone in DNA-puff formation and DNA synthesis in the polytene chromosomes of *Sciara coprophila. Proc. Nat. Acad. Sci.* **61**, 971–8. (160)
Crouse, J. V. & Keyl, H. G. (1968). Extra replications in the 'DNA-puffs' of *Sciara coprophila. Chromosoma* **25**, 357–64. (134)
Cullis, C. A. (1973). DNA differences between flax genotrophs. *Nature* **243**, 515–16. (161)
Cullis, C. A. (1975). Environmentally induced changes in flax. In *Modification of the Information Content of Plant Cells*, ed. by R. Markham, D. R. Davies and R. W. Horne, pp. 27–36. Amsterdam: North-Holland Publ. Co. (160, 161)
Curtis, H. J. (1966). *Biological Mechanisms of Aging.* Springfield: C. C. Thomas Publisher. (140, 141)
Curtis, H. J. (1971). Genetic factors in aging. *Adv. Genet.* **16**, 305–24. (140, 141)
Czeika, G. (1956). Strukturveränderungen endopolyploider Ruhekerne im Zusammenhang mit wechselnder Bündelung der Tochterchromosomen und karyologisch-anatomische Untersuchungen an Sukkulenten. *Öster. bot. Z.* **103**, 536–66. (126)
Da Cunha, A. B., Pavan, C., Morgante, J. S. & Garrido, M. C. (1969). Studies on cytology and differentiation in Sciaridae. II. DNA redundancy in salivary gland cells of *Hybosciara fragilis* (Diptera, Sciaridae). *Genetics* Suppl. **61**, 335–49. (158)

Da Cunha, A. B., Riess, R. W., Biesele, J. J., Morgante, J. S., Pavan, C. & Garrido, M. C. (1973). Studies in cytology and differentiation in Sciaridae. VI. Cytoplasmic transfer in *Hybosciara fragilis* Morgante (Diptera, Sciaridae). *Caryologia* **26**, 549–61. (158)

D'Amato, F. (1948*a*). Cytological consequences of decapitation in onion roots. *Experientia* **4**, 388. (125)

D'Amato, F. (1948*b*). Sull'attivita colchicino-mitotica e su altri effetti citologici del 2,4-diclorofenossiacetato di sodio. *Accad. Naz. Lincei, Cl. Sci. Fis. Mat. Nat. VIII* **4**, 450–8. (125)

D'Amato, F. (1949). Biotipi cariologici ed embrionia avventizia in *Nothoscordum fragrans* Kunth. *Caryologia* **1**, 378–80. (76)

D'Amato, F. (1950). Differenziazione istologica per endopoliploidia nelle radici di alcune Monocotiledoni. *Caryologia* **3**, 11–26. (113)

D'Amato, F. (1952*a*). New evidence on endopolyploidy in differentiated plant tissues. *Caryologia* **4**, 121–44. (126)

D'Amato, F. (1952*b*). Polyploidy in the differentiation and function of tissues and cells in plants. A critical examination of the literature. *Caryologia* **4**, 311–58. (115, 119, 120, 125, 136)

D'Amato, F. (1954). A brief discussion on 'endomitosis'. *Caryologia* **6**, 341–4. (123)

D'Amato, F. (1959). C-mitosis and experimental polyploidy in plants. *Genetica Agraria* **11**, 17–26. (223)

D'Amato, F. (1964*a*). Nuclear changes and their relationships to histological differentiation. *Caryologia* **17**, 317–25. (113, 167)

D'Amato, F. (1964*b*). Cytological and genetic aspects of aging. In *Genetics Today, Proc. XI. Internatl. Congr. Genet.*, pp. 285–95. Oxford: Pergamon Press. (140)

D'Amato, F. (1965*a*). Chimera formation in mutagen-treated seeds and diplontic selection. In *The use of induced mutations in plant breeding, Radiat. Bot.* **5** *Suppl.*, 303–16. (105, 106)

D'Amato, F. (1965*b*). Endopolyploidy as a factor in plant tissue development. In *Proc. Internatl. Conference on Plant Tissue Culture*, ed. by P. R. White and A. R. Grove, pp. 449–62. Berkeley: McCutchan Publ. Corp. (113, 120)

D'Amato, F. (1971). *Genetica Vegetale.* Torino: Boringhieri. (6)

D'Amato, F. (1972*a*). Morphogenetic aspects of the development of meristems in seed embryos. In *The Dynamics of Meristem Cell Populations*, ed. by M. W. Miller and C. C. Kuehnert, pp. 149–63. New York: Plenum Publ. Corp. (50, 51, 52, 53)

D'Amato, F. (1972*b*). Significato teorico e pratico delle colture *in vitro* di tessuti e cellule vegetali. *Accad. Naz. Lincei, Problemi Attuali di Scienza e Cultura, no. 173*, 55 pages. (224)

D'Amato, F. (1975*a*). The problem of genetic stability in plant tissue and cell cultures. In *Crop Genetic Resources for Today and Tomorrow*, ed. by O. Frankel and J. C. Hawkes, pp. 333–48. Cambridge: University Press. (49, 224)

D'Amato, F. (1975*b*). Recent findings on the organization of apical meristems with single apical cell. *Giorn. Bot. Ital.* **109**, 321–34. (46)

D'Amato, F. (1977). Cytogenetics of differentiation in tissue and cell cultures. In *Applied and Fundamental Aspects of Plant Cell, Tissue and Organ Culture*, ed. by J. Reinert and Y. P. S. Bajaj, pp. 343–57. Berlin: Springer-Verlag. (203, 224, 225, 226, 227).

D'Amato, F. & Avanzi, M. G. (1948). Reazioni di natura auxinica ed effeti rizogeni in *Allium cepa* L. Studio cito-istologico sperimentale. *Nuovo Giorn. Bot. Ital.* **55**, 161–213. (39, 125)

D'Amato, F. & Avanzi, S. (1965). DNA content, DNA synthesis and mitosis in the root apical cell of *Marsilea strigosa. Caryologia* **18**, 383–94. (42)

D'Amato, F. & Avanzi, S. (1968). The shoot apical cell of *Equisetum arvenvense*, a quiescent cell. *Caryologia* **21**, 83–9. (46)

D'Amato, F., Devreux, M. & Scarascia-Mugnozza, G. T. (1965). The DNA content of the nuclei of the pollen grains of tobacco and barley. *Caryologia* **18**, 377–82. (13, 14)

D'Amato, F. & Hoffmann-Ostenhof, O. (1956). Metabolism and spontaneous mutations in plants. *Adv. Genet.* **8**, 1–28. (142)

Daneholt, B. & Edström, J. E. (1967). The content of deoxyribonucleic acid in individual polytene chromosomes of *Chironomus tentans. Cytogenetics* **6**, 350–6. (131)

Daneholt, B. & Hosick, H. (1973). Evidence for transport of 75S RNA from a discrete chromosomal region via nuclear sap to cytoplasm in *Chironomus tentans. Proc. Nat. Acad. Sci.* **70**, 442–6. (176)

Darlington, C. D. & La Cour, L. (1938). Differential reactivity of the chromosomes. *Ann. Bot.* **2**, 615–25. (187)

Darlington, C. D. & La Cour, L. F. (1940). Nucleic acid starvation of chromosomes in *Trillium. J. Genet.* **40**, 185–213. (187)

Darlington, C. D. & Thomas, P. T. (1941). Morbid mitosis and the activity of inert chromosomes in *Sorghum. Proc. Roy. Soc. B* **130**, 127–50. (109)

Darlington, C. D. & Upcott, M. B. (1941). The genetics of inert chromosomes in *Zea mays. J. Genet.* **41**, 275–96. (193)

Darzynkievicz, Z., Bolund, L. & Ringertz, N. R. (1969). Nucleoprotein changes and initiation of RNA synthesis in PHA stimulated lymphocytes. *Exptl. Cell Res.* **56**, 418–24. (212)

Das, N. K. & Alfert, M. (1968). Cytochemical studies on the concurrent synthesis of DNA and histone in primary spermatocytes of *Urechis caupo. Exptl. Cell Res.* **49**, 51–8. (32)

Davidson, D. (1966). The onset of mitosis and DNA synthesis in roots of germinating beans. *Amer. J. Bot.* **53**, 491–5. (50)

Davidson, E. H. (1968). *Gene Activity in Early Development.* New York: Academic Press. (166)

Davidson, R. L. (1973). *Somatic Cell Hybridization: Studies on Genetics and Development.* Reading, Mass.: Addison Wesley. (227)

De Litardière, R. (1923). Les anomalies de la caryocinèse somatique chez le *Spinacia oleracea. Rev. Gén. Bot.* **35**, 369–81. (136)

Dermen, H. (1941). Intranuclear polyploidy in bean induced by naphtalene acetic acid. *J. Heredity* **32**, 133–8. (125)

Dermen, H. (1947). Periclinal cytochimeras and histogenesis in cranberry. *Amer. J. Bot.* **34**, 32–43. (94, 100, 101, 103)

Dermen, H. (1951). Ontogeny of tissues in stem and leaf of cytochimeral apples. *Amer. J. Bot.* **38**, 753–60. (100, 101)

Dermen, H. (1953). Pattern of tetraploidy in the flower and fruit of a cytochimeral apple. *J. Hered.* **44**, 31–9. (100, 103)

Dermen, H. (1960). Nature of plant sports. *Am. Hort. Magaz.*, July, 123–73. (98)

Dermen, H. (1965). Colchiploidy and histological instability in triploid apple and pear. *Amer. J. Bot.* **52**, 353–9. (98, 100)

Dermen, H. & Stewart, R. N. (1973). Ontogenetic study of the floral organs of peach (*Prunus persica*) utilizing cytochimeral plants. *Amer. J. Bot.* **60**, 283–91. (103)

Dev, V. G., Warburton, D., Miller, O. J., Miller, D. A., Erlanger, B. F. & Beiser, S. M. (1972). Consistent pattern of binding of antiadenosine antibodies to human metaphase chromosomes. *Exptl. Cell Res.* **74**, 288–93. (187)

Dewey, D. L. & Howard, A. (1963). Cell dynamics in the bean root tip. *Radiat. Bot.* **3**, 259–63. (113)

Dickson, E., Boyd, J. B. & Laird, C. D. (1971). Sequence diversity of polytene chromosome DNA from *Drosophila hydei. J. Mol. Biol.* **61**, 615–27. (133)

Dobzhansky, T. (1936). The persistence of the chromosome pattern in successive cell divisions in *Drosophila pseudoobscura. J. Exp. Zool.* **74**, 119–35. (130)

Doljanski, F. (1960). The growth of the liver with special reference to mammals. *Int. Rev. Cytol.* **10**, 207–41. (116)

Dommergues, P., Heslot, H., Gillot, J. & Martin, C. (1967). L'induction de mutations chez les Rosiers. In *Induced Mutations and Their Utilization*, ed. by H. Stubbe, pp. 319–48. Berlin: Akademie Verlag. (100)

Donner, R., Chyle, P. & Sanerova, H. (1969). Malformation syndrome in *Gallus domesticus* associated with triploidy. *J. Hered.* **60**, 113–5. (20)

Drescher, W. & Rothenbuhler, W. C. (1963). Gynandromorph production by chilling. *J. Hered.* **54**, 195–201. (83)

Dumont, J. N. & Yamada, T. (1972). Dedifferentiation of iris epithelial cells. *Develop. Biol.* **29**, 385–401. (218)

Dumont, J. N., Yamada, T. & Cone, M. V. (1970). Alteration of nucleolar ultrastructure in iris epithelial cells during initiation of Wolffian lens regeneration. *J. Exp. Zool.* **174**, 187–203. (215)

Duncan, R. E. & Ross, J. G. (1950). The nucleus in differentiation and development. III. Nuclei of maize endosperm. *J. Hered.* **41**, 259–68. (127)

Dunn, H. O., Kenney, R. M. & Lein, D. H. (1968). XX/XY chimerism in a bovine true hermaphrodite and insight into the understanding of freemartinism. *Cytogenetics* **7**, 390–402. (87)

Dunn, H. O., McEntee, K. & Hansel, W. (1970). Diploid–triploid chimerism in a bovine true hermaphrodite. *Cytogenetics* **9**, 245–59. (87, 88)

DuPraw, E. J. (1970). *DNA and Chromosomes*. New York: Holt, Rinehart and Winston, Inc. (29)

Durrant, A. (1962). The environmental induction of heritable change in flax. *Heredity* **17**, 27–61. (160)

Durrant, A. (1971). Induction and growth of flax genotrophs. *Heredity* **27**, 277–98. (160)

Durrant, A. (1972). Studies on reversion of induced plant weight changes in flax by outcrossing. *Heredity* **29**, 71–81. (160)

Durrant, A. & Jones, T. W. A. (1971). Reversion of induced changes in amount of nuclear DNA in *Linum*. *Heredity* **27**, 431–9. (162)

Dutt, M. K. (1970). DNA determination in the nuclei of the liver and kidney of the indian buffalo and the pig. *Proc. Zool. Soc., Calcutta* **23**, 147–53. (118)

Dutt, M. K. (1971). Microspectrophotometric estimation of the amount of DNA in the liver of some rodents. *Proc. Zool. Soc., Calcutta* **24**, 139–47. (118)

Dwidevi, R. S. & Naylor, J. M. (1968). Influence of apical dominance on the nuclear proteins in cells of the lateral bud meristem in *Tradescantia paludosa*. *Can. J. Bot.* **46**, 289–98. (54)

Dyer, A. F. (1967). The maintenance of structural heterozygosity in *Nothoscordum fragrans* Kunth. *Caryologia* **20**, 287–308. (76)

Ebstein, B. S. (1969). The distribution of DNA within the nucleoli of the amphibian oocyte as demonstrated but tritiated actinomycin D radioautography. *J. Cell Sci.* **5**, 27–44. (144)

Eckhardt, R. A. & Gall, J. G. (1971). Satellite DNA associated with heterochromatin in *Rhynchosciara*. *Chromosoma* **32**, 407–27. (24, 158, 185)

Edström, J. E. & Tangway, R. (1974). Cytoplasmic ribonucleic acids with messenger characteristics in salivary gland cells of *Chironomus tentans*. *J. Mol. Biol.* **84**, 569–83. (174)

Eicher, E. M. (1970). X-autosome translocation in the mouse: total inactivation versus partial inactivation of the X chromosome. *Adv. Genet.* **15**, 175–259. (192, 193)

Einset, J. (1952). Spontaneous polyploidy in cultivated apples. *Proc. Amer. Soc. Hort. Sci.* **59**, 291–302. (100)

Einset, J. & Pratt, C. (1954). Giant 'sports' of grapes. *Proc. Amer. Soc. Hort. Sci.* **63**, 251–6. (100, 103)

Eisenberg, S. & Yamada, T. (1966). A study of DNA synthesis during the transformation of the iris into lens in the lentectomized newt. *J. Exptl. Zool.* **162**, 353–68. (218)

Ellis, J. R., Marshall, R., Normand, I. C. S. & Penrose, L. S. (1963). A girl with triploid cells. *Nature* **198**, 411. (87)

Ellison, J. R. & Barr, H. J. (1972). Quinacrine fluorescence of specific chromosome regions. *Chromosoma* **36**, 375–90. (187)

Emmerich, H. (1972). Ecdysone binding proteins in nuclei and chromatin from *Drosophila* salivary glands. *J. Gen. Comp. Endocr.* **19**, 543–51. (200)

Emsweller, S. L. & Jones, H. A. (1945). Further studies on the chiasmata of the *Allium cepa*×*A. fistulosum* hybrid and its derivatives. *Amer. J. Bot.* **32**, 370–9. (65, 69)

Enzenberg, U. (1961). Beiträge zur Karyologie des Endosperms. *Öster. bot. Z.* **108**, 245–85. (120, 125)

Epifanova, O. I. & Terskikh, V. V. (1969). On the resting periods in the cell life cycle. *Cell Tissue Kinet.* **2**, 75–93. (111, 112, 113, 130)

Epstein, C. J. (1967). Cell size, nuclear content, and the development of polyploidy in the mammalian liver. *Proc. Nat. Acad. Sci.* **57**, 327–34. (116, 117)

Erbrich, P. (1965). Über Endopolyploidie und Kernstrukturen in Endospermhanstorien. *Öster. bot. Z.* **112**, 197–262. (125, 127, 128)

Erickson, R. O. (1964). Synchronous cell and nuclear division in tissues of the higher plants. In *Synchrony in Cell Division and Growth*, ed. by E. Zeuthen, pp. 11–37. New York: Interscience Publishers. (35)

Esau, K. (1971). The sieve element and its immediate environment: thoughts on research of the past fifty years. *J. Indian Bot. Soc.* **50 A**, 115–29. (140)

Esau, K. (1973). Comparative structure of companion cells and phloem parenchyma cells in *Mimosa pudica* L. *Ann. Bot.* **37**, 625–32. (140)

Esser, K. & Künen, R. (1967). *Genetics of fungi.* Berlin: Springer-Verlag. (58)

Evans, D. & Birnstiel, M. (1968). Localization of amplified ribosomal DNA in the oocyte of *Xenopus laevis. Biochim. Biophys. Acta* **166**, 274–6. (144)

Evans, G. M. (1968). Nuclear changes in flax. *Heredity* **23**, 25–38. (160)

Evans, G. M., Durrant, A. & Rees, H. (1966). Associated nuclear changes in the induction of flax genotrophs. *Nature* **212**, 697–9. (160, 161)

Evans, G. M., Rees, H., Snell, C. L. & Sun, S. (1972). The relation between nuclear DNA amount and the duration of the mitotic cycle. In *Chromosomes Today* **3**, ed. by C. D. Darlington and K. R. Lewis, pp. 24–31. Edinburgh: Longman Group Ltd. (34)

Evans, H. G. & Savage, J. R. K. (1963). The relation between DNA synthesis and chromosome structure as resolved by X-ray damage. *J. Cell Biol.* **18**, 525–40. (49)

Evert, R. F., Davis, J. D., Tucker, C. M. & Alfieri, F. J. (1970). On the occurrence of nuclei in mature sieve elements. *Planta* **95**, 281–96. (139)

Fahmy, O. G. (1952). The cytology and genetics of *Drosophila subobscura*. VI. Maturation, fertilization and cleavage in normal eggs and in the presence of *cross-over suppressor* gene. *J. Genet.* **50**, 486–506. (65, 69)

Fakan, S., Turner, G. N., Pagano, J. S. & Hancock, H. (1972). Sites of replication of chromosomal DNA in an eukaryotic cell. *Proc. Nat. Acad. Sci.* **69**, 2300–5. (28)

Feinbrun, N. & Klein, S. (1962). ^{3}H-thymidine incorporation into cell nuclei in germinating lettuce seeds. *Plant & Cell Physiol.* **3**, 407–13. (49)

Feldmann, J. (1972). Les problèmes actuels de l'alternance de générations chez les Algues. *Soc. Bot. Fr. Mém.* 1972, 7–38. (2, 6)

Feldman, M. (1966). The effect of chromosomes VB, VD and VA on chromosomal pairing in *Triticum aestivum. Proc. Nat. Acad. Sci.* **55**, 1447–53. (63)

Fenzl, E. & Tschermak-Woess, E. (1954). Untersuchungen zur karyologischen Anatomie der Achse der Angiospermen. *Öster. bot. Z.* **101**, 140–64. (113, 125)

Fernandes, A. (1943). Sur l'origine des chromosomes surnuméraires hétérochromatiques chez *Narcissus bulbocodium* L. *Bol. Soc. Broter*, **17**, 251–6. (194)

Fernandes, A. (1949). Le problème de l'hétérochromatinisation chez *Narcissus bulbocodium* L. *Bol. Soc. Broter.* **23**, 5–69. (193, 194)

Ferrier, P., Ferrier, S., Stalder, G., Bühler, E., Bamatter, F. & Klein, D. (1964). Congenital asymmetry associated with diploid–triploid mosaicism and large satellites. *Lancet i*: 80–2. (87)

Filner, P. (1965). Semiconservative replication of DNA in higher plant cells. *Exptl. Cell Res.* **39**, 33–9. (31)

Flamm, W. G., Bernheim, N. J. & Brubacker, P. E. (1971). Density gradient analysis of newly replicated DNA from synchronized mouse limphoma cells. *Exptl. Cell Res.* **64**, 97–104. (31)

Flavell, R. B. & Walker, G. W. R. (1973). The occurrence and role of DNA synthesis during meiosis in wheat and rye. *Exptl. Cell Res.* **77**, 15–24. (61)

Focke, W. O. (1881). *Die Pflanzenmischlinge, ein Beitrag zur Biologie der Gewächse.* Berlin: Borntraeger. (72)

Ford, P. J. & Southern, E. M. (1973). Different sequences for 5s RNA in kidney cells and ovaries of *Xenopus laevis. Nature New Biol.* **241**, 7–12. (179)

Forer, A. (1969). Chromosome movements during cell-division. In *Handbook of Molecular Cytology*, ed. by A. Lima-de-Faria, pp. 553–601. Amsterdam: North-Holland Publ. Co. (34)

Fossati-Tallard, J. (1967). Etude microfluorométrique de la synthèse des acides désoxyribonucleiques dans le cycle mitotique. *Z. wiss. Mikroskop, mikroskop. Techn.* **68**, 1–21. (28)

Fox, D. P. (1969). DNA values in somatic tissues of *Dermestes*. I. Abdominal fat body and testis wall of the adult. *Chromosoma* **28**, 445–56. (162)

Fox, D. P. (1970). A non-doubling DNA series in somatic tissues of the locust *Schistocerca gregaria* and *Locusta migratoria* (Linn.). *Chromosoma* **29**, 446–61. (162)

Fox, D. P. (1971). The replicative status of heterochromatin and euchromatin in two somatic tissues of *Dermestes maculatus* (Dermestidae, Coleoptera). *Chromosoma* **33**, 183–95. (162)

Fox, D. P., Hewitt, G. M. & Hall, D. J. (1974). DNA replication and RNA transcription of euchromatic and heterochromatic chromosome regions during grasshopper meiosis. *Chromosoma* **45**, 43–62. (194)

Frenster, J. H. (1965). Nuclear polyanions as derepressors of synthesis of ribonucleic acid. *Nature* **206**, 680–3. (195, 196)

Frenster, J. H. (1969). Biochemistry and molecular biophysics of heterochromatin and euchromatin. In *Handbook of Molecular Cytology*, ed. by A. Lima-de-Faria, pp. 251–76. Amsterdam: North-Holland Publ. Co. (25)

Frenster, J. H. (1974). Ultrastructure and function of heterochromatin and euchromatin. In *The Cell Nucleus*, ed. by H. Busch, pp. 565–80. New York: Academic Press. (185, 187)

Friedberg, S. H. & Davidson, D. (1970). Duration of S phase and cell cycle in diploid and tetraploid cells of mixoploid meristems. *Exptl. Cell Res.* **61**, 216–8. (34)

Fukui, N. (1971). Factors regulating thymidine kinase in regenerating liver. *J. Biochem.* **69**, 1075–82. (213)

Gabriel, A. (1970). L'implantation d'un greffon marqué permet de suivre la régeneration de planaires normales ou irradiées *in toto. Ann. Biol.* **9**, 519–26. (207)

Gall, J. G. (1954). Lampbrush chromosomes from oocyte nuclei of the newt. *J. Morph.* **94**, 283–352. (58)

Gall, J. G. (1968). Differential synthesis of the genes for ribosomal RNA during amphibian oogenesis. *Proc. Nat. Acad. Sci.* **60**, 553–60. (144, 169)
Gall, J. G. (1969). The genes for ribosomal RNA during oögenesis. *Genetics*, Suppl. **61**, 121–32. (146, 153)
Gall, J. G. (1973). Repetitive DNA in *Drosophila*. In *Molecular Cytogenetics*, ed. by B. A. Hamkalo and J. Papaconstantinou, pp. 59–74. New York: Plenum Publ. Corp. (24)
Gall, J. G., Cohen, E. H. & Polan, M. L. (1971). Repetitive DNA sequences in *Drosophila* chromosomes. *Chromosoma* **33**, 319–44. (24, 132, 133, 185)
Gall, J. G., Macgregor, H. C. & Kidston, M. E. (1969). Gene amplification in the oocytes of dytiscid water bettles. *Chromosoma* **26**, 169–87. (148, 149)
Gall, J. G. & Pardue, M. L. (1969). Formation and detection of RNA–DNA hybrid molecules in cytological preparations. *Proc. Nat. Acad. Sci.* **63**, 378–83. (145, 146, 168, 169)
Gall, J. G. & Rochaix, J.-D. (1974). The amplified ribosomal DNA of dytiscid beetles. *Proc. Nat. Acad. Sci.* **71**, 1819–23. (149)
Gallien, C. L. (1970). Recherches sur la greffe nucléaire interspécifique dans le genre *Pleurodeles* (Amphibien-Urodèle). *Ann. Embr. Morph.* **3**, 145–92. (207, 219)
Gallwitz, D. & Mueller, G. C. (1969). Histone synthesis *in vitro* on HeLa cells microsomes. The nature of the coupling to deoxyribonucleic acid synthesis. *J. Biol. Chem.* **244**, 5947–52. (32)
Gambarini, A. & Lara, F. J. S. (1974). Under-replication of ribosomal cistrons in polytene chromosomes of *Rhynchosciara. J. Cell Biol.* **62**, 215–22. (134)
Gambarini, A. G. & Meneghini, R. (1972). Ribosomal RNA genes in salivary gland and ovary of *Rhynchosciara angelae. J. Cell Biol.* 54, 421–6. (134, 158)
Gardner, R. L. & Munro, A. J. (1974). Successful construction of chimaeric rabbit. *Nature* **250**, 146–7. (91)
Gaul, H. (1954). Asynapsis und ihre Bedeutung für die Genomanalyse. *Z. indukt. Abst. Vererbl.* **86**, 69–100. (66)
Gaul, H. (1959). Über Chimärenbildung in Gerstenpflanzen nach Röntgenbestrahlung von Samen. *Flora* **147**, 207–41. (104)
Gaul, H. (1961). Studies on diplontic selection after X-irradiation of barley seeds. In *Effects of Ionizing Radiations on Seeds*, pp 117–36. Vienna: IAEA. (104)
Gautheret, R. (1939). Sur la possibilité de réaliser la culture indefinie des tissus de tubercules de carotte. *C.R. Acad. Sci.* **208**, 218–20. (224)
Gay, H., Das, C. C., Forward, K. & Kaufmann, B. P. (1970). DNA content of mitotically active condensed chromosomes of *Drosophila melanogaster. Chromosoma* **32**, 213–23. (136)
Geitler, L. (1937). Die Analyse des Kernbaus und der Kernteilung der Wasserläufer *Gerris lateralis* und *Gerris lacustris* (Hemiptera, Heteroptera) und die Somadifferenzierung. *Z. Zellforsch.* **26**, 641–72. (120)
Geitler, L. (1938). Über den Bau des Ruhekerns mit besonderer Berüchsichtigung der Heteropteren und Dipteren. *Biol. Zentralbl.* **58**, 152–79. (122)
Geitler, L. (1939*a*). Die Entstehung der polyploiden Somakerne der Heteropteren durch Chromosomenteilung ohne Kernteilung. *Chromosoma* **1**, 1–22. (120, 122)
Geitler, L. (1939*b*). Das Heterochromatin der Geschlechtschromosomen bei Heteropteren. *Chromosoma* **1**, 197–229. (122)
Geitler, L. (1948). Ergebuisse und Probleme der Endomitoseforschung. *Öster. bot. Z.* **95**, 277–99. (120, 123)
Geitler, L. (1952). Karyologische Anatomie. *Scientia* **87**, 216–19. (125)
Geitler, L. (1953). *Endomitose und endomitotische Polyploidisierung. Protoplasmatologia, VIC.* Vienna: Springer-Verlag. (120, 122, 123, 125)
Gelfant, S. (1962). Initiation of mitosis in relation to the cell division cycle. *Exptl. Cell Res.* **26**, 395–403. (47, 48, 111)

Gelfant, S. (1963). A new theory on the mechanism of cell division. *Symp. Int. Soc. Cell Biol.* **2**, 229–59. (48, 111)
Gelfant, S. (1966). Patterns of cell division: the demonstration of discrete cell populations. In *Methods in Cell Physiology* **2**, ed. by D. M. Prescott, pp. 359–95. New York: Academic Press. (48, 111, 112)
Gentcheff, A. & Gustafsson, Å. (1939). The double chromosome reproduction in *Spinacia* and its causes. I and II. *Hereditas* **25**, 349–58; 371–86. (136)
Georgiev, G. P. (1969 *a*). On the structural organization of operon and the regulation of RNA synthesis in animal cells. *J. Theor. Biol.* **25**, 473–90. (184)
Georgiev, G. P. (1969 *b*). Histones and the control of gene action. *Ann. Rev. Genet.* **3**, 155–80. (195)
Gerbi, S. (1971). Localization and characterization of the ribosomal RNA cistrons in *Sciara coprophila. J. Mol. Biol.* **58**, 499–511. (158)
Gerner, E. W. & Humphrey, R. M. (1973). The cell-cycle phase synthesis of non-histone proteins in mammalian cells. *Biochim. Biophys. Acta* **331**, 117–27. (32)
Giardina, A. (1901). Origine dell'oocite e delle cellule nutrici nel *Dytiscus. Int. Mschr. Anat. Physiol.* **18**, 418–84. (148)
Gibson, I. & Hewitt, G. M. (1970). Isolation of DNA from B chromosomes in grasshoppers. *Nature* **225**, 67–8. (193)
Gierer, A., Berking, S., Bode, H., David, C. N., Flick, K., Hansmann, G., Schaller, H. & Trenkner, E. (1972). Regeneration of hydra from reaggregated cells. *Nature New Biol.* **239**, 98–101. (204, 206)
Gifford, E. M. Jr. (1960). Incorporation of H^3-thymidine into shoot and root apices of *Ceratopteris thalictroides. Amer. J. Bot.* **47**, 834–7. (42)
Gifford, E. M. Jr. & Nitsch, J. P. (1969). Response of tobacco pith nuclei to growth substances. *Planta* **85**, 1–10. (226)
Gilbert, C. W. & Lajtha, L. G. (1965). The importance of cell population kinetics in determining response to irradiation of normal and malignant tissue. In *Cellular Radiation Biology*, pp. 474–97. Baltimore: Williams and Wilkins Co. (112)
Gilmour, R. S. & Paul, J. (1973). Tissue-specific transcription of the globin gene in isolated chromatin. *Proc. Nat. Acad. Sci.* **70**, 3440–2. (198, 199)
Gläss, E. (1957). Das Problem der Genomsonderung in den Mitosen unbehandelter Rattenlebern. *Chromosoma* **8**, 468–92. (139)
Gläss, E. (1958). Aneuploiden Chromosomenzahlen in den Mitosen der Leber verschiedenalter Ratten. *Chromosoma* **9**, 269–85. (139)
Golikowa, M. N. (1965). Der Aufbau des kernapparates und die Verteilung der Nuleinsäuren und Protein bei *Nyctotherus cordifornis* Stein. *Arch. Protistenk.* **108**, 191-216. (130)
Goss, R. J. (1969). *Principles of regeneration.* New York: Academic Press. (204, 205, 210)
Gosselin, A. (1940). Action sur la mitose des végétaux de deux alcaloides puriques. *C.R. Acad. Sci.* **210**, 544–46. (119)
Gottschalk, W. (1958). Über Abregulierungsvorgänge bei künstlich ergestellten hochpolyploiden Pflanzen. *Z. Vererbl.* **89**, 204–15. (139)
Gottschalk, W. & Baquar, S. R. (1971). Desynapsis in *Pisum sativum* induced through gene mutation. *Can. J. Genet. Cytol.* **13**, 138–43. (65, 67, 68)
Gottschalk, W. & Konvicka, O. (1970). Die Meiosis einer partiell sterilen Mutante von *Brassica oleracea* var. *capitata. Cytologia* **36**, 269–80. (65, 68)
Gottschalk, W. & Villalobos-Pietrini, R. (1965). The influence of mutant genes on chiasmata formation in *Pisum sativum. Cytologia* **30**, 88–97. (65)
Gowen, J. W. (1933). Meiosis as a genetic character in *Drosophila melanogaster. J. Exp. Zool.* **65**, 83–106. (65, 70)
Grafl, I. (1939). Kernwachstum durch Chromosomenvermehrung als regelmässiger Vorgang bei pflanzlichen Gewebedifferenzierung. *Chromosoma* **1**, 265–75. (123)

Grafl, I. (1941). Über das Wachstum der Antipodenkerne von *Caltha palustris. Chromosoma* **2**, 1–11. (119)

Graham, C. F. (1966). The effect of cell size and DNA content on the cellular regulation of DNA synthesis in haploid and diploid embryos. *Exptl. Cell Res.* **43**, 13–19. (34)

Graham, C. F., Arms, K. & Gurdon, J. B. (1966). The induction of DNA synthesis by frog egg cytoplasm. *Develop. Biol.* **14**, 349–81. (222)

Gregor, D., Reinert, J. & Matsumoto, H. (1974). Changes in chromosomal proteins from embryo induced carrot cells. *Plant & Cell Physiol.* **15**, 875–81. (227)

Grell, R. F., Bank, H. & Gassner, G. (1972). Meiotic exchange without the synaptinemal complex. *Nature New Biol.* **240**, 155–7. (59)

Grell, S. M. (1946). Cytological studies in *Culex*. I. Somatic reduction divisions. *Genetics* **31**, 60–76. (137)

Gremigni, V. & Puccinelli, I. (1977). A contribution to the problem of the origin of the blastema cells in planarians. A karyological and ultrastructural investigation. *J. Exptl. Zool.*, **199**, 57–72. (208)

Grierson, D. & Loening, U. E. (1972). Distinct transcription products of ribosomal genes in two different tissues. *Nature New Biol.* **235**, 80–2. (171)

Grierson, D., Rogers, M. E., Sartirana, M. L. & Loening, U. E. (1970). The synthesis of ribosomal RNA in different organisms; structure and evolution of the rRNA precursor. *Cold Spring Harbor Symp. Quant. Biol.* **35**, 589–98. (172)

Grigliatti, T. A., White, B. N., Tener, G. M., Kaufman, T. C., Holden, J. J. & Suzuki, D. T. (1974). Studies on the transfer RNA genes of *Drosophila. Cold Spring Harbor Symp. Quant. Biol.* **38**, 461–74. (173)

Grossbach, U. (1969). Chromosomen-Aktivität und biochemische Zelldifferenzierung in den Speicheldrüsen von *Camptochironomus. Chromosoma* **28**, 136–87. (176)

Guevara, M. & Basile, R. (1973). DNA and RNA puffs in *Rhynchosciara. Caryologia* **26**, 275–95. (157)

Guillé, E. & Quetier, F. (1973). Heterochromatic, redundant and metabolic DNAs: a new hypothesis about their structure and function. *Progr. Biophys. Mol. Biol.* **27**, 121–42. (184)

Gurdon, J. B. (1962). Adult frogs derived from the nuclei of single somatic cells. *Develop. Biol.* **4**, 256–73. (219)

Gurdon, J. B. (1968). Changes in somatic cell nuclei inserted into growing and maturing amphibian oocytes. *J. Embryol. Exp. Morph.* **20**, 401–14. (222)

Gurdon, J. B. (1974). *The Control of Gene Expression in Animal Development.* Oxford: Clarendon Press. (165, 166, 179, 180, 203, 218, 219, 220, 221, 227)

Gurdon, J. B. & Laskey, R. A. (1970). The transplantation of nuclei from single cultured cells into enucleate frogs' eggs. *J. Embryol. Exp. Morph.* **24**, 227–48. (219)

Gurley, L. R., Irvin, J. L. & Hoolbrook, D. J. (1964). Inhibition of DNA polymerase by histone. *Biochem. Biophys. Res. Commun.* **14**, 527–32. (226)

Gustafsson, Å. (1946). Apomixis in higher plants. I. The mechanisms of apomixis. *Lunds Univ. Arsskr. N.F. Avd. 2* **42**, 1–66. (70, 72, 73, 75)

Gustafsson, Å. (1947). Apomixis in higher plants. II. The causal aspects of apomixis. *Lunds Univ. Arsskr. N.F. Avd. 2* **43**, 71–178. (70, 73, 75)

Guyenot, E. & Danon, M. (1953). Chromosomes et ovocytes de Batraciens. *Rev. Suisse Zool.* **60**, 1–129. (143)

Haccius, B. & Lakshmann, K. K. (1969). Adventiv-Embryonen-Embyoide-Adventiv-Knospen. Ein Beitrag zur Klärung der Begriffe. *Oster. bot. Z.* **116**, 145–58. (222)

Hadorn, E., Ruch, F. & Staub, M. (1964). Zum DNS-Gehalt in Speicheldrüsenkernen mit 'übergrossen Riesenchromosomen' von *Drosophila melanogaster. Experientia* **20**, 566–7. (132)

Hagerup, O. (1944). On fertilization, polyploidy and haploidy in *Orchis maculatus. Dansk. Bot. Ark.* **11**, 1–26. (72)

Hagerup, O. (1947). The spontaneous formation of haploid, polypoloid and aneuploid embryos in some orchids. *Danske Vidensk. Selsk* **20**, 1–22. (72)
Haggis, A. J. (1966). Deoxyribonucleic acid in germinal vesicles of oocytes of *Rana pipiens. Science* **154**, 670–1. (144)
Hahn, H. P. (1970). Structural and functional changes in nucleoprotein during aging of the cell. *Gerontologia* **16**, 116–29. (142)
Håkansson, A. (1951). Parthenogenesis in *Allium. Bot. Notiser* 1951, 143–79. (73)
Håkansson, A. & Levan, A. (1957). Endo-duplicational meiosis in *Allium odorum. Hereditas* **43**, 179–200. (73, 75, 79)
Halperin, W. (1970). Embryos from somatic plant cells. In *Control Mechanisms in the Expression of Cellular Phenotypes*, ed. by H. A. Padykula, pp. 169–91. New York: Academic Press. (222)
Hamerton, J. L., Richardson, B. J., Gee, P. A., Allen, W. R. & Short, R. V. (1971). Non-random X chromosome expression in female mules and hinnies. *Nature* **232**, 312–15. (192)
Hanstein, J. (1868). Die Scheitelzellgruppe in Vegetationspunkt der Phanerogamen. *Festschr. Niederrhein Ges. Natur-und Heilkunde* 1868, 109–34. (38)
Hardin, J. A., Einem, G. E. & Lindsay, D. T. (1967). Simultaneous synthesis of histone and DNA in synchronously dividing *Tetrahymena pyriformis. J. Cell Biol.* **32**, 709–17. (32)
Harford, A. G. (1974). Ribosomal gene magnification in *Drosophila*: a chromosomal change. *Genetics* **78**, 887–96. (156)
Harkness, R. D. (1957). Regeneration of liver. *Brit. Med. Bull.* **13**, 87–93. (211, 212)
Hay, E. D. (1962). Cytological studies of dedifferentiation and differentiation in regenerating amphibian limbs. In *Regeneration*, ed. by D. Rudnick, pp. 177–210. New York: The Ronal Press Co. (209)
Hay, E. D. (1966). *Regeneration.* New York: Holt, Rinehart and Winston. (204)
Hay, E. D. (1968). Dedifferentiation and metaplasia in vertebrate and invertebrate regeneration. In *The Stability of the Differentiated State*, ed. by H. Ursprung, pp. 85–108. Berlin: Springer-Verlag. (206, 207, 209, 211)
Hay, E. D. & Coward, S. J. (1975). Fine structure studies on the planarian *Dugesia*. I. Nature of the 'neoblast' and other cell types in noninjured worms. *J. Ultrastruc. Res.* **50**, 1–21. (207)
Hay, E. D. & Fischmann, D. A. (1961). Origin of the blastema in regenerating limbs of the newt, *Triturus viridescens*. An autoradiographic study using tritiated thymidine to follow cell proliferation and migration. *Develop. Biol.* **3**, 26–59. (208, 210)
Haynes, J. & Burnett, A. L. (1963). Dedifferentiation and redifferentiation of cells in *Hydra viridis. Science* **142**, 1481–3. (204, 206, 209)
Hearst, J. E., Cech, T. R., Marx, K. A., Rosenfeld, A. & Allen, J. R. (1974). Characterization of the rapidly renaturing sequences in the main CsCl density bands of *Drosophila*, mouse and human DNA. *Cold Spring Harbor Symp. Quant. Biol.* **38**, 329–39. (186)
Heitz, E. (1928). Das Heterochromatin der Moose. *Jb. wiss. Bot.* **69**, 762–818. (184)
Heitz, E. (1929). Heterochromatin, Chromozentren, Chromomeren. *Ber. deutsch. bot. Ges.* **47**, 274–84. (123, 184)
Heitz, E. (1931). Nucleolen und Chromosomen in der Gattung *Vicia. Planta* **15**, 495–505. (168)
Heitz, E. (1933). Die somatische Heteropyknose bei *Drosophila melanogaster* und ihre genetische Bedeutung. *Z. Zellforsch.* **20**, 237–87. (132, 184)
Heitz, E. (1934). Über α und β Heterochromatin sowie Konstanz und Bau der Chromomeren by *Drosophila. Biol. Zentralbl.* **54**, 588–609. (132)
Heitz, E. & Bauer, H. (1933). Beweise für die Chromosomennatur der Kernschleifen in den Knäuelkernen von *Bibio hortulanus* L. *Z. Zellforsch.* **17**, 67–82. (130)
Helmsing, P. J. (1972). Induced accumulation of non-histone proteins in polytene nuclei of

Drosophila hydei. II. Accumulation of proteins in polytene nuclei and chromatin of different larval tissues. *Cell Differentiation* **1**, 19–24. (200)

Helmsing, P. J. & Berendes, H. D. (1971). Induced accumulation of nonhistone proteins in polytene nuclei of *Drosophila hydei*. *J. Cell Biol.* **50**, 893–6. (200)

Henderson, A. S., Warburton, D. & Atwood, K. C. (1972). Location of ribosomal DNA in the human chromosome complement. *Proc. Nat. Acad. Sci.* **69**, 3394–8. (169)

Henderson, A. S., Warburton, D. & Atwood, K. C. (1974). Localization of rDNA in the chimpanzee (*Pan troglodytes*) chromosome complement. *Chromosoma* **46**, 435–41. (169)

Henderson, S. A. (1970). The time and place of meiotic crossing-over. *Ann. Rev. Genet.* **4**, 295–324. (58)

Hennen, S. (1970). Influence of spermine and reduced temperature on the ability of transplanted nuclei to promote normal development in eggs of *Rana pipiens*. *Proc. Nat. Acad. Sci.* **66**, 630–7. (219, 220)

Hennig, W., Hennig, H. & Stein, H. (1970). Repeated sequences in the DNA of *Drosophila* and their localization in giant chromosomes. *Chromosoma* **32**, 31–63. (133)

Hennig, W. & Meer, B. (1971). Reduced polyteny of ribosomal RNA cistrons in giant chromosomes of *Drosophila hydei*. *Nature New Biol.* **233**, 10–12. (133)

Herreros, B. & Giannelli, F. (1967). Spatial distribution of old and new chromatid subunits and frequency of chromatid exchanges in induced human lymphocyte endoreduplication. *Nature* **126**, 286–7. (128, 129)

Hesemann, C. U. (1971). Untersuchungen zur Pollenentwicklung und Pollenschlauchbildung bei höheren Pflanzen. I. Quantitative Bestimmungen des DNS-Gehalts generativer und vegetativer Kerne in Pollenkörner und Pollenschläuchen von *Petunia-hybrida*-Mutanten. *Theor. Appl. Genet.* **41**, 338–51. (13)

Hesemann, C. U. (1973). Untersuchungen zur Pollenentwicklung und Pollenschlauchbildung bei höheren Pflanzen. III. Replication bei vegetativen und Sperma-kernen in reifen Pollenkörner von Gerste. *Theor. Appl. Genet.* **43**, 232–41. (13)

Heslop-Harrison, J. (1966). Cytoplasmic connections between angiosperm meiocytes. *Ann. Bot.* **30**, 221–30. (36)

Heslop-Harrison, J. (1968). Synchronous pollen mitosis and the formation of the generative cell in massulate Orchids. *J. Cell Sci.* **3**, 457–66. (12, 36)

Hesse, M. (1968). Karyologische Anatomie der Zoocecidien und ihre Kernstrukturen. *Öster. bot. Z.* **115**, 34–83. (125, 127)

Hewitt, G. M. (1974). The integration of supernumerary chromosomes into the orthopteran genome. *Cold Spring Harbor Symp. Quant. Biol.* **38**, 183–94. (193, 194)

Hewitt, G. M. & John, B. (1968). Parallel polymorphism for supernumerary segments in *Corthippus parallelus*. I. British populations. *Chromosoma* **25**, 319–42. (194)

Heyden, H. W. & Zachau, H. G. (1971). Characterization of RNA in fractions of calf thymus chromatin. *Biochim. Biophys. Acta* **232**, 651–60. (25)

Hinegardner, R. T., Rao, B. & Feldman, D. E. (1964). The DNA synthetic period during the early development of the sea urchin egg. *Exptl. Cell Res.* **36**, 127–60. (15, 36)

Holliday, R. & Pugh, J. E. (1975). DNA modification mechanisms and gene activity during development. *Nature* **187**, 226–32. (167, 184)

Holmes, D. S., Mayfield, J. E., Sander, G. & Bonner, J. (1972). Chromosomal RNA: its properties. *Science* **177**, 72–4. (25)

Holt, T. K. H. (1971). Local protein accumulation during gene activation. II. Interferometric measurement of the amount of solid material in temperature induced puffs of *Drosophila hydei*. *Chromosoma* **32**, 428–35. (200)

Holzer, K. (1952). Untersuchungen zur karyologischen Anatomie der Wurzel. *Öster. bot. Z.* **99**, 118–55. (113, 125)

Hotta, Y., Ito, M. & Stern, H. (1966). Synthesis of DNA during meiosis. *Proc. Nat. Acad. Sci.* **56**, 1184–91. (59)

Hotta, Y., Parchman, L. G. & Stern, H. (1968). Protein synthesis during meiosis. *Proc. Nat. Acad. Sci.* **60**, 575–82. (61)

Hotta, Y. & Stern, H. (1971*a*). Analysis of DNA synthesis during meiotic prophase in *Lilium. J. Mol. Biol.* **55**, 337–55. (59, 60)

Hotta, Y. & Stern, H. (1971*b*). A DNA binding protein in meiotic cells of *Lilium. Develop. Biol.* **26**, 87–99. (61)

Hotta, Y. & Stern, H. (1971*c*). Meiotic protein in spermatocytes of mammals. *Nature New Biol.* **234**, 83–6. (61)

Hourcade, D., Dressler, D. & Wolfson, J. (1973). The amplification of ribosomal RNA genes involves a rolling circle intermediate. *Proc. Nat. Acad. Sci.* **70**, 2926–30. (146, 147)

Hourcade, D., Dressler, D. & Wolfson, J. (1974). The nucleolus and the rolling circle. *Cold Spring Harbor Symp. Quant. Biol.* **38**, 537–50. (146, 147)

Howard, A. & Pelc, R. (1953). Synthesis of deoxyribonucleic acid in normal and irradiated cells and its relation to chromosome breakage. *Heredity* **6** Suppl., 261–73. (27, 34)

Howard, H. W. (1942). Heteroauxin and the production of tetraploid shoots by the callus method in *Brassica oleracea. J. Genet.* **44**, 1–9. (223)

Howard, H. W. (1961). An octoploid potato from eye excision experiments. *J. Hered.* **52**, 191–2. (223)

Howard, H. W. (1967). Experiments on X-ray irradiation of potatoes: analysis of chimeras and differentiation effects. In *Induced Mutations and Their Utilization*, ed. by H. Stubbe, pp. 311–17. Berlin: Akademie Verlag. (97)

Howell, S. H. & Stern, H. (1971). The appearance of DNA breakage and repair activities in the synchronous meiotic cycle of *Lilium. J. Mol. Biol.* **55**, 357–78. (60)

Hsu, T. C., Arrighi, F. E. & Saunders, G. F. (1972). Compositional heterogeneity of human heterochromatin. *Proc. Nat. Acad. Sci.* **69**, 1464–6. (186)

Hsu, T. C., Dewey, W. C. & Humphrey, R. M. (1962). Radiosensitivity of cells of Chinese hamster *in vitro* in relation to the cell cycle. *Exptl. Cell Res.* **27**, 441–52. (49)

Hsu, T. C. & Moorehead, P. S. (1956). Chromosome anomalies in human neoplasms with special reference to the mechanisms of polyploidization and aneuploidization in the HeLa strain. *Ann. N.Y. Acad. Sci.* **63**, 1083–94. (123)

Huang, R. C. & Bonner, J. (1962). Histone, a suppressor of chromosomal RNA synthesis. *Proc. Nat. Acad. Sci.* **48**, 1216–22. (195)

Huberman, J. A. & Riggs, A. D. (1968). On the mechanism of DNA replication in mammalian chromosomes. *J. Mol. Biol.* **32**, 327–41. (29)

Hughes-Schrader, S. (1948). Cytology of Coccids (Coccoidea-Homoptera). *Adv. Genet.* **2**, 127–203. (64, 123, 187)

Hughes-Schrader, S. & Ris, H. (1941). The diffuse spindle attachment of coccids as verified by the mitotic behaviour of induced fragments. *J. Exp. Zool.* **87**, 429–56. (63)

Hughes-Schrader, S. & Schrader, F. (1961). The kinetochore of the Hemiptera. *Chromosoma* **12**, 327–50. (63)

Humphrey, R. R. & Fankhauser, G. (1957). The origin of spontaneous and experimental haploids in the Mexican axolotl (*Siredon* or *Ambystoma mexicanum*). *J. Exp. Zool.* **134**, 427–47. (21)

Hunter, H. F. (1967). The effects of delayed insemination on fertilization and early cleavage in the pig. *J. Reprod. Fertil.* **13**, 133–47. (20)

Huskins, C. L. (1947). The subdivision of the chromosomes and their multiplication in nondividing tissues: possible interpretation in terms of gene structure and gene action. *Amer. Nat.* **81**, 401–34. (123, 125)

Huskins, C. L. (1948). Segregation and reduction in somatic tissues. I. Initial observations on *Allium cepa*. *J. Heredity* **39**, 311–25. (137)

Huskins, C. L. & Cheng, K. C. (1950). Segregation and reduction in somatic tissues. IV. Reductional groupings induced in *Allium cepa* by low temperature. *J. Hered.* **41**, 13–18. (137)

Huskins, C. L. & Steinitz, L. (1948*a*). The nucleus in differentiation and development. I. Heterochromatic bodies in energic nuclei of *Rhoeo* roots. *J. Hered.* **39**, 35–43. (125)

Huskins, C. L. & Steinitz, L. (1948*b*). The nucleus in differentiation and development. II. Induced mitosis in differentiated tissues of *Rhoeo* roots. *J. Hered.* **39**, 67–77. (125)

Ingle, J., Pearson, G. G. & Sinclair, J. (1973). Species distribution and properties of nuclear satellite DNA in higher plants. *Nature New Biol.* **242**, 193–7. (161)

Ingle, J. & Timmis, J. N. (1975). A role for differential replication of DNA in development. In: *Modification of the Information Content of Plant Cells*, ed. by R. Markham, D. R. Davies, D. A. Hopwood and R. W. Horne, pp. 37–52. Amsterdam: North-Holland Publ. Co. (162)

Ingle, J., Timmis, J. N. & Sinclair, J. (1975). The relationship between satellite deoxyribonucleic acid, ribosomal ribonucleic acid gene redundancy, and genome size in plants. *Plant Physiol.* **55**, 496–501. (24, 168)

Innocenti, A. M. (1973). Aspects ultrastructuraux des premiers stades de la différenciation nucléaire du métaxylème chez les racines de l'*Allium cepa*. *C.R. Acad. Sci.* **277**, 2153–6. (153)

Innocenti, A. M. & Avanzi, S. (1971). Some cytological aspects of the differentiation of metaxylem in the root of *Allium cepa*. *Caryologia* **24**, 283–92. (153)

Ito, M., Hotta, Y. & Stern, H. (1967). Studies of meiosis *in vitro*. II. Effect of inhibiting DNA synthesis during meiotic prophase on chromosome structure and behavior. *Develop. Biol.* **16**, 54–77. (60)

Izawa, M., Allfrey, V. G. & Mirsky, A. E. (1963). Composition of the nucleus and chromosomes in the lampbrush stage of the newt oocyte. *Proc. Nat. Acad. Sci.* **50**, 811–17. (144)

Jacob, J., Gillies, K., Macleod, D. & Jones, K. W. (1974). Molecular hybridization of mouse satellite DNA-complementary RNA in ultrathin sections prepared for electron microscopy. *J. Cell Sci.* **14**, 253–61. (186)

Jacobj, W. (1925). Über das rhythmische Wachstum der Zellen durch Verdoppelung ihrer Volumens. *Arch. Entw. Mech. Organ.* **106**, 124–92. (27)

Jacqmard, A., Miksche, J. P. & Bernier, G. (1972). Quantitative study of nucleic acids and proteins in the shoot apex of *Sinapis alba* during transition from the vegetative to the reproductive condition. *Amer. J. Bot.* **59**, 714–21. (107, 108)

Jakob, K. M. & Bovey, F. (1969). Early nucleic acid and protein syntheses and mitoses in the primary root tips of germinating *Vicia faba*. *Exptl. Cell Res.* **54**, 118–26. (50)

Jalouzot, R. (1969). Differenciation nucleaire et cytoplasmique du grain de pollen de *Lilium candidum*. *Exptl. Cell Res.* **55**, 1–8. (166)

Janczewski, E. (1874). Das Spitzenwachstum der Phanerogamenwurzeln. *Bot. Ztg.* **32**, 113–27. (38)

Jaworska, H., Avanzi, S. & Lima-de-Faria, A. (1973). Amplification of ribosomal DNA in *Acheta*. VIII. Binding of H^3-actinomycin to DNA in the nucleus and cytoplasm. *Hereditas* **74**, 205–10. (31, 152)

Jaworska, H. & Lima-de-Faria, A. (1973). Amplification of ribosomal DNA in *Acheta*. VI. Ultrastructure of two types of nucleolar components associated with ribosomal DNA. *Hereditas* **74**, 169–86. (152)

Jeanny, J.-C. (1973). Etude cytophotométrique des acides nucléiques et des histones des cellules cartilagineuses activées au cours de la régénération du membre de *Desmognathus fuscus* (Amphibien, Urodèle, Pléthodontidé). *Ann. Embryol. Morphogén.* **6**, 25–41. (210)

Jeanny, J.-C. & Gontcharoff, M. (1974). Synthèse de l'ADN aucours des phases précoces de la régénération des membres postérieurs de *Desmognathus fuscus*. *C.R. Acad. Sci.* **278 D**, 1513–16. (210)

Jensen, W. A. (1958). The nucleic acid and protein content of root tip cells of *Vicia faba* and *Allium cepa*. *Exptl. Cell Res.* **14**, 575–83. (40)

Joarder, I. O., Al-Saheal, Y., Begum, J. & Durrant, A. (1974). Environments inducing changes in amount of DNA in flax. *Heredity* **34**, 247–53. (162)

John, B. & Lewis, K. R. (1965). *The Meiotic System. Protoplasmatologia VI, Fl.* Vienna: Springer-Verlag. (58, 64)

John, B. & Lewis, K. R. (1968). *The Chromosome Complement. Protoplasmatologia, VI A.* Vienna: Springer-Verlag. (87, 123, 134)

John, B. & Lewis, K. R. (1969). *The Chromosome Cycle. Protoplasmatologia VI B.* Vienna: Springer-Verlag. (26)

John, H. A., Birnstiel, M. L. & Jones, K. W. (1969). RNA–DNA hybrids at a cytological level. *Nature* **223**, 582–7. (145, 168, 169)

Johnson, J. D., Douvas, A. S. & Bonner, J. (1974). Chromosomal proteins. *Int. Rev. Cytol.* Suppl. **4**, 273–361. (24, 25, 32, 195, 196)

Johnson, R., Chrisp, C. & Strehler, B. (1972). Selective loss of ribosomal RNA genes during the aging of post-mitotic tissues. *Mech. Age. Dev.* **1**, 183–98. (141, 142)

Johri, B. M. (1963). Female gametophytes. In *Recent Advances in the Embryology of Angiosperms*, ed. by P. Maheshwari, pp. 69–103. Delhi: International Society of Plant Morphologists. (9)

Jones, K. W. (1970). Chromosomal and nuclear location of mouse satellite DNA in individual cells. *Nature* **255**, 912–15. (24, 185)

Jones, K. W., Prosser, J., Corneo, G. & Ginelli, E. (1973). The chromosomal location of human satellite DNA III. *Chromosoma* **42**, 445–51. (186)

Jones, K. W. & Robertson, F. W. (1970). Localization of reiterated nucleotide sequences in *Drosophila* and mouse by *in situ* hybridization of complementary RNA. *Chromosoma* **31**, 331–45. (24)

Jones, R. B. & Irvin, J. L. (1972). Effect of hydrocortisone on the synthesis of DNA and histones and the acetylation of histones in regenerating liver. *Arch. Biochem. Biophys.* **152**, 828–38. (212)

Jones, R. N. (1975). B-chromosome systems in flowering plants and animal species. *Int. Rev. Cytol.* **40**, 1–100. (193, 194)

Jørgensen, C. A. & Crane, M. B. (1927). Formation and morphology of potato chimeras. *J. Genet.* **18**, 247–74. (93, 95)

Juel, O. (1898). Parthenogenesis bei *Antennaria alpina* (L.). R. Br. *Bot. Zentralbl.* **74**, 369–72. (75)

Kahn, E. B. & Simpsson, S. B. Jr. (1974). Satellite cells in mature, uninjured skeletal muscle of the lizard tail. *Develop. Biol.* **37**, 219–23. (211)

Kalt, M. R. & Gall, J. G. (1974). Observations on early germ cell development and premeiotic ribosomal DNA amplification in *Xenopus laevis*. *J. Cell Biol.* **62**, 460–72. (146)

Kambysellis, M. P. & Wheeler, M. B. (1972). Banded polytene chromosomes in pericardial cells of *Drosophila*. *J. Hered.* **63**, 214–15. (130)

Karlson, P. (1965). Biochemical studies of ecdysone control of chromosomal activity. *J. Cell Comp. Physiol.* **66**, 69–76. (199)

Kato, K. (1968). Cytochemistry and fine structure of elimination chromatin in *Dytiscidae*. *Exptl. Cell Res.* **52**, 507–22. (148)

Kaukis, K. & Reitz, L. P. (1955). Ontogeny of the *Sorghum* inflorescence as revealed by seedling mutants. *Amer. J. Bot.* **42**, 660–3. (104)

Kavenoff, R. & Zimm, B. H. (1973). Chromosome-sized DNA molecules from *Drosophila*. *Chromosoma* **41**, 1–27. (29)

Kedes, L. & Birnstiel, M. (1971). Reiteration and clustering of DNA sequences complementary to histone messenger RNA. *Nature New Biol.* **230**, 165–9. (173)

Keep, E. (1971). Nucleolar suppression, its inheritance and association with taxonomy and sex in the genus *Ribes*. *Heredity* **26**, 443–52. (182, 183)

Kezer, J., cited in Peacock, W. (1965). Chromosome replication. *Nat. Cancer Inst. Mono.* **18**, 101–31. (144)

Kezer, J. & Macgregor, H. C. (1971). A fresh look at meiosis and centromeric heterochromatin in the red-backed Salamander, *Plethodon cinereus cinereus* (Green). *Chromosoma* **33**, 146–66. (56, 58, 62)

Kihlman, B. A. & Levan, A. (1949). The cytological effect of caffeine. *Hereditas* **35**, 109–11. (119)

Kimber, G. & Riley, R. (1963). Haploid Angiosperms. *Bot. Rev.* **29**, 480–531. (21, 71)

King, R. L. & Beams, H. W. (1934). Somatic synapsis in *Chironomus* with special reference to the individuality of the chromosomes. *J. Morph.* **56**, 577–86. (130)

Kiortsis, V. & Trampusch, H. A. L. Ed. (1965). *Regeneration in Animals and Related Problems*. Amsterdam: North-Holland Publ. Co. (204)

Klinger, H. P. & Schwarzacher, H. G. (1962). XY/XXY and sex chromatin positive cell distribution in a 60 mm human fetus. *Cytogenetics* **1**, 266–90. (86, 87)

Koch, J. & Cruceanu, A. (1971). Hormone-induced gene amplification in somatic cells. *Hoppe-Seyler Z. Physiol. Chem.* **352**, 137–42. (153)

Kohler, P. O., Grimley, P. M. & O'Malley, B. W. (1968). Protein synthesis: differential stimulation of cell-specific proteins in epithelial cells of chick oviduct. *Science* **160**, 86–7. (200)

Kolodny, G. M. & Gross, P. R. (1969). Changes in patterns of protein synthesis during the mammalian cell cycle. *Exptl. Cell Res.* **56**, 117–21. (32)

Konar, R. N., Thomas, E. & Street, H. E. (1972). Origin and structure of embryoids arising from epidermal cells of the stem of *Ranunculus sceleratus* L. *J. Cell Sci.* **11**, 77–93. (222)

Kram, R., Botchan, M. & Hearst, J. E. (1972). Arrangement of the highly reiterated DNA sequences in the centric heterochromatin of *Drosophila melanogaster*. Evidence for interspersed spacer DNA. *J. Mol. Biol.* **64**, 103–17. (186)

Krider, H. M. & Plaut, W. (1972). Studies on nucleolar RNA synthesis in *Drosophila melanogaster*. I. The relationship between number of nucleolar organizers and rate of synthesis. *J. Cell Sci.* **11**, 675–87. (181)

Kriegstein, H. J. & Hogness, D. S. (1974). Mechanism of DNA replication in *Drosophila* chromosomes: structure of replication forks and evidence for bidirectionality. *Proc. Nat. Acad. Sci.* **71**, 135–9. (29)

Kusanagi, A. (1964). RNA synthetic activity in the mitotic nuclei. *Japan. J. Genet.* **39**, 254–8. (33)

La Cour, L. F. (1949). Nuclear differentiation in the pollen grain. *Heredity* **3**, 319–37. (166)

La Cour, L. F. (1953). The *Luzula* system analyzed by X-rays. *Heredity* **6**, Suppl. Vol., 77–81. (64)

Laird, C. D. (1971). Chromatid structure: Relationship between DNA content and nucleotide sequence diversity. *Chromosoma* **32**, 378–406. (186)

Lajtha, L. G. (1963). On the concept of the cell cycle. *J. Cell. Comp. Physiol.* **67**, Suppl. **1**, 142–5. (111)

Lajtha, L. G. (1964). Recent studies in erythroid differentiation and proliferation. *Medicine* **43**, 625–33. (111)

Lakhotia, S. C. (1974). EM autoradiographic studies on polytene nuclei of *Drosophila melanogaster*. III. Localization of nonreplicating chromatin in the chromocentre heterochromatin. *Chromosoma* **46**, 145–59. (133)

Lambert, B. (1972). Repeated DNA sequences in a Balbiani ring. *J. Mol. Biol.* **72**, 65–75. (174)

Lambert, B. & Edström, J. E. (1974). Balbiani ring nucleotide sequences in cytoplasmic 75S RNA of *Chironomus tentans* salivary gland cells. *Mol. Biol. Rep.* **1**, 457–64. (174)

Lambert, B., Wieslander, L., Daneholt, B., Egyhazi, E. & Ringborg, U. (1972). *In situ* demonstration of DNA hybridizing with chromosomal and nuclear sap RNA in *Chironomus tentans. J. Cell Biol.* **53**, 407–18. (174)

Langlet, O. (1927). Zur Kenntniss der polysomatischen Zellkerne in Wurzelmeristem. *Svensk. Bot. Tids.* **21**, 397–422. (136)

Lark, K. G., Consigli, R. & Toliver, A. (1971). DNA replication in chinese hamster cells: evidence for a single replication fork per replicon. *J. Mol. Biol.* **58**, 873–5. (29)

Laskey, R. A. & Gurdon, J. B. (1970). Genetic content of adult somatic cells tested by nuclear transplantation from cultured cells. *Nature* **228**, 1332–4. (219)

Lauber, H. (1947). Untersuchungen über das Wachstum der Früchte einiger Angiospermen unter endomitotische Polyploidisierung. *Öster. bot. Z.* **94**, 30–60. (125)

Leighton, T. J., Dill, B. C., Stock, J. J. & Phillips, C. (1971). Absence of histones from the chromosomal proteins of fungi. *Proc. Nat. Acad. Sci.* **68**, 677–80. (23)

Lejeune, J., Salmon, Ch., Berger, R., Réthoré, M. O., Rossier, A. & Job. J. C. (1967). Chimère 46, XX/69, XXY. *Ann. Génét.* **10**, 188–92. (87)

Lepori, N. G. (1949). Ricerche sulla ovogenesi e sulla fecondazione nella planaria *Polycelis nigra* Ehrenberg con particolare riguardo all'ufficio del nucleo spermatico. *Caryologia* **1**, 280–95. (79)

Lepori, N. G. (1950). Il ciclo cromosomico, con poliploidia, endomitosi e ginogenesi, in popolazioni italiane di *Polycelis nigra* Ehrenberg. *Caryologia* **2**, 301–24. (79)

Leshem, B. & Clowes, F. A. L. (1972). Rates of mitosis in shoot apices of potato at the beginning and end of dormancy. *Ann. Bot.* **36**, 687–91. (46)

Levan, A. (1938). The effect of colchicine on root mitoses in *Allium. Hereditas* **24**, 471–86. (122)

Levan, A. (1939). Cytological phenomena connected with the root swelling caused by growth substances. *Hereditas* **25**, 87–96. (124)

Levan, A. & Emsweller, S. L. (1938). Structural hybridity in *Nothoscordum fragrans. J. Hered.* **29**, 291–4. (76)

Levan, A. & Hauschka, T. S. (1953). Endomitotic reduplication mechanisms in ascites tumors of the mouse. *J. Nat. Cancer. Inst.* **14**, 1–43. (120, 123)

Levan, A. & Hsu, T. C. (1961). Repeated endoreduplication in a mouse cell. *Hereditas* **47**, 69–71. (123)

Lewin, B. (1974*a*). *Gene Expression. Vol. I: Bacterial Genomes.* London: John Wiley and Sons. (184)

Lewin, B. (1974*b*). *Gene Expression. Vol. II: Eucaryotic Chromosomes.* London: John Wiley and Sons. (192)

Lewis, E. B. (1948). Location of *c3G* in the salivary gland chromosomes. *D.I.S.* **22**, 72–3. (65, 69)

Lewis, E. B. (1950). The phenomenon of position effect. *Adv. Genet.* **3**, 73–115. (191)

Lewis, E. B. & Gencarella, W. (1952). Claret and non-disjunction in *Drosophila melanogaster. Rec. Genet. Soc. Amer.* **21**, 44–5. (65, 69, 70)

Liapunova, N. A. & Babadjanian, D. P. (1973). Quantitative study of histone in meiocytes. I. Investigation of the histone amount in cricket spermatogenesis by interference microscopy. *Chromosoma* **40**, 387–99. (56, 61, 62)

Lima-de-Faria, A. (1959). Differential uptake of tritiated thymidine into hetero- and euchromatin in *Melanoplus* and *Secale. J. Biophys. Biochem. Cytol.* **6**, 457–65. (30)

Lima-de-Faria, A. (1969). DNA replication and gene amplification in heterochromatin. In *Handbook of Molecular Cytology*, ed. by A. Lima-de-Faria, pp. 277–325. Amsterdam: North-Holland Publ. Co. (30, 31)

Lima-de-Faria, A. (1974). The molecular organization of the chromosomes of *Acheta* involved in ribosomal DNA amplification. *Cold Spring Harbor Symp. Quant. Biol.* **38**, 559–71. (150, 152)

Lima-de-Faria, A., Birnstiel, M. & Jaworska, H. (1969). Amplification of ribosomal cistrons in the heterochromatin of *Acheta. Genetics* Suppl. **61**, 145–59. (149)

Lima-de-Faria, A., Daskaloff, S. & Enell, A. (1973*a*). Amplification of ribosomal DNA in *Acheta*. I. The number of chromomeres involved in the amplification process. *Hereditas* **73**, 99–118. (152)

Lima-de-Faria, A., German, J., Ghatnekar, M., McGovern, J. & Anderson, L. (1968). DNA synthesis in the meiotic chromosomes of man. *Hereditas* **56**, 398–9. (60)

Lima-de-Faria, A., Gustafsson, T. & Jaworska, H. (1973*b*). Amplification of ribosomal DNA in *Acheta*. II. The number of nucleotide pairs of the chromosomes and chromomeres involved in amplification. *Hereditas* **73**, 119–42. (152)

Lima-de-Faria, A. & Jaworska, H. (1964). Haplo-diploid chimeras in *Haplopappus gracilis. Hereditas* **52**, 119–22. (138)

Lima-de-Faria, A., Jaworska, H., Gustafsson, T. & Daskaloff, S. (1973*c*). Amplification of ribosomal DNA in *Acheta*. III. The release of DNA copies from chromomeres. *Hereditas* **73**, 163–84. (151, 152)

Lima-de-Faria, A. & Moses, M. J. (1966). Ultrastructure and cytochemistry of metabolic DNA in *Tipula. J. Cell Biol.* **30**, 177–92. (148)

Lima-de-Faria, A., Nilsson, B., Cave, D., Puga, A. & Jaworska, H. (1968). Tritium labelling and cytochemistry of extra DNA in *Acheta. Chromosoma* **25**, 1–20. (149)

Lima-de-Faria, A., Pero, R., Avanzi, S., Durante, M., Ståhle, U., D'Amato, F. & Granström, H. (1975). Relation between ribosomal RNA genes and the DNA satellites of *Phaseolus coccineus. Hereditas* **79**, 5–20. (128, 134, 158, 159)

Lin, M. (1955). Chromosome control of nuclear composition in maize. *Chromosoma* **7**, 340–70. (168)

Linskens, H. F. (1969). Fertilization mechanisms in higher plants. In *Fertilization*, ed. by C. B. Metz and A. Monroy, pp. 189–253. New York: Academic Press. (9, 19, 20, 71)

Lipp, C. (1953). Über Kernwachstum, Endomitosen und Funktionszyklen in den trichogenen Zellen von *Corixa punctata. Chromosoma* **5**, 454–86. (122)

Lison, L. & Pasteels, J. (1951). Etudes histophotométriques sur la teneur en acide désoxyribonucléique des noyaux au cours du développement embryonnaire chez l'oursin *Paracentrotus lividus. Arch. Biol. Liège* **62**, 2–43. (36)

Longwell, A. R. & Svihla, G. (1960). Specific chromosomal control of the nucleolus and of the cytoplasm in wheat. *Exptl. Cell Res.* **20**, 294–312. (156, 168)

López-Sáez, F., Giménez-Martin, G. & González-Fernández, A. (1965). Nuclear fusion in somatic cells: a new mechanism of polyploidization. *Phyton* **22**, 1–5. (119)

Lorz, A. (1937). Cytological investigations on five chenopodiaceous genera with special emphasis on chromosome morphology and somatic doubling in *Spinacia. Cytologia* **8**, 241–70. (136)

Lorz, A. (1947). Supernumerary chromonemal reproduction, polytene chromosomes, endomitosis, multiple chromosome complexes, polynemy. *Bot. Rev.* **13**, 597–624. (120, 136)

Lowell, R. D. & Burnett, A. L. (1969). Regeneration of complete hydra from isolated epidermal explants. *Biol. Bull.* **137**, 312–20. (204)

Lund, H. A. (1956). Growth hormones in the styles and ovaries of tobacco responsible for fruit development. *Amer. J. Bot.* **43**, 562–8. (71)

Lyon, M. F. (1961). Gene action in the X-chromosome of the mouse (*Mus musculus* L.). *Nature* **190**, 372–3. (191)

Lyon, M. F. (1972). X chromosome inactivation and developmental patterns in mammals. *Biol. Rev.* **47**, 1–35. (192, 193)

McCarthy, B. J., Nishiura, J. T., Doenecke, D., Nasser, D. S. & Johnson, C. B. (1974).

Transcription and chromosome structure. *Cold Spring Harbor Symp. Quant. Biol.* **38**, 763–71. (195, 196)
McClintock, B. (1934). The relation of a particular chromosome element to the development of the nucleoli in *Zea mays. Z. Zellforsch.* **21**, 294–328. (168, 171)
McFeeley, R. A., Hare, W. C. D. & Biggers, J. D. (1967). Chromosome studies in 14 cases of intersex in domestic mammals. *Cytogenetics* **6**, 242–53. (87)
Macgregor, H. C. (1968). Nucleolar DNA in oocytes of *Xenopus laevis. J. Cell Sci.* **3**, 437–44. (169)
Macgregor, H. C. (1973). Amplification, polytenisation and nucleolus organizers. *Nature* **246**, 81–2. (145)
Macgregor, H. C. & Kezer, J. (1970). Gene amplification in oocytes with 8 germinal vesicles from the tailed frog *Ascaphus truei* Stejneger. *Chromosoma* **29**, 189–206. (148)
Macgregor, H. C. & Kezer, J. (1971). The chromosomal localization of a heavy satellite DNA in the testis of *Plethodon c. cinereus. Chromosoma* **33**, 167–82. (186)
Macgregor, H. C. & Kezer, J. (1973). The nucleolar organizer of *Plethodon cinereus cinereus* (Green). I. Location of the nucleolar organizer by *in situ* nucleic acid hybridization. *Chromosoma* **42**, 415–26. (145)
McLaren, A. (1969). Recent studies on developmental regulation in vertebrates. In *Handbook of Molecular Cytology*, ed. by A. Lima-de-Faria, pp. 639–55. Amsterdam: North-Holland Publ. Co. (92)
McLeish, J. & Sunderland, N. (1961). Measurements of deoxyribosenucleic acid (DNA) in higher plants by Feulgen photometry and chemical methods. *Exptl. Cell Res.* **24**, 527–40. (114)
Maheshwari, P. (1950). *An Introduction to the Embryology of Angiosperms.* Toronto: McGraw Hill Book Co. (9)
Mangold, O. & Seidel, F. (1927). Homoplastische und heteroplastische Verschmelzung ganzer Tritonkeime. *Arch. Entwicklungsmech. Organ.* **111**, 593–665. (91)
Manthriratna, M. A. P. P. & Hayward, M. D. (1973). Pollen development and variation in the genus *Lolium.* 3. Pollen histochemistry. *Z. Pflanzenzüchtg.* **70**, 11–21. (13)
Margulis, L. (1973). Colchicine sensitive microtubules. *Int. Rev. Cytol.* **34**, 333–61. (32, 34)
Marquardt, H. & Gläss, E. (1957). Die Chromosomenzahlen in den Leberzellen von Ratten verschiedenen Alters. *Chromosoma* **8**, 617–36. (139)
Martens, P. (1954*a*). Alternance de phases et cycles de développement chez les végétaux. *Acad. R. Belg., Bull. Cl. Sci. V*, **40**, 508–17. (2)
Martens, P. (1954*b*). Evolution et cycles de développement chez les Champignons. *Rev. Gén. Bot.* **61**, 1–15. (2)
Martini, G. & Bozzini, A. (1966). Radiation induced asynaptic mutations in *durum* wheat (*Triticum durum* Desf.). *Chromosoma* **20**, 251–66. (65, 66)
Mascarenhas, J. P. (1966). Pollen tube growth and ribonucleic acid synthesis by vegetative and generative nuclei of *Tradescantia. Amer. J. Bot.* **53**, 563–9. (13)
Matsubayashi, I. & Yamaguchi, H. (1971). First cell cycle in the shoot apical cells of germinating rice seeds. *Radioisotopes* **20**, 29–32. (50)
Matthysse, A. G. & Torrey, J. G. (1967). DNA synthesis in relation to polyploid mitosis in excised pea root segments cultured *in vitro. Exptl. Cell Res.* **48**, 484–98. (114)
Mechelke, R. (1953). Reversible Strukturmodifikationen der Speicheldrüsenchromosomen von *Acricotopus lucidus. Chromosoma* **5**, 511–43. (176)
Melchers, G. & Labib, G. (1974). Somatic hybridization of plants by fusion of protoplasts. I. Selection of light resistant hybrids of 'haploid' light sensitive varieties of tobacco. *Mol. Gen. Genet.* **135**, 277–94. (228)
Metz, C. W. (1916). Chromosome studies on the Diptera. II. The paired association of chromosomes in the Diptera and its significance. *J. Exp. Zool.* **21**, 213–79. (130)
Metz, C. W. (1942). Mosaic salivary glands in *Sciara. Amer. Nat.* **86**, 623–30. (138)

Metz, C. W. & Schmuck, M. L. (1931). Differences between the chromosome groups of soma and germ-line in *Sciara. Proc. Nat. Acad. Sci.* **15**, 862–6. (110)

Meyer, J. (1915). Die *Crataegomespili* von Bronvaux. *Z. indukt. Abst. Vererbl.* **13**, 193–233. (94)

Miassod, R., Cecchini, J. P., Becerra De Lares, L. & Richard, J. (1973). Maturation of ribosomal RNA's in suspensions of higher plant cells. A DNA-RNA hybridization study. *FEBS Letters* **35**, 71–5. (172)

Michaelis, P. (1966). The proof of cytoplasmic inheritance in *Epilobium* (a historical survey as example for the necessary proceeding). *Nucleus* **9**, 1–16. (58)

Michaelis, P. (1967). Untersuchungen zur Entwicklungsgeschichte des Epilobium-sprosses mit Hilfe genetisch markierter Zelldeszendenzen. *Bot. Jb.* **86**, 50–112. (107)

Michaux, N. (1970). Détermination par cytophotométrie de la quantité d'ADN contenue dans le noyau de la cellule apicale des méristèmes jeunes et adultes du *Pteris cretica* L. *C.R. Acad. Sci.* **271**, 656–9. (44)

Miller, O. L. Jr. (1963). Cytological studies in asynaptic maize. *Genetics* **48**, 1445–66. (65, 66)

Miller, O. L. Jr. (1964). Extra-chromosomal nucleolar DNA in amphibian oocytes. *J. Cell Biol.* **23**, 60A. (144)

Miller, O. L. Jr. (1966). Structure and composition of peripheral nucleoli of salamander oöcytes. *Nat. Cancer Inst. Mono.* **23**, 55–66. (144)

Miller, O. L. Jr. & Beatty, B. R. (1969). Visualization of nucleolar genes. *Science* **164**, 955–7. (144, 171)

Miller, O. L. Jr., Beatty, B. R., Hamkalo, B. A. & Thomas, C. A. Jr. (1970). Electron microscopic visualization of transcription. *Cold Spring Harbor Symp. Quant. Biol.* **35**, 505–12. (174)

Miller, O. L. Jr. & Hamkalo, B. A. (1973). Visualization of RNA synthesis on chromosomes. *Int. Rev. Cytol.* **33**, 1–25. (171)

Mintz, B. (1964). Formation of genetically mosaic mouse embryos, and early development of 'lethal (t^{12}/t^{12})'-normal mosaics. *J. Exp. Zool.* **157**, 273–92. (91)

Mintz, B. (1965). Experimental genetic mosaicism in the mouse. In *Preimplantation Stages of Pregnancy*, ed by G. E. W. Wolstenholme and M. O'Connor, pp. 194–207. London: J. and A. Churchill. (91)

Mintz, B. (1967). Gene control of mammalian pigmentary differentiation. Clonal origin of melanocytes. *Proc. Nat. Acad. Sci.* **58**, 344–51. (92)

Mintz, B. (1968). Hermaphroditism, sex chromosomal mosaicism and germ cell selection in allophenic mice. *J. Animal. Sci.* **27**, Suppl. **1**, 51–60. (91)

Mitchell, J. P. (1967). DNA synthesis during the early division-cycle of Jerusalem artichoke callus cultures. *Ann. Bot.* **31**, 427–35. (114)

Mitchison, J. M. (1971). *The Biology of the Cell Cycle.* Cambridge: University Press. (26, 28)

Mitra, S. (1958). Effects of X-rays on chromosomes of *Lilium longiflorum* during meiosis. *Genetics* **43**, 771–89. (56)

Mittwoch, U. (1967). *Sex Chromosomes.* New York: Academic Press. (67, 87, 191)

Moens, P. B. (1964). A new interpretation of meiotic prophase in *Lycopersicum esculentum* (tomato). *Chromosoma* **15**, 231–42. (58, 63)

Moens, P. E. (1969). The fine structure of meiotic chromosome polarization and pairing in *Locusta migratoria* spermatocytes. *Chromosoma* **28**, 1–25. (57)

Moens, P. B. (1973). Mechanisms of chromosome synapsis at meiotic prophase. *Int. Rev. Cytol.* **35**, 117–34. (58)

Mohan, J. & Flavell, R. B. (1974). Ribosomal RNA cistron multiplicity and nucleolar organizers in hexaploid wheat. *Genetics* **76**, 33–44. (156, 157)

Mohan, J. & Ritossa, F. M. (1970). Regulation of ribosomal RNA synthesis and its bearing on the bobbed phenotype in *Drosophila melanogaster. Develop. Biol.* **22**, 495–512. (182)

Montgomery, J. R. & Coward, S. J. (1974). On the minimal size of a planarian capable of regeneration. *Trans. Amer. Micros. Soc.* **93**, 386–91. (206)

Morgan, T. H. & Bridges, C. B. (1919). Contributions to the genetics of *Drosophila melanogaster.* The origin of gynandromorphs. *Carnegie Instn. Wash. Publ.* **278**, 1–122. (85)

Morzlock, F. V. & Stocum, D. L. (1971). Patterns of RNA synthesis in regenerating limbs of the adult newt, *Triturus viridescens. Develop. Biol.* **24**, 106–18. (210)

Morzlock, F. V. & Stocum, D. L. (1972). Neural control of RNA synthesis in regenerating limb of the adult newt *Triturus viridescens. Wm. Roux Arch.* **171**, 170–80. (210)

Moses, M. J. (1956). Chromosome structures in crayfish spermatocytes. *J. Biophys. Biochem. Cytol.* **2**, 215–17. (58)

Moses, M. J. (1968). The synaptinemal complex. *Ann. Rev. Genet.* **2**, 363–412. (58)

Mufti, S. A. & Simpson, S. B. Jr. (1972). Tail regeneration following autotomy in the adult salamander *Desmognathus fuscus. J. Morph.* **136**, 297–306. (211)

Muggleton-Harris, A. L. & Pezzella, K. (1972). The ability of the lens cell nucleus to promote complete embryonic development and its applications to ophthalmic gerontology. *Exp. Geront.* **7**, 427–31. (219)

Mukherjee, A. B. & Cohen, M. M. (1968). DNA synthesis during meiotic prophase in male mice. *Nature* **219**, 489–90. (60)

Mukherjee, B. B. & Sinha, A. K. (1964). Single X hypothesis: cytological evidence for random inactivation of X chromosomes in female mule complement. *Proc. Nat. Acad. Sci.* **51**, 252–9. (192)

Muldal, S. (1952). The chromosomes of the Earthworms. I. The evolution of polyploidy. *Heredity* **6**, 55–76. (79)

Mulder, M. P., Van Duyn, P. & Gloor, J. H. (1968). The replicative organization of DNA in polytene chromosomes of *Drosophila hydei. Genetica* **39**, 385–428. (132)

Müntzing, A. (1974). Accessory chromosomes. *Ann. Rev. Genet.* **8**, 243–66. (193, 194)

Müntzing, A. & Akdik, S. (1948). Cytological disturbances in the first inbred generations of rye. *Hereditas* **34**, 485–509. (65, 69)

Murashige, T. (1974). Plant propagation through tissue culture. *Ann. Rev. Plant Physiol.* **25**, 135–66. (203, 222, 224)

Mystkowska, E. T. & Tarkowski, A. K. (1968). Observations on CBA-p/CBA-T6T6 mouse chimeras. *J. Embryol. Exp. Morph.* **20**, 33–52. (91, 92)

Mystkowska, E. T. & Tarkowski, A. K. (1970). Behaviour of germ cells and sexual differentiation in late embryonic and early post-natal mouse chimeras. *J. Embryol. Exp. Morph.* **23**, 395–405. (92)

Nadal, C. & Zajdela, F. (1966*a*). Polyploidie somatique dans le foie de rat. I. Le rôle des cellules binucléées dans la genèse des cellules polyploides. *Exptl. Cell Res.* **42**, 99–116. (118, 119, 213)

Nadal, C. & Zajdela, F. (1966*b*). Polyploidie somatique dans le foie de rat. II. Le rôle de l'hypophyse et de la carence protéique. *Exptl. Cell Res.* **42**, 117–29. (118, 213)

Nagl, W. (1962*a*). Über Endopolyploidie, Restitutionskernbildung und Kernstrukturen im Suspensor von Angiospermen und einer Gymnosperme. *Öster. bot. Z.* **109**, 431–94. (120, 128)

Nagl, W. (1962*b*). 4096-Ploidie und 'Riesenchromosomen' im Suspensor von *Phaseolus coccineus. Naturwiss.* **49**, 261–2. (120, 128)

Nagl, W. (1965*a*). Die SAT-Riesenchromosomen der Kerne des Suspensors von *Phaseolus coccineus* und ihr Verhalten während der Endomitose. *Chromosoma* **16**, 511–20. (128)

Nagl. W. (1965*b*). Karyologische Anatomie der Samenanlage von *Pinus silvestris. Öster. bot. Z.* **112**, 359–70. (123, 136)

Nagl, W. (1967*a*). Die Riesenchromosomen von *Phaseolus coccineus* L.: Bau, Strukturmodifikationen, Zusätliche Nukleolen und Vergleich mit den mitotischen Chromosomen. *Öster. bot. Z.* **114**, 171–82. (128)

Nagl, W. (1967*b*). Microphotometrische DNS-Messungen an Interphase- und Ruhe-kernen sowie Mitosen in der Samenanlage von *Pinus silvestris*. *Z. Pflanzenphysiol.* **56**, 40–56. (136)

Nagl, W. (1969*a*). Correlation of structure and RNA synthesis in the nucleolus organizing polytene chromosomes of *Phaseolus vulgaris*. *Chromosoma* **28**, 85–91. (128)

Nagl, W. (1969*b*). Puffing of polytene chromosomes in a plant (*Phaseolus vulgaris*). *Naturwiss.* **56**, 221–2. (128)

Nagl, W. (1969*c*). Banded polytene chromosomes in the legume *Phaseolus vulgaris*. *Nature* **221**, 70–1. (128)

Nagl, W. (1970*a*). The mitotic and endomitotic nuclear cycle in *Allium carinatum*. II. Relation between DNA replication and chromatin structure. *Caryologia* **23**, 71–8. (123)

Nagl, W. (1970*b*). Inhibition of polytene chromosome formation in *Phaseolus* by polyploid mitoses. *Cytologia* **35**, 252–8. (123)

Nagl, W. (1970*c*). Temperature dependent functional structures in the polytene chromosomes of *Phaseolus* with special reference to the nucleolus organizers. *J. Cell Sci.* **6**, 87–107. (128)

Nagl, W. (1970*d*). Differentielle RNS Synthese an pflanzlichen Riesenchromosomen. *Ber. deutsch. bot. Ges.* **83**, 301–9. (128)

Nagl, W. (1972*a*). Giant sex chromatin in endopolyploid trophoblast nuclei of the rat. *Experientia* **28**, 217–18. (128)

Nagl, W. (1972*b*). Evidence of DNA amplification in the orchid *Cymbidium in vitro*. *Cytobios* **5**, 145–54. (164)

Nagl, W. (1973). The angiosperm suspensor and the mammalian trophoblast: organs with similar cell structure and function? *Soc. Bot. Fr., Mém. 1973, Coll. Morphologie*, 289–302. (128)

Nagl, W. (1974). Role of heterochromatin in the control of cell cycle duration. *Nature* **249**, 53–4. (35)

Nagl, W., Hendon, J. & Rücker, W. (1972). DNA amplification in *Cymbidium* protocorms *in vitro* as it relates to cytodifferentiation and hormone treatment. *Cell Differentiation* **1**, 229–37. (164)

Nagl, W. & Rücker, W. (1972). Beziehungen zwischen Morphogenese und nuklearem DNS-Gehalt bei aseptischen Kulturen von *Cymbidium* nach Wuchsstoffbehandlung. *Z. Pflanzenphysiol.* **67**, 120–34. (164)

Narbel, M. (1946). La cytologie de la parthénogénèse chez *Apterona helix* Sieb. (Lepidoptera: Psychides). *Rev. Suisse Zool.* **53**, 625–81. (80)

Narbel-Hofstetter, M. (1950). La cytologie de la parthénogénèse chez *Solenobia* sp. (*lichenella* L.?) (Lépidoptères, Psychides). *Chromosoma* **4**, 56–90. (79)

Narbel-Hofstetter, M. (1961). Cytologie comparée de l'espèce parthénogénétique *Luffia ferchaultella* Steph. et de l'espèce bisexuée *L. lapidella* Goeze (Lépid. Psychidae). *Chromosoma* **12**, 505–52. (79)

Narbel-Hofstetter, M. (1964). *Les Alterations de la Meiose chez les Animaux Parthénogénétiques. Protoplasmatologia VI*-F2. Vienna: Springer-Verlag. (70, 79, 80, 81)

Navashin, M. (1933). Alter der Samen als Ursache von Chromosomenmutationen. *Planta* **20**, 233–43. (142)

Navashin, M. (1934). Chromosome alterations caused by hybridization and their bearing upon certain general genetic problems. *Cytologia* **5**, 169–203. (182, 183)

Naylor, E. & Johnson, B. (1937). A histological study of vegetative reproduction in *Saintpaulia ionantha*. *Amer. J. Bot.* **24**, 673–8. (224)

Nelson-Rees, W. A. (1962). The effects of radiation damaged heterochromatic chromosomes on male fertility in the mealy bug *Planococcus citri*. *Genetics* **47**, 661–83. (189)

Nes, N. (1966). Diploid-triploid chimerism in a true hermaphrodite mink (*Mustela vison*). *Hereditas* **56**, 159–70. (87, 88)

Nicklas, R. B. (1971). Mitosis. *Adv. Cell Biol.* **3**, 225–97. (34)

Nicklas, R. B. & Jaqua, R. A. (1965). X chromosome DNA replication: developmental shift from synchrony to asynchrony. *Science* **147**, 1041–3. (31)

Nieman, R. H. (1965). Expansion of bean leaves and its suppression by salinity. *Plant Physiology* **42**, 946–52. (126)

Nigon, V. (1949). Les modalités de la reproduction et le déterminisme du sexe chez quelques Nématodes libres. *Ann. Sci. Nat.* **11**, 1–132. (80)

Nobécourt, P. (1939). Sur la perennité et l'augmentation de volume des cultures des tissus végétaux. *C.R. Soc. Biol.* **130**, 1270–1. (224)

Nordenskiöld, H. (1962). Studies of meiosis in *Luzula purpurea*. *Hereditas* **48**, 503–19. (64)

Normandin, D. K. (1960). Regeneration of Hydra from the entoderm. *Science* **132**, 678. (204)

Nöthiger, R. (1972). The larval development of imaginal disks. In *The Biology of Imaginal Disks*, ed. by H. Ursprung and R. Nöthiger, pp. 1–34. Berlin: Springer-Verlag. (4, 46, 47, 91)

Nougarède, A. (1965). *Organisation et fonctionnement du méristeme apical des végétaux vasculaires*. Paris: Masson et Cie. (46)

Nougarède, A. (1967). Experimental cytology of the shoot apical cells during vegetative growth and flowering. *Int. Rev. Cytol.* **21**, 203–351. (38, 46)

Nur, U. (1963). Meiotic parthenogenesis and heterochromatinization in a soft scale, *Pulvinaria hydrangeae* (Coccoidea: Homoptera). *Chromosoma* **14**, 123–39. (81)

Nur, U. (1966). Nonreplication of heterochromatic chromosomes in a mealy bug, *Planococcus citri* (Coccoidea: Homoptera). *Chromosoma* **19**, 439–48. (122)

Nur, U. (1967). Reversal of heterochromatization and the activity of the paternal chromosome set in the male mealy bug. *Genetics* **56**, 375–89. (189)

Nur, U. (1968). Endomitosis in the mealy bug, *Planococcus citri* (Homoptera: Coccoidea). *Chromosoma* **24**, 202–9. (122)

Nur, U. (1971). Parthenogenesis in coccids (Homoptera). *Amer. Zool.* **11**, 301–8. (77, 79, 80, 81)

Nur, U. (1972). Diploid arrhenotoky and automictic telytoky in soft scale insects (Lecaniidae: Coccoidea: Homoptera). *Chromosoma* **39**, 381–401. (77, 81)

Nuti-Ronchi, V., Avanzi, S. & D'Amato, F. (1965). Chromosome endoreduplication (endopolyploidy) in pea root meristems induced by 8-azaguanine. *Caryologia* **18**, 599–617. (129)

Nuti-Ronchi, V., Bennici, A. & Martini, G. (1973). Nuclear fragmentation in dedifferentiating cells of *Nicotiana glauca* pith tissue grown *in vitro*. *Cell Differentiation* **2**, 77–85. (164)

Nygren, A. (1957). *Poa timoleontis* Heldr., a new diploid species of the section Bolbophorum A. & Gr. with accessory chromosomes only in the meiosis. *Kungl. Lantbr. Ann.* **23**, 489–95. (109)

Oberpriller, J. (1967). A radioautographic analysis of the potency of blastemal cells in the adult newt, *Diemictylus viridescens*. *Growth* **31**, 251–96. (209)

Oehlert, W., Lauf, P. & Seemayer, N. (1962). Autoradiographische Untersuchungen über den Generationszyclus der Zellen des Ehrlichschen Aszites-Carcinoms. *Naturwiss.* **49**, 137. (111)

Ohno, S. (1974). Conservation of ancient linkage groups in evolution and some insight into the genetic regulatory mechanisms of X-inactivation. *Cold Spring Harbor Symp. Quant. Biol.* **38**, 155–64. (193)

Ohno, S., Kittrell, W. A., Christian, L. C., Stenius, C. & Witt, G. (1963). An adult triploid chicken (*Gallus domesticus*) with a left ovotestis. *Cytogenetics* **2**, 42–9. (20)

Okabe, S. (1932). Parthenogenesis in *Ixeris dentata*. *Bot. Mag.* **46**, 518–23. (73)

Okamoto, M. (1962). Identification of the chromosomes of common wheat belonging to the A and B genomes. *Can. J. Genet. Cytol.* **4**, 31–7. (66)

O'Malley, B. W., Spelsberg, T. C., Schrader, W. T., Chytil, F. & Steggles, A. W. (1972). Mechanisms of interaction of a hormone-receptor complex with the genome of an eukaryotic target cell. *Nature* **235**, 141–4. (200, 201)

Omodeo, P. (1952). Cariologia dei *Lumbricidae. Caryologia* **4**, 173–275. (79)

Oppenheim, A. & Wahrman, J. (1973). DNA-membrane association during the mitotic cycle of *Physarum polycephalum. Exptl. Cell Res.* **79**, 287–94. (28)

Oprescu, St. & Thibault, C. (1965). Duplication de l'ADN dans les oeufs de lapine après la fécondation. *Ann. Biol. Anim. Bioch. Biophys.* **5**, 151–6. (15)

Östergren, G. & Fröst, S. (1962). Elimination of accessory chromosomes from the roots in *Haplopappus gracilis. Hereditas* **48**, 363–6. (109)

Painter, T. S. (1933). A new method for the study of chromosome rearrangements and the plotting of chromosome maps. *Science* **78**, 575–6. (130)

Painter, T. S. (1934). A new method for the study of chromosome aberrations and the plotting of chromosome maps. *Genetics* **19**, 175–88. (130)

Painter, T. S. & Taylor, A. N. (1942). Nucleic acid storage in the toad's egg. *Proc. Nat. Acad. Sci.* **28**, 311–17. (143)

Papaconstantinou, J. (1969). Mechanisms of translation of genetic information during cellular differentiation. *Excerpta Medica Int. Congr. Ser.* **204**, 42–52. (165)

Parchman, L. G. & Stern, H. (1969). The inhibition of protein synthesis in meiotic cells and its effect on chromosome behaviour. *Chromosoma* **26**, 298–311. (61)

Pardue, M. L. (1974). Localization of repeated DNA sequences in *Xenopus* chromosomes. *Cold Spring Harbor Symp. Quant. Biol.* **38**, 475–82. (24, 169, 185)

Pardue, M. L., Brown, D. D. & Birnstiel, M. L. (1973). Localization of the genes for 5S ribosomal RNA in *Xenopus laevis. Chromosoma* **42**, 191–203. (172)

Pardue, M. L. & Gall, J. G. (1970). Chromosomal localization of mouse satellite DNA. *Science* **168**, 1356–8. (24, 31, 185)

Pardue, M. L., Gerbi, S. A., Eckhardt, R. A. & Gall, J. G. (1970). Cytological localization of DNA complementary to ribosomal RNA in polytene chromosomes of *Diptera. Chromosoma* **29**, 268–90. (158, 169)

Parenti, R., Guillé, E., Grisvard, J., Durante, M., Giorgi, L. & Buiatti, M. (1973). Transient DNA satellite in dedifferentiating pith tissue. *Nature New Biol.* **246**, 237–9. (164)

Partanen, C. R. (1959). Quantitative chromosomal changes and differentiation in plants. In *Developmental Cytology*, ed. by D. Rudnick, pp. 21–45. New York: The Ronald Press. (114)

Partanen, C. R. (1965). The chromosomal basis for cellular differentiation. *Amer. J. Bot.* **52**, 204–9. (120)

Parthasarathy, M. V. & Tomlinson, P. B. (1967). Anatomical features of metaphloem in stems of *Sabal, Cocos* and two other palms. *Amer. J. Bot.* **54**, 1143–51. (139)

Pätau, K. (1950). A correlation between separation of the two chromosome groups in somatic reduction and their degree of homologous segregation. *Genetics* **35**, 128. (137)

Pätau, K. & Das, N. K. (1961). The relation of DNA synthesis and mitosis in tobacco pith tissue cultured *in vitro. Chromosoma* **11**, 553–72. (114, 123)

Pätau, K., Das, N. K. & Skoog, F. (1957). Induction of DNA synthesis by kinetin and indoleacetic acid in excised tobacco pith tissue. *Physiol. Plant.* **10**, 949–66. (114)

Pätau, K. & Swift, H. (1953). The DNA content (Feulgen) of nuclei during mitosis in a root tip of onion. *Chromosoma* **6**, 149–69. (27)

Patel, N. & Holoubek, V. (1973). RNA associated with nonhistone chromosomal proteins of dog liver. *Biochem. Biophys. Res. Com.* **54**, 524–30. (25)

Paul, J. (1972). General theory of chromosome structure and gene activation in eukaryotes. *Nature* **238**, 444–6. (184)

Paul, J. & Gilmour, R. S. (1968). Organ specific restriction of transcription in mammalian chromatin. *J. Mol. Biol.* **34**, 305–16. (197)

Pavan, C. & Da Cunha, A. B. (1968). Chromosome activities in normal and in infected cells of Sciaridae. *Nucleus*, Suppl. **12**, 183–96. (131)

Pavan, C. & Da Cunha, A. B. (1969). Chromosomal activities in *Rhynchosciara* and other Sciaridae. *Ann. Rev. Genet.* **3**, 425–50. (130, 157)

Pearson, G. G., Timmis, J. N. & Ingle, J. (1974). The differential replication of DNA during plant development. *Chromosoma* **45**, 281–94. (162, 163)

Pedersen, K. J. (1972). Studies on regeneration blastemas of the planarian *Dugesia tigrina* with special reference to differentiation of the muscle-connective filament system. *Wm. Roux. Arch.* **169**, 134–69. (206)

Pelling, C. (1972). RNA synthesis in giant chromosomal puffs and the mode of puffing. *FEBS Symp.* **24**, 77–89. (175, 176)

Pera, F. (1969). Die Entstehung haploider und triploider Zellen durch multipolare Mitosen. *Verh. anat. Ges.* **64**, 53–5. (138)

Pera, F. (1970). *Mechanismen der Polyploidisierung und der somatischen Reduktion.* Berlin: Springer-Verlag. (30, 115, 138)

Pera, F. & Rainer, B. (1973). Studies of multipolar mitoses in euploid tissue cultures. I. Somatic reduction to exact haploid and triploid chromosome sets. *Chromosoma* **42**, 71–86. (138)

Perkins, J. M., Eglington, E. G. & Jinks, J. L. (1971). The nature of the inheritance of permanently induced changes in *Nicotiana rustica. Heredity* **27**, 441–57. (160)

Perkowska, E., Macgregor, H. C. & Birnstiel, M. L. (1968). Gene amplification in the oöcyte nucleus of mutant and wild-type *Xenopus laevis. Nature* **217**, 649–50. (145)

Pero, R., Lima-de-Faria, A., Ståhle, U., Granström, H. & Ghatnekar, R. (1973). Amplification of ribosomal DNA in *Acheta*. IV. The number of cistrons for 28S and 18S ribosomal RNA. *Hereditas* **73**, 195–210. (149)

Perry, R. P. (1962). The cellular sites of synthesis of ribosomal and 4s RNA. *Proc. Nat. Acad. Sci.* **48**, 2179–86. (168)

Pfeiffer, S. E. (1968). RNA synthesis in synchronously growing populations of HeLa S_3 cells. II. Rate of synthesis in individual RNA fractions. *J. Cell Physiol.* **71**, 95–104. (28)

Phillips, H. L. Jr. & Torrey, J. G. (1971). The quiescent center in cultured roots of *Convolvulus. Plant Physiol.* **48**, 213–18. (39, 41)

Phillips, R. L., Kleese, R. A. & Wang, S. S. (1971). The nucleolus organizer region of maize (*Zea mays* L.): chromosomal site of DNA complementary to ribosomal RNA. *Chromosoma* **36**, 79–88. (169, 171)

Phillips, R. L., Weber, D. E., Kleese, R. A. & Wang, S. S. (1974). The nucleolus organizer region of maize (*Zea mays* L.): tests for ribosomal gene compensation or magnification. *Genetics* **77**, 285–97. (156)

Picheral, B. (1962). Capacités des noyaux de cellules endodermiques embryonnaires à organiser un germe viable chez l'Urodèle *Pleurodeles Walltii* Michah. *C.R. Acad. Sci.* **255**, 2509–11. (219)

Piesco, N. P. & Alvarez, M. R. (1972). Nuclear cytochemical changes in onion roots stimulated by kinetin. *Exptl. Cell Res.* **73**, 129–39. (212)

Pilet, P. E. & Lance-Nougarède, A. (1965). Quelques caracteristiques structurales et physiologiques du meristème radiculaire du *Lens culinaris. Bull. Soc. Fr. Physiol. Végét.* **11**, 187–201. (41)

Pogo, B. G. T., Pogo, A. O., Allfrey, V. G. & Mirsky, A. E. (1968). Changing pattern of histone acetylation and RNA synthesis in regeneration of the liver. *Proc. Nat. Acad. Sci.* **59**, 1337–44. (212)

Popham, R. A. (1963). Developmental studies of flowering. *Brookhaven Symp. Biol.* **16**, 138–56. (96, 107)

Porter, K. R. (1963). Diversity at the subcellular level and its significance. In *The Nature of Biological Diversity*, ed. by J. M. Allen, pp. 121–63. New York: McGraw-Hill Book Co. (23)

Postlethwait, J. H. & Schneiderman, H. A. (1973). Developmental genetics of *Drosophila* imaginal discs. *Ann. Rev. Genet.* **7**, 381–433. (91)

Prabhoo, N. R. (1961). A note on the giant chromosomes in the salivary glands of *Womersleya* sp. (Collembola: Insecta). *Bull. Entomol. Madras* **2**, 21–2. (130)

Prakken, R. (1943). Studies of asynapsis in rye. *Hereditas* **29**, 475–95. (65, 67)

Pratt, C., Ourecky, D. K. & Einset, J. (1967). Variation in apple cytochimeras. *Amer. J. Bot.* **54**, 1295–1301. (100, 101, 102, 103)

Prensky, W., Steffensen, D. M. & Hughes, W. L. (1973). The use of iodinated RNA for gene localization. *Proc. Nat. Acad. Sci.* **70**, 1860–4. (169)

Prescott, D. M. (1970). Structure and replication of eukaryotic chromosomes. *Adv. Cell Biol.* **1**, 57–117. (31)

Quastler, H. (1963). The analysis of cell population kinetics. In *Cell Proliferation*, ed. by L. F. Lamerton and R. J. M. Fry, pp. 18–34. Oxford: Blackwell Scientific Publications. (111, 112)

Quastler, H. & Sherman, F. G. (1959). Cell population kinetics in the intestinal epithelium of the mouse. *Exptl. Cell Res.* **17**, 420–38. (34, 111)

Rae, P. M. M. (1970). Chromosome distribution of rapidly reannealing DNA in *Drosophila melanogaster. Proc. Nat. Acad. Sci.* **67**, 1018–20. (24, 186)

Rae, P. M. M. & Franke, W. W. (1972). The interphase distribution of satellite DNA-containing heterochromatin in mouse nuclei. *Chromosoma* **39**, 443–56. (24)

Ramirez, S. A. & Sinclair, J. H. (1975). Ribosomal gene localization and distribution (arrangement) within the nucleolar organizer region of *Zea mays. Genetics* **80**, 505–18. (171)

Randolph, L. F. (1928). Types of supernumerary chromosomes in maize. *Anat. Rec.* **41**, 102. (193)

Randolph, L. F. (1941). Genetic characteristics of the B-chromosomes in maize. *Genetics* **26**, 608–31. (193)

Rao, P. T. & Johnson, R. T. (1970). Mammalian cell fusion: studies on the regulation of DNA synthesis and mitosis. *Nature* **225**, 159–64. (32)

Raper, J. R. & Flexer, A. S. (1970). The road to diploidy with emphasis on a detour. *Symp. Soc. Gen. Microbiol.* **20**, 401–32. (2, 15, 17, 115)

Rasch, E. & Woodard, J. W. (1959). Basic proteins of plant nuclei during normal and pathological cell growth. *J. Biophys. Biochem. Cytol.* **6**, 263–76. (13)

Rasch, E. M. (1966). Developmental changes in patterns of DNA synthesis by polytene chromosomes of *Sciara coprophila. J. Cell Biol.* **31**, 91A. (131)

Rasch, E. M. (1970*a*). Two-wavelength cytophotometry of *Sciara* salivary gland chromosomes. In *Introduction to Quantitative Cytochemistry*, ed. by G. L. Wied and G. F. Bahr, pp. 335–55. New York: Academic Press. (134)

Rasch, E. M. (1970*b*). DNA cytophotometry of salivary gland nuclei and other tissue systems in dipteran larvae. In *Introduction to Quantitative Cytochemistry*, ed. by G. L. Wied and G. F. Bahr, pp. 357–97. New York: Academic Press. (131, 132)

Rees, H. (1955). Genotypic control of chromosome behaviour in rye. I. Inbred lines. *Heredity* **9**, 93–116. (65, 69)

Rees, H. (1961). Genotypic control of chromosome form and behaviour. *Bot. Res.* **27**, 288–318. (64, 69)

Rees, H. & Thompson, J. B. (1956). Genotypic control of chromosome behaviour in rye. III. Chiasma frequency in homozygotes and heterozygotes. *Heredity* **10**, 409–24. (65, 69)

Reese, D. H., Puccia, E. & Yamada, T. (1969). Activation of ribosomal RNA synthesis in initiation of Wolffian lens regeneration. *J. Exp. Zool.* **170**, 259–68. (217)

Reinert, J. (1973). Aspects of organization. Organogenesis and embryogenesis. In *Plant Tissue and Cell Culture*, ed. by H. E. Street, pp. 338–355. Oxford: Blackwell Sci. Public. (203, 222, 224)

Reitberger, A. (1964). Lineare Anordnung der Chromosomen im Kern des Spermatozoids des Lebermooses *Spaerocarpus donnellii. Naturwiss.* **51**, 395–6. (15)

Resch, A. (1958). Weitere Untersuchungen über das Phloem von *Vicia faba. Planta* **52**, 121–43. (140)

Resende, F. (1967). General principles of sexual and asexual reproduction and life cycles. In *Encyclopedia of Plant Physiology* **18**, ed. by W. Ruhland, pp. 257–81. Berlin: Springer-Verlag. (2)

Resende, F. & Catarino, F. M. (1963). The effect of photoperiodic action and organ age on chromonematic replication in kalanchoideae. *Portug. Acta Biol.* (A) **18**, 1–12. (126)

Resende, F., Linskens, H. F. & Catarino, F. M. (1964). Growth regulators and protein content, their correlation with endopolyploidy. *Rev. Biologia* **4**, 101–12. (126)

Reyer, R. W. (1971). DNA synthesis and the incorporation of labeled iris cells into the lens during lens regeneration in adult newts. *Develop. Biol.* **24**, 533–58. (218)

Ribbert, D. (1967). Die Polytänchromosomen der Borstenbildungszellen von *Calliphora erytrocephala* unter besonderer Berücksichtigung der Geschlechtsgebundenen Heterozygotie und des Puffmusters während der Metamorphose. *Chromosoma* **21**, 296–344. (179)

Ribbert, D. (1972). Relation of puffing to bristle and footpad differentiation in *Calliphora* and *Sarcophaga*. In *Developmental Studies on Giant Chromosomes*, ed. by W. Beermann, pp. 153–79. Berlin: Springer-Verlag. (176, 177, 178, 179)

Ribbert, D. & Bier, K. (1969). Multiple nucleoli and enhanced nucleolar activity in the nurse cells of the insect ovary. *Chromosoma* **27**, 178–97. (148)

Richardson, B. J., Czuppon, A. B. & Sharman, G. B. (1971). Inheritance of glucose-6-phosphate dehydrogenase variation in kangaroos. *Nature New Biol.* **230**, 154–5. (192)

Rieffel, S. M. & Crouse, H. (1966). The elimination and differentiation of chromosomes in the germ line of *Sciara. Chromosoma* **19**, 231–76. (110)

Riggs, A. D. (1975). X inactivation, differentiation and DNA methylation. *Cytogenet. Cell Genet.* **14**, 9–25. (193)

Riley, R. (1966). The genetic regulation of meiotic behaviour in wheat and its relatives. In *Proc. 2nd. Int. Wheat Genetics Symp., Hereditas* Suppl. **2**, 395–408. (65, 66)

Riley, R. (1967). Theoretical and practical aspects of chromosome pairing. *Genetica Agraria* **21**, 111–28. (65, 66, 67)

Riley, R. & Bennett, M. D. (1971). Meiotic DNA synthesis. *Nature* **230**, 182–5. (60)

Riley, R. & Chapman, V. (1958). Genetic control of the cytologically diploid behaviour of the hexaploid wheat. *Nature* **182**, 713–15. (65, 66, 67)

Riley, R. & Chapman, V. (1963). The effects of the deficiency of chromosome V (5B) of *Triticum aestivum* on the meiosis of synthetic amphiploids. *Heredity* **18**, 473–84. (65, 66)

Riley, R. & Kempanna, C. (1963). The homoeologous nature of the nonhomologous meiotic pairing in *Triticum aestivum* deficient for chromosome V (5B). *Heredity* **18**, 287–306. (65, 66)

Riley, R. & Law, C. N. (1965). Genetic variation in chromosome pairing. *Adv. Genet.* **13**, 57–114. (64, 65, 66, 69)

Rimpau, J. & Flavell, R. B. (1975). Characterisation of rye B chromosome DNA by DNA/DNA hybridisation. *Chromosoma* **52**, 207–17. (193)

Ris, H. & Kubai, D. F. (1970). Chromosome structure. *Ann. Rev. Genet.* **4**, 263–94. (31)

Ritossa, F. M. (1968). Unstable redundancy of genes for ribosomal RNA. *Proc. Nat. Acad. Sci.* **60**, 509–16. (154, 155)

Ritossa, F. M. (1972). Procedure for magnification of lethal deletions of genes for ribosomal RNA. *Nature New Biol.* **240**, 109–11. (155)

Ritossa, F. M. (1973). Crossing-over between X and Y chromosomes during ribosomal DNA magnification in *Drosophila melanogaster. Proc. Nat. Acad. Sci.* **70**, 1950–4. (155)

Ritossa, F. M., Atwood, K. C. & Spiegelman, S. (1966*a*). On the redundancy of DNA complementary to amino acid transfer RNA and its absence from the nucleolar organizer region of *Drosophila melanogaster. Genetics* **51**, 663–76. (173)

Ritossa, F. M., Atwood, K. C. & Spiegelman, S. (1966*b*). A molecular explanation of the bobbed mutants of *Drosophila* as partial deficiencies of 'ribosomal' DNA. *Genetics* **54**, 819–34. (154, 181)

Ritossa, F. M. & Scala, G. (1969). Equilibrium variations in the redundancy of rDNA in *Drosophila melanogaster. Genetics* Suppl. **61**, 305–17. (155, 181)

Ritossa, F., Scalenghe, F., Di Turi, N. & Contini, A. M. (1974). The cell stage of X-Y recombination during rDNA magnification in *Drosophila. Cold Spring Harbor Symp. Quant. Biol.* **38**, 483–90. (155)

Ritossa, F. M. & Spiegelman, S. (1965). Localization of DNA complementary to ribosomal RNA in the nucleolus organizer region of *Drosophila melanogaster. Proc. Nat. Acad. Sci.* **53**, 737–45. (133, 154, 168)

Röbbelen, G. & Nirula, S. (1965). Der Replikationszustand der chromosomalen DNS in ruhenden Samen von *Haplopappus gracilis. Naturwiss.* **52**, 649–50. (50)

Roberts, P. A., Kimball, R. F. & Pavan, C. (1967). Response of *Rhynchosciara* chromosomes to microsporidian infection. Increased polyteny and generalized puffing. *Exptl. Cell. Res.* **47**, 408–22. (131)

Robinson, R. & Darrow, G. M. (1929). A pink *Poinsettia* chimera. *J. Hered.* **20**, 335–7. (96)

Rodman, T. C. (1967). Control of polytenic replication in dipteran larvae. I. Increased number of cycles in a mutant strain of *Drosophila melanogaster. J. Cell Physiol.* **70**, 179–86. (132)

Rodman, T. C. (1969). Morphology and replication of intranucleolar DNA in polytene nuclei, *J. Cell Biol.* **42**, 575–82. (169)

Rosenberg, O. (1906). Über die Embryobildung in der Gattung *Hieracium. Sv. Bot. Tidskr.* **28**, 143–70. (75)

Rossen, J. M. & Westergaard, M. (1966). Studies on the mechanism of crossing over. II. Meiosis and the time of meiotic chromosome replication in the ascomycete *Neottiella rutilans* (Fr.) Dennis. *C.R. Trav. Lab. Carlsberg* **35**, 233–66. (15, 55, 56, 57, 58)

Rudkin, G. T. (1965). The structure and function of heterochromatin. In *Genetics Today* **2**, ed. by S. J. Geerts, pp. 359–74. Oxford: Pergamon Press. (132)

Rudkin, G. T. (1969). Non replicating DNA in *Drosophila. Genetics*, Suppl. **61**, 227–38. (132)

Rudkin, G. T. (1972). Replication in polytene chromosomes. In *Developmental Studies on Giant Chromosomes*, ed. by W. Beermann, pp. 59–85. Berlin: Springer-Verlag. (130, 131)

Rudkin, G. T. & Corlette, S. L. (1957). Disproportionate synthesis of DNA in a polytene chromosome region. *Proc. Nat. Acad. Sci.* **43**, 964–8. (134)

Russell, L. B. (1961). Genetics of mammalian sex chromosomes. *Science* **133**, 1795–1803. (191)

Russell, L. B. (1963). Mammalian X-chromosome action: Inactivation limited in spread and region of origin. *Science* **140**, 976–8. (191)

Russell, L. B. (1964). Genetic and functional mosaicism in the mouse. In *The Role of Chromosomes in Development*, ed. by M. Locke, pp. 153–81. New York: Academic Press. (89)

Russell, L. B. & Montgomery, C. S. (1970). Comparative studies on X-autosome translocation in the mouse. II. Inactivation of autosomal loci, segregation, and mapping of autosomal break points in five T (X; 1)'s. *Genetics* **64**, 281–312. (191, 192)

Russell, W. L. (1951). X-ray induced mutations in mice. *Cold Spring Harbor Symp. Quant. Biol.* **16**, 327–36. (89)

Rutishauser, A. (1967). *Fortpflanzungsmodus und Meiose apomiktischer Blütenpflanzen. Protoplasmatologia VI F3.* Vienna: Springer-Verlag. (70, 73, 75)

Ryffel, G., Hagenbüchle, O. & Weber, R. (1973). Unchanged number of rRNA genes in liver and tail muscle of *Xenopus* larvae during thyroxine induced metamorphosis. *Cell Differentiation* **2**, 191–8. (153)

Sagawa, Y. & Mehlquist, G. A. L. (1957). The mechanism responsible for some X-ray induced changes in flower color of the carnation, *Dianthus caryophyllus. Amer. J. Bot.* **44**, 397–403. (98)

Satina, S. (1944). Periclinal chimeras in *Datura* in relation to development and structure (A) of the style and stigma (B) of calix and corolla. *Amer. J. Bot.* **31**, 493–502. (100, 103)

Satina, S. (1945). Periclinal chimeras in *Datura* in relation to the development and structure of the ovule. *Amer. J. Bot.* **32**, 72–81. (100, 103)

Satina, S. & Blakeslee, A. F. (1935). Cytological effects of a gene in *Datura* which causes dyad formation in sporogenesis. *Bot. Gaz.* **96**, 521–32. (73)

Satina, S. & Blakeslee, A. F. (1941). Periclinal chimeras in *Datura stramonium* in relation to development of leaf and flower. *Amer. J. Bot.* **18**, 862–71. (100, 101, 103)

Satina, S. & Blakeslee, A. F. (1943). Periclinal chimeras in *Datura* in relation to the development of the carpel. *Amer. J. Bot.* **30**, 453–62. (100, 103)

Satina, S., Blakeslee, A. F. & Avery, A. (1940). Demonstration of the three germ layers in the shoot apex of *Datura* by means of induced polyploidy in periclinal chimeras. *Amer. J. Bot.* **27**, 895–905. (94, 100)

Sauerland, H. (1956). Quantitative Untersuchungen von Röntgeneffekten nach Bestrahlung verschiedener Meiosistadien bei *Lilium candidum* L. *Chromosoma* **7**, 627–54. (56)

Sax, K. (1935). The effects of temperature on nuclear differentiation in microspore development. *J. Arnold Arb.* **16**, 301–10. (166)

Sax, K. (1962). Aspects of aging in plants. *Ann. Rev. Plant Physiol.* **13**, 489–506. (142)

Schieder, O. (1974). Selektion einer somatischen Hybride nach Fusion von Protoplasten auxotropher Mutanten von *Sphaerocarpos donnellii* Aust. *Z. Pflanzenphysiol.* **74**, 357–65. (228)

Schimke, R. T., Rhoads, R. E., Palacios, R. & Sullivan, D. (1973). Ovalbumin mRNA, complementary DNA and hormone regulation in chick oviduct. *Karolinska Symp. Reproductive Endocrinol.* **6**, 357–79. (180)

Schmidt, A. (1924). Histologische Studien an phanerogamen Vegetations punkten. *Bot. Arch.* **8**, 345–404. (96)

Schnedl, W. (1974). Banding patterns in chromosomes. *Int. Rev. Cytol. Suppl.* **4**, 237–72. (31)

Schrader, F. (1921). The chromosomes of *Pseudococcus nipae. Biol. Bull.* **40**, 259–70. (187)

Schrader, F. & Leuchtenberger, C. (1950). A cytochemical analysis of the functional interrelationships of various cell structures in *Arvelius albopunctatus* (De Geer). *Exptl. Cell Res.* **1**, 421–52. (135)

Schreck, R. R., Warburton, D., Miller, O. J., Beiser, S. M. & Erlanger, B. F. (1973). Chromosome structure as revealed by a combined chemical and immunochemical procedure. *Proc. Nat. Acad. Sci.* **70**, 804–7. (187)

Schreiber, E. (1935). Über Kultur und Geschlechtsbestimmung von *Dictyota dichotoma. Planta* **24**, 266–75. (6)

Schreiber, G. (1949). Statistical and physiological studies on the interphase growth of the nucleus. *Biol. Bull.* **97**, 187–205. (27)

Schreiber, G. (1950). Some basic facts concerning the variability of nuclear size. *Ann. Acad. Bras. Ciencias* **22**, 151–60. (27)

Schwarzacher, H. G. & Pera, F. (1969). Multipolar mitosis and somatic segregation in cell cultures of *Microtus agrestis.* In *Comparative Mammalian Cytogenetics*, ed. by K. Benirscke, pp. 186–90. New York: Springer-Verlag. (138)

Schwarzacher, H. G. & Schnedl, W. (1965). Autoradiographische Untersuchungen über die Endoreduplikation. *Naturwiss.* **53**, 23. (128)

Schwarzacher, H. G. & Schnedl, W. (1966). Position of labelled chromatids in diplochromosomes of endo-reduplicated cells after uptake of tritiated thymidine. *Nature* **209**, 107–8. (128)

Schweizer, D. & Nagl, W. (1976). Heterochromatin diversity in *Cymbidium* and its relationship to differential DNA replication. *Exptl. Cell Res.* **98**, 411–23. (164)

Scott, N. S. & Ingle, J. (1973). The genes for cytoplasmic ribosomal ribonucleic acid in higher plants. *Plant Physiol.* **51**, 677–84. (171)

Sears, E. R. (1954). The aneuploids of common wheat. *Missouri Univ. Agr. Exp. Sta. Res. Bull.* no. *572.* (65, 66)

Sears, E. R. (1958). The aneuploids of common wheat. *Proc. 1st Int. Wheat Genet. Symp.*, pp. 221–8. Winnipeg: Public Press Ltd. (65, 66)

Sears, E. R. & Okamoto, M. (1959). Intergenomic chromosome relationships in hexaploid wheat. *Proc. X Int. Congr. Genet., Montreal* **2**, 258–9. Toronto: University of Toronto Press. (65, 66, 67)

Seiler, J. (1963). Untersuchungen über die Parthenogenese bei *Solenobia triquetrella* F.R. (Lepidoptera, Psychidae). IV Mitteilung. *Z. Vererb.* **94**, 29–66. (81)

Seiler, J. & Schäffer, K. (1960). Untersuchungen über die Entstehung der Parthenogenese bei *Solenobia triquetrella* F.R. (Lepidoptera, Psychidae). II. Mitteilung. *Chromosoma* **11**, 29–102. (80, 81)

Semple, J. C. (1972). Behavior of B-chromosomes in *Xanthisma texanum* D.C.: a nonrandom phenomenon. *Science* **175**, 666. (109)

Serra, J. A. & Picciochi, G. C. (1961). The course of mitosis in polycentric chromosomes. *Rev. Portug. Zool. Biol. Ger.* **3**, 95–114. (109)

Shah, V. C., Lakhotia, S. C. & Rao, S. R. V. (1973). Nature of heterochromatin. *J. Sci. Industr. Res.* **32**, 467–80. (185)

Sharman, G. B. (1971). Late DNA replication in the paternally derived X chromosome of female kangaroos. *Nature* **230**, 231–2. (192)

Sheridan, W. F. (1975). Plant regeneration and chromosome stability in tissue cultures. In *Genetic Manipulations with Plant Material*, ed. by L. Ledoux, pp. 263–95. New York: Plenum Publ. Corp. (225, 226)

Sheridan, W. F. & Stern, H. (1967). Histones of meiosis. *Exptl. Cell Res.* **45**, 323–35. (56, 62)

Sherman, F. G., Quastler, H. & Wimber, D. R. (1961). Cell population kinetics in the ear epidermis of mice. *Exptl. Cell Res.* **25**, 114–9. (47)

Sherman, M. I., McLaren A. & Walker, P. M. B. (1972). Mechanism of accumulation of DNA in giant cells of mouse trophoblast. *Nature New Biol.* **238**, 175–6. (128)

Sibatani, A. (1971). Difference in the proportion of DNA specific to ribosomal RNA between adults and larvae of *Drosophila melanogaster. Mol. Gen. Genet.* **114**, 177–80. (133)

Signoret, J. (1973). Obtention experimentale d'embryons non zygotiques chez les vertebrés: bilan et perspectives. *Soc. bot. Fr., Memoires 1973, Coll. Morphologie*, 195–200. (218)

Signoret, J., Briggs, R. & Humphrey, R. R. (1962). Nuclear transplantation in the Axolotl. *Develop. Biol.* **4**, 134–64. (219)

Signoret, J. & Picheral, B. (1962). Transplantation de noyaux chez *Pleurodeles walltii* Michah. *C.R. Acad. Sci.* **254**, 1150–1. (219)

Simpson, S. B. Jr. (1970). Studies on regeneration of the lizard's tail. *Amer. Zool.* **10**, 157–65. (211)

Sirlin, J. L. (1972). *Biology of RNA*, New York: Academic Press. (25, 28, 32, 165)

Skoog, F. (1971). Aspects of growth factor interactions in morphogenesis of tobacco tissue cultures. In *Coll. Internat. C.N.R.S. no. 193, Les Cultures de Tissus de Plantes*, pp. 115–35. Paris: C.N.R.S. (226)

Smith, L. (1936). Cytogenetic studies in *Triticum monococcum L. and T. aegilopoides* Bal. *Univ. Missouri Agr. Exp. Sta. Res. Bull.* **248**, 1–38. (65, 66)
Smith, L. D. (1965). Transplantation of the nuclei of primordial germ cells into enucleated eggs of *Rana pipiens. Proc. Nat. Acad. Sci.* **54**, 101–7. (219)
Smyth, D. R. & Stern, H. (1973). Repeated DNA synthesized during pachytene in *Lilium henryi. Nature New Biol.* **245**, 94–6. (60)
Snoad, B. (1955). Somatic instability of chromosome number in *Hymenocallis palathinum. Heredity* **9**, 129–34. (137)
Sossountzov, L. (1965). Etude morphologique et histologique du sporophyte de la Fougère aquatique *Marsilea drummondii* A. Br. cultivée *in vitro*; structure et functionnement de l'apex. I. Le sporophyte normal. *Rev. Cytol. Biol. Vég.* **28**, 175–211. (46)
Sossountzov, L. (1969). Incorporation de précurseurs tritiés des acides nucléiques dans les méristèmes apicaux de la Fougère aquatique *Marsilea drummondii* A. *Br. Rev. Gen. Bot.* **76**, 109–56. (45, 46)
Southern, E. M. (1970). Base sequence and evolution of guinea pig α-satellite DNA. *Nature* **227**, 794–8. (24)
Southern, E. M. (1975). Long range periodicities in mouse satellite DNA. *J. Mol. Biol.* **94**, 51–69. (24)
Soyer, M. O. (1971). Structure des noyaux des *Blastodinium* (dinoflagellés parasites). Division et condensation chromosomique. *Chromosoma* **33**, 70–114. (23)
Sparrow, A. H., Sparrow, R. C. & Schairer, L. A. (1960). The use of X-rays to induce somatic mutations in *Saintpaulia. Afr. Violet Mag.* **13**, 32–7. (224)
Spear, B. B. (1974). The genes for ribosomal RNA in diploid and polytene chromosomes of *Drosophila melanogaster. Chromosoma* **48**, 159–79. (133)
Spear, B. B. & Gall, J. G. (1973). Independent control of ribosomal gene replication in polytene chromosomes of *Drosophila melanogaster. Proc. Nat. Acad. Sci.* **70**, 1359–63. (133, 156)
Spelsberg, T. C. & Hnilica, L. S. (1970). Deoxyribonucleoproteins and the tissue specific restriction of DNA in chromatin. *Biochem. J.* **120**, 435–7. (197, 198)
Spieler, R. A. (1963). Genic control of chromosome loss and non-disjunction in *Drosophila melanogaster. Genetics* **48**, 73–90. (65, 69, 70)
Stack, S. M. & Brown, W. V. (1969). Somatic pairing, reduction and recombination: an evolutionary hypothesis. *Nature* **222**, 1275–6. (63)
Stange, L. (1965). Plant cell differentiation. *Ann. Rev. Plant Physiol.* **16**, 119–40. (120, 166)
St Aubin, P. M. G. & Bucher, N. L. R. (1952). Study of binucleate cell counts in resting and regenerating rat liver employing a mechanical method for separation of liver cells. *Anat. Rec.* **112**, 797–810. (115)
Stebbins, G. L. & Jain, S. K. (1960). Developmental studies of cell differentiation in the epidermis of monocotyledons. I. *Allium, Rhoeo* and *Commelina. Develop. Biol.* **2**, 409–26. (167)
Stedman, E. & Stedman, E. (1950). Cell specificity and histones. *Nature* **166**, 780–1. (194)
Steeves, T. & Sussex, I. (1972). *Patterns in Plant Development.* Englewood Cliffs: Prentice-Hall. (94, 96)
Steffen, K. (1963). Fertilization. In *Recent Advances in Embryology of Angiosperms*, ed. by P. Maheshwari, pp. 105–143. Delhi: International Society of Plant Morphologists. (12)
Steffensen, D. (1959). A comparative view of the chromosome. *Brookhaven Symp. Biol.* **12**, 103–24. (25)
Steffensen, D. M. (1966). Synthesis of ribosomal RNA during growth and division in *Lilium. Exptl. Cell Res.* **44**, 1–12. (166)
Steffensen, D. M. & Duffey, P. (1974). Localization of 5S ribosomal RNA genes on human chromosome 1. *Nature* **252**, 741–3. (172)

Steffensen, D. M. & Wimber, D. E. (1971). Localization of tRNA genes in the salivary chromosomes of *Drosophila* by RNA:DNA hybridization. *Genetics* **69**, 163–78. (173)
Stein, G. & Baserga, R. (1970). Continued synthesis of non-histone chromosomal proteins during mitosis. *Biochem. Biophys. Res. Comm.* **41**, 715–22. (32)
Stein, G., Park, W., Thrall, C., Mans, R. & Stein, J. (1975). Regulation of cell cycle stage-specific transcription of histone genes from chromatin by non-histone chromosomal proteins. *Nature* **257**, 764–7. (199)
Stein, G. S., Spelsberg, T. C. & L. J. Kleinsmith (1974). Nonhistone chromosomal proteins and gene regulation. *Science* **183**, 817–24. (196, 197, 200, 201)
Stein, O. L. & Quastler, H. (1963). The use of tritiated thymidine in the study of tissue activation during germination in *Zea mays. Amer. J. Bot.* **50**, 1006–11. (50)
Stern, C. (1929). Untersuchungen über Aberrationen des Y-chromosoms von *Drosophila melanogaster. Z. indukt. Abst. Vererbl.* **51**, 253–353. (185)
Stern, C. (1936). Somatic crossing over and segregation in *Drosophila melanogaster. Genetics* **21**, 625–730. (90)
Stern, C. (1960). A mosaic of *Drosophila* consisting of 1X, 2X and 3X tissue and its probable origin by mitotic non-disjunction. *Nature* **186**, 179–80. (85)
Stern, C. (1968). *Genetic Mosaics and Other Essays.* Cambridge, Mass.: Harvard University Press. (82, 83, 84, 87, 90, 91)
Stern, H. & Hotta, Y. (1973). Biochemical controls of meiosis. *Ann. Rev. Genet.* **7**, 37–66. (59)
Steward, F. C. (1968). *Growth and Organization in Plants.* Reading, Mass.: Addison-Wesley. (204).
Steward, F. C., Blakely, L. M., Kent, A. T. & Mapes, M. O. (1963). Growth and organization in free cell cultures. *Brookhaven Symp. Biol.* **16**, 73–88. (223)
Stewart, R. N. & Burk, L. G. (1970). Independence of tissues derived from apical layers in ontogeny of the tobacco leaf and ovary. *Amer. J. Bot.* **57**, 101–16. (100)
Stewart, R. N., Mayer, F. G. & Dermen, H. (1972). *Camellia*+'Daisy Eagleson', a graft chimera of *Camellia sasanqua* and *C. japonica. Amer. J. Bot.* **59**, 515–24. (92)
Stocker, A. J. & Pavan, C. (1974). The influence of ecdysterone on gene amplification, DNA synthesis and puff formation in the salivary gland chromosomes of *Rhynchosciara hollaenderi. Chromosoma* **45**, 295–319. (160)
Strasburger, E. (1907). Über die Individualität der Chromosomen und die Propfhybriden-Frage. *Jb. wiss. Bot.* **44**, 482–550. (92)
Strasburger, E. (1909). Meine Stellungnahme zur Frage der Propfbastarde. *Ber. deutsch. bot. Ges.* **27**, 511–28. (94)
Strokov, A. A., Bogdanov, Y. F. & Reznickova, S. A. (1973). A quantitative study of histones from *Lilium* microsporocytes. *Chromosoma* **43**, 247–60. (56, 62)
Sturtevant, A. H. (1929). The claret mutant type of *Drosophila simulans*: a study of chromosome elimination and of cell lineage. *Z. wiss. Zool.* **135**, 323–56. (85)
Sullivan, D., Palacios, R., Starnezer, J., Taylor, J. M., Faras, A. J., Kiely, M. L., Summers, N. M., Bishop, J. M. & Schimke, R. T. (1973). Synthesis of a deoxyribonucleic sequence complementary to ovalbumin messenger ribonucleic acid and quantification of ovalbumin genes. *J. Biol. Chem.* **248**, 7530–9. (180)
Sunderland, N., Collins, G. B. & Dunwell, J. M. (1974). The role of nuclear fusion in pollen embryogenesis of *Datura innoxia* Mill. *Planta* **117**, 227–41. (225)
Suomalainen, E. (1950). Parthenogenesis in animals. *Adv. Genet.* **3**, 193–253. (70, 76, 77)
Suzuki, Y., Gage, L. P. & Brown, D. D. (1972). The genes for silk fibroin in *Bombyx mori. J. Mol. Biol.* **70**, 637–49. (174, 180)
Swartz, F. J. (1956). The development in the human liver of multiple deoxyribose nucleic acid (DNA) classes and their relationship to the age of the individual. *Chromosoma* **8**, 53–72. (115)

Swift, H. (1962). Nucleic acids and cell morphology in dipteran salivary glands. In *The Molecular Control of Cellular Activity*, ed. by M. Allen, pp. 73–125. New York: McGraw-Hill. (130, 131, 134)

Swift, H. (1974). The organization of genetic material in eukaryotes: progress and prospects. *Cold Spring Harbor Symp. Quant. Biol.* **38**, 963–79. (24, 195)

Swift, H. & Kleinfeld, R. (1953). DNA in grasshopper spermatogenesis, oogenesis and cleavage. *Physiol. Zool.* **26**, 301–11. (15)

Tanaka, R. (1969*a*). Deheterochromatinization of the chromosomes in *Spiranthes sinensis. Japan. J. Genet.* **44**, 291–6. (189, 190)

Tanaka, R. (1969*b*). Speciation and karyotypes in *Spiranthes sinensis. J. Sci. Hiroshima University, B* Div. *2*, **12**, 165–97. (189, 190)

Tanaka, T. (1927). Bizzarria – a clear case of periclinal chimera. *J. Genet.* **18**, 77–85. (95)

Tardent, P. (1963). Regeneration in the hydrozoa. *Biol. Rev.* **38**, 293–333. (204, 205, 206)

Tarkowski, A. K. (1961). Mouse chimeras developed from fused eggs. *Nature* **190**, 857–60. (91, 92)

Tarkowski, A. K. (1963). Studies on mouse chimeras developed from eggs fused *in vitro. Nat. Cancer Inst. Mono.* **11**, 51–71. (91, 92)

Tarkowski, A. K. (1964). Patterns of pigmentation in experimentally produced mouse chimaerae. *J. Embryol. Exp. Morph.* **12**, 575–85. (91, 92)

Tartof, K. D. (1971). Increasing the multiplicity of ribosomal RNA genes in *Drosophila melanogaster. Science* **171**, 294–7. (156)

Tartof, K. D. (1973). Regulation of ribosomal RNA gene multiplicity in *Drosophila melanogaster. Genetics* **73**, 57–71. (156, 168)

Tartof, K. D. (1974). Unequal mitotic sister chromatid exchange and disproportionate replication as mechanisms regulating ribosomal RNA gene redundancy. *Cold Spring Harbor Symp. Quant. Biol.* **38**, 491–500. (155, 156)

Tartof, K. D. & Perry, R. (1970). The 5S RNA genes of *Drosophila melanogaster. J. Mol. Biol.* **51**, 171–83. (172)

Taylor, J. H. (1958). The mode of chromosome duplication in *Crepis capillaris. Exptl. Cell Res.* **15**, 350–7. (28)

Taylor, J. H. (1965). Distribution of tritium-labelled DNA among chromosomes during meiosis. I. Spermatogenesis in the grasshopper. *J. Cell Biol.* **25**, 57–67. (58)

Taylor, J. H. (1967). Meiosis. In *Encyclopedia of Plant Physiology* **18**, ed. by W. Ruhland, pp. 444–67. Berlin: Springer-Verlag. (56)

Taylor, J. H., Woods, P. S. & Hughes, W. L. (1957). The organization and duplication of chromosomes as revealed by autoradiographic studies using tritium-labelled thymidine. *Proc. Nat. Acad. Sci.* **43**, 122–8. (31)

Teissere, M., Penon, P., Van Huystee, R. B., Azou, Y. & Picard, J. (1975). Hormonal control of transcription in higher plants. *Biochim. Biophys. Acta* **402**, 391–402. (202)

Thomas, D. R. (1967). Quiescent centre in excised tomato roots. *Nature* **214**, 739. (39)

Thompson, J. & Clowes, F. A. L. (1968). The quiescent centre and rates of mitosis in the root meristem of *Allium sativum. Ann. Bot.* **32**, 1–13. (39)

Thomsen, M. (1927). Studien über die Parthenogenese bei einigen Cocciden und Aleurodiden. *Z. Zellforsch.* **5**, 1–116. (80)

Thornton, C. S. (1938). The histogenesis of muscle in the regenerating forelimb of larval *Ambystoma punctatum. J. Morphol.* **62**, 17–47. (209)

Tidwell, T., Allfrey, V. G. & Mirsky, A. E. (1968). The methylation of histone during regeneration of the liver. *J. Biol. Chem.* **243**, 707–15. (213)

Timmis, J. N. & Ingle, J. (1973). Environmentally induced changes in rRNA gene redundancy. *Nature New Biol.* **244**, 235–6. (160, 161)

Timmis, J. N. & Ingle, J. (1974). The nature of the variable DNA associated with environmental induction in flax. *Heredity* **33**, 339–46. (161)

Timmis, J. N., Sinclair, J. & Ingle, J. (1972). Ribosomal RNA genes in euploids and aneuploids of hyacinth. *Cell Differentiation* **1**, 335–9. (156)

Tobia, A. M., Schildkraut, C. M. & Maio, J. J. (1970). Deoxyribonucleic acid replication in synchronized mammalian cells. I. Time of synthesis of molecules of different average guanine+cytosine content. *J. Mol. Biol.* **54**, 499–515. (31)

Tobler, H. (1975). Occurrence and developmental significance of gene amplification. In *The Biochemistry of Animal Development*, ed. by R. Weber, pp. 91–143. New York: Academic Press. (143, 153)

Tobler, H., Smith, K. D. & Ursprung, H. (1972). Molecular aspects of chromatin elimination in *Ascaris lumbricoides. Develop. Biol.* **27**, 190–203. (110)

Tobler, H., Zulauf, E. & Kuhn, O. (1974). Ribosomal RNA genes in germ line and somatic cells of *Ascaris lumbricoides. Develop. Biol.* **41**, 218–23. (110)

Tocchini-Valentini, G. P., Mahdavi, V., Brown, R. & Crippa, M. (1974). The synthesis of amplified ribosomal DNA. *Cold Spring Harb. Symp. Quant. Biol.* **38**, 551–8. (146)

Tomkins, G. M. & Martin, D. W. (1970). Hormones and gene expression. *Ann. Rev. Genet.* **4**, 91–106. (200)

Torrey, J. G. (1963). Cellular patterns in developing root. *Symp. Soc. Exp. Biol.* **17**, 285–314. (41)

Torrey, J. G. (1965). Physiological bases of organization and development in the root. In *Encyclopedia of Plant Physiol. 16/1*, ed. by W. Ruhland, pp. 1256–1327. Berlin: Springer-Verlag. (38)

Torrey, J. G. (1966). The initiation of organized development in plants. *Adv. Morphogen.* **5**, 39–91. (203)

Torrey, J. G. (1972). On the initiation of organization in the root apex. In *The Dynamics of Meristem Cell Population*, ed. by M. W. Miller and C. C. Kuehnert, pp. 1–13. New York: Plenum Publ. Corp. (39, 41)

Trampusch, H. A. L. (1972). Cytodifferentiation. *Acta Morph. Neerl. Scand.* **10**, 155–163. (209)

Trampusch, H. A. L. & Harrebomée, A. E. (1965). Dedifferentiation a prerequisite of regeneration. In *Regeneration in Animals and Related Problems*, ed. by V. Kiortsis and H. A. L. Trampusch, pp. 341–76. Amsterdam: North-Holland Publ. Co. (209)

Trampusch, H. A. L. & Harrebomée, A. E. (1969). Dedifferentiation and the interconvertibility of the different cell-types in the amphibian extremity. *Acta Embryol. Exp. 1969*, 35–69. (209)

Trendelenburg, M. (1974). Morphology of ribosomal RNA cistrons in oocytes of the water beetle, *Dytiscus marginalis* L. *Chromosoma* **48**, 119–35. (149)

Trendelenburg, M. F., Scheer, U. & Franke, W. W. (1973). Structural organization of the transcription of ribosomal DNA in oocytes of the house cricket. *Nature New Biol.* **245**, 167–70. (149)

Troy, M. R. & Wimber, D. E. (1968). Evidence for a constancy of the DNA synthetic period between diploid–polyploid groups in plants. *Exptl. Cell Res.* **53**, 145–54. (34)

Tschermak-Woess, E. (1956). Karyologische Pflanzenanatomie. *Protoplasma* **46**, 798–834. (113, 120, 125)

Tschermak-Woess, E. (1957). Über das regelmässige Auftreten von 'Riesenchromosomen' im Chalazalhaustorium von *Rhinanthus. Chromosoma* **8**, 523–44. (128)

Tschermak-Woess, E. (1959). Die DNS-Reproduktion in ihrer Beziehung zum endomitotischen Strukturwechsel. *Chromosoma* **10**, 497–503. (123)

Tschermak-Woess, E. (1963). *Strukturtypen der Ruhekerne von Pflanzen und Tieren. Protoplasmatologia, V/1.* Vienna: Springer-Verlag. (120, 128)

Tschermak-Woess, E. (1971). Endomitose. In *Handbuch der allegemeinen Pathologie*, ed. by H. W. Altmann, pp. 569–625. Berlin: Springer-Verlag. (120)

Tschermak-Woess, E. & Dolezal, R. (1953). Durch Seitenwurzelbildung induzierte und spontane Mitosen in den Dauergeweben der Wurzel. *Öster. Bot. Z.* **100**, 358–402. (125)

Tschermak-Woess, E. & Enzenberg-Kunz, U. (1965). Die Struktur der hoch endopolyploiden Kerne im Endosperm von *Zea mays,* das auffallende Verhalten ihrer Nucleolen und ihr Endopolyploidiegrad. *Planta* **64**, 149–69. (127)

Tschermak-Woess, E. & Hasitschka, G. (1953). Veräuderungen der Kernstruktur während der Endomitose, rhytmisches Kernwachstum und verschiedenes Heterochromatin bei Angiospermen. *Chromosoma* **5**, 574–614. (123)

Turcotte, E. L. & Feaster, C. V. (1967). Semigamy in Pima cotton. *J. Hered.* **58**, 55–7. (75)

Turkington, R. W. (1971). Hormonal regulation of cell proliferation and differentiation. In *Developmental Aspects of the Cell Cycle*, ed. by I. L. Cameron, G. M. Padilla and A. M. Zimmerman, pp. 315–55. New York: Academic Press. (200)

Ullman, J. S., Lima-de-Faria, A., Jaworska, H. & Bryngelsson, T. (1973). Amplification of ribosomal DNA in *Acheta.* V. Hybridization of RNA complementary to ribosomal DNA with pachytene chromosomes. *Hereditas* **74**, 13–24. (152)

Urbani, E. (1969). Cytochemical and ultrastructural studies of oogenesis in the Dytiscidae. *Mon. Zool. Ital. N.S.* **3**, 55–87. (148)

Urbani, E. & Russo-Caia, S. (1964). Osservazioni citochimiche e autoradiografiche sul metabolismo degli acidi nucleici nella oogenesi di *Dytiscus marginalis* L. *Rend. Ist. Sci. Univ. Camerino* **5**, 19–50. (148)

Vaarama, A. (1949). Spindle abnormalities and variation in chromosome number in *Ribes nigrum. Hereditas* **35**, 135–62. (137)

Vaarama, A. (1950). Cases of asyndesis in *Matricaria inodora* and *Hyosciamus niger. Hereditas* **36**, 342–62. (65, 66)

Van de Flierdt, K. (1975). No multistrandedness in mitotic chromosomes of *Drosophila melanogaster. Chromosoma* **50**, 431–4. (136)

Van Parijs, R. & Vandendriessche, L. (1966). Changes of the DNA content of nuclei during the process of cell elongation in plants. I. The formation of polytene chromosomes. *Arch. Physiol. Bioch.* **74**, 579–86. (123)

Van't Hof, J. (1965*a*). Discrepancies in mitotic cycle time when measured with tritiated thymidine and colchicine. *Exptl. Cell Res.* **37**, 292–9. (34)

Van't Hof, J. (1965*b*). Relationship between mitotic cycle duration, S period duration and the average rate of DNA synthesis in the root meristem cells of several plants. *Exptl. Cell Res.* **39**, 48–58. (34, 35)

Van't Hof, J. (1966). Comparative cell population kinetics of tritiated thymidine labeled diploid and colchicine-induced tetraploid cells in the same tissue of *Pisum. Exptl. Cell Res.* **41**, 274–88. (34)

Van't Hof, J. (1968). Control of cell progression through the mitotic cycle by carbohydrate provision. I. Regulation of cell division. *J. Cell Biol.* **37**, 773–8. (53)

Van't Hof, J. (1974). Control of the cell cycle in higher plants. In *Cell cycle controls*, ed. by G. M. Padilla, I. L. Cameron and A. Zimmerman, pp. 77–85. New York: Academic Press. (53)

Van't Hof, J. & Kovacs, C. J. (1972). Mitotic cycle regulation in the meristem of cultured roots. In *The Dynamics of Meristem Cell Populations*, ed. by M. W. Miller and C. C. Kuehnert, pp. 15–33. New York: Plenum Publ. Corp. (53)

Van't Hof, J. & Sparrow, A. H. (1963). A relationship between DNA content, nuclear volume and minimum mitotic cycle time. *Proc. Nat. Acad. Sci.* **49**, 897–902. (34, 35)

Vasil, I. K. (1971). Physiology and ultrastructure of callus formation. In *Coll. Internat. C.N.R.S. no. 193, Les Cultures de Tissus de Plantes*, pp. 103–12. Paris: C.N.R.S. (226)

Vasil, I. K. & Vasil, V. (1972). Totipotency and embryogenesis in plant cell and tissue cultures. *In Vitro* **8**, 117–27. (203, 222, 224)

Vincent, W. S., Halvorson, H. O., Chen, H.-R. & Shin, D. (1969). A comparison of ribosomal gene amplification in uni- and multinucleolate oocytes. *Exptl. Cell Res.* **57**, 240–50. (152)

Vosa, C. G. (1970). Heterochromatin recognition with fluorochromes. *Chromosoma* **30**, 366–72. (187)

Vosa, C. G. (1973). Heterochromatin and chromosome structure. In *Chromosome identification, Nobel symposium* **23**, ed. by T. Casperson and L. Zech, pp. 152–5. New York: Academic Press. (185, 187)

Vosa, C. G. & Marchi, O. (1972). Quinacrine fluorescence and Giemsa staining in plants. *Nature New Biol.* **237**, 191–2. (187)

Wada, B. (1976). Mitotic cell studies based on *in vivo* observations. VIII. The evolution of mitotic spindles in eukaryota: a negation of the breakdown of the nuclear membrane. *Cytologia* **41**, 153–75. (33)

Wald, H. (1936). Cytologic studies on the abnormal development of the eggs of the claret mutant type of *Drosophila simulans. Genetics* **21**, 364–81. (65, 69, 70)

Walen, K. H. (1965). Spatial relationships in the replication of chromosomal DNA. *Genetics* **51**, 915–29. (128)

Walker, P. M. B. (1971). 'Repetitive' DNA in higher organisms. *Progr. Biophys. Mol. Biol.* **23**, 145–90. (24)

Walker, P. M. B. & Yates, H. B. (1952). Nuclear components of dividing cells. *Proc. Roy. Soc. B* **140**, 274–99. (27)

Wallace, H. R. & Birnstiel, M. L. (1966). Ribosomal cistrons and the nucleolus organiser. *Biochim. Biophys. Acta* **114**, 296–310. (168)

Walters, M. S. (1970). Evidence on the time of chromosome pairing from the preleptotene spiral stage in *Lilium longiflorum* 'Croft'. *Chromosoma* **29**, 375–418. (56)

Wang, T. Y. & Nyberg, L. M. (1974). Androgen receptors in the nonhistone protein fractions of prostatic chromatin. *Int. Rev. Cytol.* **39**, 1–33. (196, 197, 200)

Warmke, A. E. (1954). Apomixis in *Panicum maximum. Amer. J. Bot.* **41**, 5–11. (75)

Webster, P. L. & Langenauer, H. D. (1973). Experimental control of the activity of the quiescent centre in excised root tips of *Zea mays. Planta* **112**, 91–100. (39, 41, 42)

Webster, P. L. & Van't Hof, J. (1969). Dependence on energy and aerobic metabolism of initiation of DNA synthesis and mitosis by G1 and G2 cells. *Exptl. Cell Res.* **55**, 88–94. (53)

Webster, P. L. & Van't Hof, J. (1970). DNA synthesis and mitosis in meristems: requirements for RNA and protein synthesis. *Amer. J. Bot.* **57**, 130–9. (53)

Weisblum, B. (1974). Fluorescent probes of chromosomal DNA structure: Three classes of acridines. *Cold Spring Harbor Symp. Quant. Biol.* **38**, 441–9. (187)

Weisblum, B. & de Haseth, P. L. (1972). Quinacrine, a chromosome stain specific for deoxyadenilate–deoxythimydilate-rich regions in DNA. *Proc. Nat. Acad. Sci.* **69**, 629–32. (187)

Wen, W., Leon, P. E. & Hague, D. R. (1974). Multiple gene sites for 5S and 18+28S RNA on chromosomes of *Glyptotendipes barbipes* (Staeger). *J. Cell. Biol.* **62**, 132–44. (172)

Wensink, P. C. & Brown, D. D. (1971). Denaturation map of the ribosomal DNA of *Xenopus laevis. J. Mol. Biol.* **60**, 235–47. (147, 171)

Westergaard, M. (1964). Studies on the mechanism of crossing-over. I. Theoretical considerations. *C.R. Trav. Lab. Carlsberg* **34**, 359–405. (55)

Westergaard, M. & Wettstein, D. von (1968). The meiotic cycle in an ascomycete. In *Effects of Radiation on Meiotic Systems*, pp. 113–21. Vienna: IAEA. (15, 56)

Westergaard, M. & Wettstein, D. von (1972). The synaptinemal complex. *Ann. Rev. Genet.* **6**, 71–110. (15, 56, 58, 59)

Wetmore, R. H., De Maggio, A. E. & Morel, G. (1963). A morphogenetic look at the alternation of generations. *J. Indian Bot. Soc.* **42A**, 306–20. (2, 18)

Wettstein, R. & Sotelo, J. R. (1971). Electron microscope serial reconstruction of the spermatocyte I nuclei at pachytene. *J. Microsc.* **6**, 557–76. (58)

Wheatley, D. N. (1972). Binucleation in mammalian liver. Studies on the control of cytokinesis *in vivo. Exptl. Cell. Res.* **74**, 455–65. (117, 118, 213, 214)

White, M. J. D. (1973). *Animal Cytology and Evolution.* Cambridge: University Press. (67, 70, 77, 79, 80, 81, 206)

White, M. J. D., Cheney, J. & Key, K. H. L. (1963). A parthenogenetic species of grasshopper with complex structural heterozygosity (Ortoptera: Acridoidea). *Aust. J. Zool.* **11**, 1–19. (79)

White, P. R. (1939). Potentially unlimited growth of excised plant callus in an artificial nutrient. *Amer. J. Bot.* **26**, 59–64. (224)

Whitten, J. M. (1969). Coordinated development in the footpad of the fly *Sarcophaga bullata* during metamorphosis: changing puffing patterns. *Chromosoma* **26**, 215–44. (158, 179)

Wieslander, L., Lambert, B. & Egyhazi, E. (1975). Localization of 5S RNA genes in *Chironomus tentans. Chromosoma* **51**, 49–56. (172)

Williamson, J. H. & Procunier, J. D. (1975). Disproportionately-replicated, non functional rDNA in compound chromosomes of *Drosophila melanogaster. Mol. Gen. Genet.* **139**, 33–7. (156, 168)

Wilson, E. B. (1925). *The Cell in Development and Heredity.* New York: Macmillan. (72)

Wilson, G. B. & Cheng, K. C. (1949). Segregation and reduction in somatic tissues. II. The separation of homologous chromosomes in *Trillium* species. *J. Hered.* **40**, 3–6. (137)

Wilson, J. W. & Leduc, E. H. (1948). The occurrence and formation of binucleate cells and polyploid nuclei in the mouse liver. *Amer. J. Anat.* **82**, 353–92. (115)

Wimber, D. (1961). Asynchronous replication of deoxyribonucleic acid in root tip chromosomes of *Tradescantia paludosa. Exptl. Cell Res.* **23**, 402–7. (29)

Wimber, D. E., Duffey, P. A., Steffensen, D. M. & Prensky, W. (1974). Localization of the 5S RNA genes in *Zea mays* by RNA–DNA hybridization *in situ. Chromosoma* **47**, 353–9. (172)

Wimber, D. E. & Prensky, W. (1963). Autoradiography with meiotic chromosomes of the male newt (*Triturus viridescens*) using H^3-thymidine. *Genetics* **48**, 1731–8. (60)

Wimber, D. E. & Steffensen, D. M. (1970). Localization of 5S RNA genes on *Drosophila* chromosomes by RNA–DNA hybridization. *Science* **170**, 639–41. (172)

Wimber, D. E. & Steffensen, D. M. (1973). Localization of gene function. *Ann. Rev. Genet.* **7**, 205–23. (168, 169, 172, 173, 174)

Winkler, H. (1907). Über Propfbastarde und pflanzliche Chimären. *Ber. deutsch bot. Ges.* **25**, 568–76. (82, 92, 94, 95)

Winkler, H. (1908). Über Parthogenesis und Apogamie im Pflanzenreich. *Progr. Rei Bot.* **2**, 293–454. (70, 95)

Winkler, H. (1910). Über die Nachkommenschaft der *Solanum* Propfbastarde und die Chromosomenzahlen ihrer Keimzellen. *Z. Bot.* **2**, 1–38. (92, 93, 95)

Winkler, H. (1916). Über die experimentelle Erzeugung von Pflanzen mit abweichenden Chromosomenzahlen. *Z. Bot.* **8**, 417–531. (223)

Wischnitzer, S. (1973). The submicroscopic morphology of the interphase nucleus. *Int. Rev. Cytol.* **34**, 1–48. (23)

Witkus, E. R., Ferschl, J. B. & Berger, C. A. (1959). The cytology of *Xantisma texanum* D.C. V. Further observations on cytology and breeding behaviour. *Bull Torrey Bot. Club* **86**, 300–7. (109)

Witsch, H. & Flügel, A. (1952). Über Polyploidieerhöhung im Kurztag bei *Kalanchoe blossfeldiana. Z. Bot.* **40**, 281–91. (126)

Wolff, E. & Dubois, F. (1948). Sur la migration des cellules de régeneration chez les Planaires. *Rev. Suisse Zool.* **55**, 218–27. (206)

Wolff, G. (1895). Entwicklungsphysiologische Studien. I. Die Regeneration der Urodelenlinse. *Arch. Entwicklungsmech. Organ.* **1**, 380–90. (215)

Wolff, S. (1969). Strandedness of chromosomes. *Int. Rev. Cytol.* **25**, 279–95. (31)

Wolff, S. & Luippold, H. E. (1964). Chromosome splitting as revealed by combined X-ray and labeling experiments. *Exptl. Cell Res.* **34**, 548–56. (49)

Wollman, E. L., Jacob, F. & Hayes, W. (1956). Conjugation and genetic recombination in *Escherichia coli* K12. *Cold Spring Harbor Symp. Quant. Biol.* **21**, 141–62. (1)

Woodard, J. & Swift, H. (1964). The DNA content of cold treated chromosomes. *Exptl. Cell Res.* **34**, 131–7. (187)

Woodruff, L. S. & Burnett, A. L. (1965). The origin of the blastemal cells in *Dugesia tigrina. Exptl. Cell Res.* **38**, 295–305. (207)

Wunderlich, R. (1954). Über das Antherentapetum mit besonderer Berüchsichtigung seiner Kernzahl. *Öster. bot. Z.* **101**, 1–63. (120)

Yamada, T. (1967). Cellular and subcellular events in Wolffian lens regeneration. In *Current Topics in Developmental Biology*, ed. by A. A. Moscona and A. Monroy, pp. 247–83. New York: Academic Press. (215, 216)

Yamada, T. (1972). Control mechanisms in cellular metaplasia. In *Cell Differentiation, Proc. First Internat. Confer. Cell Differentiation*, ed. by R. Harris, P. Hallin and D. Viza, pp. 56–60. Copenhagen: Munksgaard. (215, 217, 218)

Yamada, T. & McDevitt, D. S. (1974). Direct evidence for transformation of differentiated iris epithelial cells into lens cells. *Develop. Biol.* **38**, 104–18. (215)

Yamada, T., Reese, D. H. & McDevitt, D. S. (1973). Transformation of iris into lens *in vitro* and its dependency on neural retina. *Differentiation* **1**, 65–82. (215)

Yamada, T. & Roesel, M. E. (1969). Activation of DNA replication in the iris epithelium by lens removal. *J. Exptl. Zool.* **171**, 425–31. (218)

Yamada, Y., Matsumoto, H. & Takahashi, E. (1968). Nucleic acid in callus formation and redifferentiation of callus in *Nicotiana tabacum. Sci. and Plant Nutrition* **14**, 35–8. (226)

Yamada, Y., Yasuda, T., Koge, M. & Sekiya, J. (1971). Biochemical aspects of the basic mechanism for dedifferentiation in the Alaska pea and tobacco. In *Coll. Internat.* C.N.R.S. no. *193, Les Cultures de Tissus de Plantes*, pp. 137–53. Paris: C.N.R.S. (226)

Yeoman, M. M. & Evans, P. K. (1967). Growth and differentiation of plant tissue cultures. II. Synchronous cell division in developing callus cultures. *Ann. Bot.* **31**, 323–32. (114)

Yun, K. B. & Naylor, J. M. (1973). Regulation of cell reproduction in bud meristems of *Tradescantia paludosa. Can. J. Bot.* **51**, 1137–45. (54)

Yunis, J. J. & Yasmineh, W. G. (1972). Model for mammalian constitutive heterochromatin. *Adv. Cell Mol. Biol.* **2**, 1–46. (185, 187)

Zalik, S. E. & Scott, V. (1973). Sequential disappearance of cell surface components during dedifferentiation. *Nature New Biol.* **244**, 212–14. (218)

Zetterberg, A. (1966). Synthesis and accumulation of nuclear and cytoplasmic proteins during interphase in mouse fibroblasts *in vitro. Exptl. Cell Res.* **42**, 500–11. (28)

Zetterberg, A. (1970). Nuclear and cytoplasmic growth during interphase in mammalian cells. In *Advances in Cell Biology*, ed. by D. M. Prescott, L. Goldstein and E. McConkey, pp. 211–32. Amsterdam: North-Holland Publ. Co. (28)

Zimmerman, S. B. & Levin, C. J. (1975). Do histones bind to a specific group of DNA sequences in chromatin? A test based on DNA ligase action in reconstituted chromatin. *Biochem. Biophys. Res. Comm.* **62**, 357–61. (25)

Zybina, E. V. (1970). Features of polyploidization pattern of trophoblast cells. *Tsitologija* **12**, 1081–94. (128)

Index